'SO USEFUL AN UNDERTAKING'

'SO USEFUL AN UNDERTAKING'

A History of the Royal Cornwall Show

1793 – 1993

by

CHRISTOPHER RIDDLE

with a Foreword by

HER ROYAL HIGHNESS THE PRINCESS ROYAL
GCVO

Published by
The Royal Cornwall Agricultural Association
The Royal Cornwall Showground
Wadebridge, Cornwall, England
PL27 7JE

© CHRISTOPHER RIDDLE, 1993

ISBN 0 9521037 0 2

This book is copyright and no part of it may be reproduced, stored in a retrieval system or transmitted, in any form or by any means without the prior permission of the publisher.

Published by
The Royal Cornwall Agricultural Association
The Royal Cornwall Showground
Wadebridge, Cornwall, England
PL27 7JE

Printed by Blackfords of Cornwall

CONTENTS

	Page
Preface and Acknowledgements	viii
List of Illustrations	vi
Foreword by Her Royal Highness The Princess Royal GCVO	xi

Chapter

1	From Little Acorns 1793–1827	1
2	An Amalgamation 1827–1857	31
3	On The Road 1857–1915	48
4	In War and Peace 1919–1939	102
5	A New Beginning 1945–1956	121
6	A New Regime 1957–1993	131

Show Venues, Attendances and Show Presidents, 1793–1993	189
Chairmen of Council, Directors of the Showyard and Secretaries of the Society/Association 1793–1993	196
An Index of Council Members 1891 to date	198
Entries in Livestock and Other Agricultural Classes 1857–1992	214
Cattle Section Results 1794–1992	219
Sheep Section Results 1794–1992	270
Pig Section Results 1794–1972	297
Horse Section Results 1794–1992	315
Miscellaneous Results 1793 onwards	355

List of Illustrations

Page

Her Royal Highness The Princess Royal GCVO	x
Poster, 1792 Meeting	2
Sir William Molesworth, Bart	4
Poster, 1793 Meeting	6
The Right Hon The Viscount Falmouth	7
Poster, 1794 Ploughing Match	10
Poster, J Liddell, Seedsman	15
Premiums and Adjudications, 1804	22
Henry Tresawna	54
Bath & West and Royal Cornwall Shows, Truro 1861	57
A Devon bull 'Narcissus'	61
Royal Cornwall Show, Truro 1871	64
Arch, Bodmin 1884	71
Arch, Tywardreath 1891	77
Shorthorn cattle at Trenethick Farm	82
Arch, Bodmin 1901	84
Shorthorn heifer at Trenethick Farm	85
Grand Parade of cattle, circa 1900	85
Stand of C H Harris, Engineers, circa 1905	87
Royal Cornwall Show, circa 1905	88
Royal Cornwall Show, Redruth 1906, general scene	89
Royal Cornwall Show, Redruth 1906, machinery stands	89
Royal Cornwall Show, Redruth 1906, main ring	90
Royal Cornwall Show, Redruth 1906, bandstand	90
Royal Cornwall Show, Redruth 1906, ambulance station	91
Royal Cornwall Show, Helston 1908	92
Arch, Blackcross 1909	93
Her Royal Highness The Princess of Wales, St Columb 1909	94
Their Royal Highnesses The Prince and Princess of Wales, St Columb 1909	95
Nature and Handicrafts exhibit, circa 1910	96
Cornwall County Farm and Dairy Co-Operative Society, circa 1910	96
Arch, Penzance 1912	97
Royal Cornwall Show, Penzance 1912	98
Unloading a pig	98
Prize bull, circa 1915	100
Sheep judging, circa 1915	100
Side saddle jumping, Camborne 1915	101
Agricultural mare and foal, Camborne 1923	106
Poultry Section, Helston 1925	109
Prize winning shire horse, Penzance 1929	112
Buttermakers, Penryn 1932	115
Arch, Bodmin 1948	123
George Johnstone, Chairman of Council	124
Their Majesties King George VI and Queen Elizabeth and Her Royal Highness Princess Margaret, Callington 1950	126
Albert Riddle, Show Secretary	132
Sir John Molesworth-St Aubyn, Chairman of Council	133
The Right Hon The Viscount Falmouth, Truro 1958	136
Her Majesty Queen Elizabeth, The Queen Mother, Liskeard 1959	137
Group photograph of Council Members, 1960	138
Her Royal Highness Princess Alexandra, Wadebridge 1960	139
The Royal Carriages, 1965	144
Devon bull, 'Burncoose No Use 5th', 1966	145
Stand of F Davey & Sons, 1966	146

List of Illustrations *(cont)*

	Page
His Royal Highness The Prince of Wales, Duke of Cornwall, 1970	148
Group photograph, Royal Cornwall cruise, 1971	150
Women's Institute—Keep Britain Tidy, 1971	150
The King's Troop	151
The building of Molesworth House, 1974	152
Les Tromps De Chasse, 1974	153
Stand of the Rare Breeds Survival Trust, 1975	154
His Royal Highness The Prince of Wales, Duke of Cornwall, 1977	154
His Royal Highness The Prince of Wales, Duke of Cornwall, 1977	155
Her Majesty The Queen and His Royal Highness The Duke of Edinburgh, Silver Jubilee 1977	156
Mary Chipperfield's Racing Camels, 1982	158
The Royal Navy Display Team, 1983	159
His Royal Highness The Prince of Wales, Duke of Cornwall, 1984	160
Her Majesty Queen Elizabeth, The Queen Mother, 1985	162
Her Majesty Queen Elizabeth, The Queen Mother, 1985	163
Albert Riddle's 30th Show celebrations, 1986	165
Presentation of gates in memory of George Eustice, 1987	166
Aerial view of the Showground, 1988	167
Her Royal Highness Princess Margaret, The Countess of Snowdon	168
Her Majesty The Queen, 1989	169
Her Majesty The Queen and His Royal Highness The Duke of Edinburgh, 1989	170
The Four Burrow Foxhounds, 1989	170
Albert and Betty Riddle, 1989	171
Senora Cecelia Gallinal de De Haedo, 1990	172
The Cornish Rugby team, 1991	172
Mr F Julian Williams and The Rt Hon The Viscount Falmouth	173
Christopher Riddle, Show Secretary	174
Her Royal Highness The Duchess of Kent, 1991	175
Ron Chapman and Thelma Riddle, Jewellers	176
Arthur Daniel and Hugh Lello, Director of Showyard and Departmental Steward of Cattle	177
Charlie Richards, Departmental Steward of Horses	177
Adrian Clifton-Griffith, Departmental Steward of Sheep	178
Donald Dunstan, Departmental Steward of the Grandstand	178
Shirley Parkyn, Office Staff	179
Roy Meagor and Mike Wilson, Office Staff	179
Myra Langdon, Office Staff	180
Ashton White, Cattle Commentator	180
Raymond Brooks-Ward, Main Ring Commentator	181
Reg Roberts and Alan Oliver, Showjumping Judge and Course Builder	181
Antony Ball, Farrier	182
Owen Pearce, Veterinary Surgeon	182
Don Waterhouse, Flower Show Chairman	183
Flower Show Committee	183
Maurice Lobb, Cage Birds Section Secretary	184
Poultry Section	184
Marion Dyer, Fur Section Secretary	185
Bees and Honey Section	185
John Robilliard, Pigeon Section Chairman	186
Sue Emrys-Jones, Canine Section Chairman	186
Bill Vernon, Milk Steward	187
Gerald Radford and Dick Parkes, Plumbers	187
Children's Creche	188

PREFACE AND ACKNOWLEDGEMENTS

When I first suggested the publication of a history of the Royal Cornwall Show, 1993 and the bicentenary seemed a very long way off. However, as is its habit, time has sped by and the preparation of this book has taken rather more hours and effort than was at first envisaged.

There have perhaps been times when the possibility of seeing the results of this project actually in print seemed remote. I like to think that the enthusiasm for the advancement of the work of the Society/Association and for the good of the agriculture of our County, which is so evident in the words of the Minutes of meetings long past, has helped to spur me on. The detailed and vivid reports of shows in the county newspapers, describing scenes, recognisable but of a more leisurely period, do much to paint the picture.

The wonderful support and interest expressed in the show, not only today but also in years gone by, has helped considerably in the tracking down of many minor, but often significant, historical details.

Prior to the start of my researches, little was known of the early years of this most 'respectable and patriotic society', and apart from the Minutes, (complete from 1839) and two original documents dating from 1803 and 1804, in the possession of the Association, there was little else to go on. I would like to take this opportunity to acknowledge the great help given by the many people who have made this book possible and I apologise to those whom I omit to mention.

In particular, I would like to thank Christine North, Colin Edwards and the staff of the County Records Office at Truro, for their advice and the use of their excellent facilities and fund of knowledge. Also I must mention Terry Knight and the staff of the Cornish Studies Library at Redruth, who opened up a wealth of information, in the form of newspaper archives, to a very inexperienced but very grateful researcher. Cornwall should think itself lucky to have such facilities available. To this list must certainly be added the Royal Cornwall Museum, whose Director, Caroline Dudley and Library and Photographic Department have been most supportive and the provision of some wonderful photographs has in particular been most appreciated.

I must also thank the Chairman of the Association, Lord Falmouth, who with his great support of all things Cornish, guides the Association so ably. His permission to photograph the painting by John Opie of his ancestor, the Cornwall Agricultural Society's first President, and also that of the Devon bull 'Narcissus', was much appreciated.

Likewise, the permission given by Lt. Col. Sir Arscott and Lady Molesworth-St. Aubyn to publish the photograph of the painting of Sir William Molesworth, Bart., the original proposer of all that exists today.

In addition to the above thanks should go to the following for the kind provision of photographs for inclusion: Doreen Cattran, Mrs. Enid Pryor, Mr. and Mrs. N. Arthur, Mr. J Rowe (Jethro), Mrs. J. Julian, Mrs. J. Pollinger, Mr. T. Irons, Mr. G. R. Lacy of Newquay, Mr. T. R. Bradley of Redruth, The Western Morning News, Bodmin Museum and Penzance and District Museum and Art Gallery.

My grateful and sincere thanks also go to Shirley Parkyn and Myra Langdon, the Association's office staff, whose untiring work has brought about the compilation of the past results and entries as printed towards the end of this work. My father, Albert Riddle, Show Secretary for so long a period, has provided much fascinating detail and together with my mother and Roy Meagor, a stalwart member of the Show's back-room team, have proof read the pages that follow.

Lastly, I would like to express my gratitude to my wife Thelma, for putting up with the many months of odd working hours and general disruption to normal life.

If you as the reader of this book gain only a fraction of the enjoyment that I have had in writing it, then I shall feel my time to have been well spent.

CHRISTOPHER RIDDLE

Wadebridge
February 1993

HER ROYAL HIGHNESS THE PRINCESS ROYAL, GCVO
President for 1993 of the Royal Cornwall Agricultural Association
Photograph by Norman Parkinson

BUCKINGHAM PALACE

It is a particular pleasure and honour to be President of the Royal Cornwall Agricultural Association in its 200th year and to welcome this account of one of the oldest Agricultural Shows in the country. The Association's Secretary is to be congratulated for compiling and publishing this edition to commemorate the Association's life.

The original aims of "holding a meeting for the Exhibition of Breeding Stock and Agricultural Implements" has been fulfilled splendidly throughout these years, no mean achievement when we consider the changes brought throughout these two centuries. The men and women of 1793 would have been astounded by what is to be exhibited at the 1993 Show.

It is significant that in this era no Shows were held during the years of the two Great Wars, in 1916, 1917 and 1918 and from 1940 to 1946, when the energies of Cornish Farmers and Growers were devoted totally to help saving the people who live in these islands from starvation and that in the years of peace the Show has continued to flourish exceedingly, demonstrating the truth of the words, "Peace hath her victories, no less renowned than War".

This book reflects the works of Farmers, Growers, Landowners, Designers, Farm Workers, Machinery Builders and Breeders of Livestock and Horses, who have silently and through achievements, often unsung, made these incredible changes in our industry and the country since 1793. The "Royal Cornwall" mirrors their achievements.

Change and the necessity to adapt to change has been the hallmark of the countryman in man's struggle for existence. Nowhere is this shown more clearly than in Cornwall.

I know that Cornishmen and women and all those who live in Cornwall are justly very proud of the 'Royal Cornwall Show' and jealous of its countrywide reputation as one of the leading County Shows in the Kingdom.

I wish the 'Show' good fortune and prosperity in carrying out the aims of its founders in the years to come.

Anne

'This truely respectable and patriotic Society'
Cornwall Gazette 1804

1. From Little Acorns 1793—1827

Cornwall, although distant from the great centres of population and commerce, indeed particularly so in the late eighteenth century, has never been far behind when a new idea or fashion has been sweeping the country. In the last decades of the eighteenth century agriculture was seeing many important changes to fundamental ideas which had remained, in most part, unchanged for centuries. Many areas were seeing the formation of agricultural societies, made up of forward thinking landowners and others interested in the improvement of stock husbandry and cultivation methods. Cornwall was not alone in being keen to see the establishment of such an organisation, however, many similar societies foundered within a few years and Cornwall can be proud that it's own Society, has reached its Bi-Centenary in such splendid fashion.

Although its official formation took place in 1793, the true roots of the organisation can be seen being spread in 1792 and indeed if the correspondence of one William Lamport of Honiton to Sir Christopher Hawkins of Trewithen in July 1786 had come to fruition, a society would have been formed even earlier. In these two letters in the Hawkins' Papers, William Lamport is actively encouraging Sir Christopher to lend his support and to enlist that of his friends, for the formation of *'The Cornish Society of Agriculture, Manufactures & Commerce'*. In the first letter dated July 19th 1786, Mr. Lamport offers, if he was *'certain of being reimbursed for my time, trouble and the expense of my journey'* to come to Cornwall to help with the setting up of the above Society and even goes so far as to lay out, in great detail, his proposed wording for a first advertisement and constitution.

It is interesting to see that the proposed wording is very similar to that actually used some seven years later although no mention of William Lamport is ever seen again. In the second letter, dated only a day later, Mr. Lamport reiterates his desire to be of help in such an important venture, although is at pains to express his concern over his expenses. It does appear that his own financial position is not that sound and perhaps the Society would have been of great help to him! However, at the end of his letter he makes mention of the possibility that he *'would take an estate, in order to realise in practice a somewhat superior mode of husbandry to what I have hitherto observed in the County'*. Perhaps Sir Christopher Hawkins felt that such comment on the state of Cornish agriculture, however true in parts, was a little unwelcome! The offer, however, to *'send down some heifers of the Barnstaple breed'* and to *'engage to supply the County with a change of grain and other*

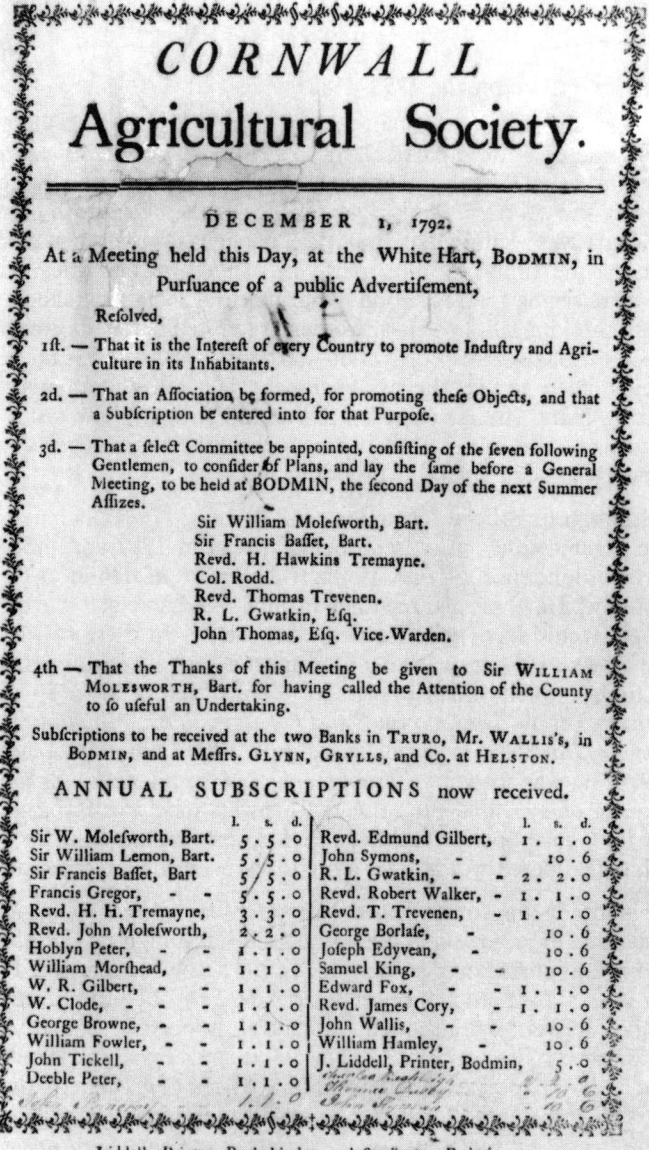

CORNWALL
Agricultural Society.

DECEMBER 1, 1792.

At a Meeting held this Day, at the White Hart, BODMIN, in Purfuance of a public Advertifement,

Refolved,

1ft. — That it is the Intereft of every Country to promote Induftry and Agriculture in its Inhabitants.

2d. — That an Affociation be formed, for promoting thefe Objects, and that a Subfcription be entered into for that Purpofe.

3d. — That a felect Committee be appointed, confifting of the feven following Gentlemen, to confider of Plans, and lay the fame before a General Meeting, to be held at BODMIN, the fecond Day of the next Summer Affizes.

    Sir William Molefworth, Bart.
    Sir Francis Baffet, Bart.
    Revd. H. Hawkins Tremayne.
    Col. Rodd.
    Revd. Thomas Treveen.
    R. L. Gwatkin, Efq.
    John Thomas, Efq. Vice-Warden.

4th — That the Thanks of this Meeting be given to Sir WILLIAM MOLESWORTH, Bart. for having called the Attention of the County to fo ufeful an Undertaking.

Subfcriptions to be received at the two Banks in TRURO, Mr. WALLIS's, in BODMIN, and at Meffrs. GLYNN, GRYLLS, and Co. at HELSTON.

ANNUAL SUBSCRIPTIONS now received.

	l. s. d.		l. s. d.
Sir W. Molefworth, Bart.	5 . 5 . 0	Revd. Edmund Gilbert,	1 . 1 . 0
Sir William Lemon, Bart.	5 . 5 . 0	John Symons, - -	10 . 6
Sir Francis Baffet, Bart	5 / 5 . 0	R. L. Gwatkin, - -	2 . 2 . 0
Francis Gregor, - -	5 . 5 . 0	Revd. Robert Walker, -	1 . 1 . 0
Revd. H. H. Tremayne,	3 . 3 . 0	Revd. T. Trevenen, -	1 . 1 . 0
Revd. John Molefworth,	2 . 2 . 0	George Borlafe, -	10 . 6
Hoblyn Peter, - -	1 . 1 . 0	Jofeph Edyvean, -	10 . 6
William Morfhead, -	1 . 1 . 0	Samuel King, - -	10 . 6
W. R. Gilbert, - -	1 . 1 . 0	Edward Fox, - -	1 . 1 . 0
W. Clode, - - -	1 . 1 . 0	Revd. James Cory, -	1 . 1 . 0
George Browne, - -	1 . 1 . 0	John Wallis, - -	10 . 6
William Fowler, - -	1 . 1 . 0	William Hamley, -	10 . 6
John Tickell, - -	1 . 1 . 0	J. Liddell, Printer, Bodmin,	5 . 0
Deeble Peter, - -	1 . 1 . 0		

Liddell, Printer, Book-binder, and Stationer, Bodmin.

Following an exploratory meeting in December of 1792, this poster was circulated announcing the intention to form an agricultural society for the county. The thanks of the meeting were given to Sir William Molesworth, Bart., for having called the attention of the County to 'so useful an undertaking'.

seeds of the best kind, (a change of seeds being of great importance for raising superior crops)' was I am sure something which would have found favour with a landowner such as Sir Christopher.

William Lamport, probably became interested in activities in Cornwall, through his involvement with the Bath & West, being recorded as present at their first Annual Meeting in December 1777. He also helped promote plans for an agricultural school and methods of raising potatoes from seed.

No further mention of this early attempt to create an agricultural organisation has been found, although it must surely have helped lay the ground for the meetings which were to follow in 1792/93.

In the winter of 1792 a more definite attempt to obtain the support for such an undertaking was made when an advertisement appeared in the Sherborne and Yeovil Mercury and General Advertiser. It is interesting to note that at this time no newspaper existed within the County and indeed was not to do so until 1801 when the Royal Cornwall Gazette was first published. Although published in Somerset, the weekly Sherborne and Yeovil Mercury, reported in detail on much that happened within Cornwall and was obviously a vital method of communication within the county. The first advertisement, which appeared on November 12th 1792, constitutes the seed for all that was to flourish over the next two centuries.

CORNWALL
AGRICULTURE and INDUSTRY
A Meeting will be held at the White Hart Inn, Bodmin, on Saturday the 1st of December next, for the purpose of taking into consideration a Plan for forming an Agricultural Society in this county: And it is earnestly hoped that all those gentlemen who wish to promote the objects of such a society will give an early attendance.
Dinner on table at two o'clock at 1s. 6d. per head.
Bodmin, November 5, 1792

The above advertisement appears to have attracted the attention of many of those active in agricultural circles within the county, as an account of this meeting was published on December 10th and a poster detailing the outcome of the meeting, a copy of which is held by the County Records Office, was printed for distribution throughout Cornwall. A series of resolutions were passed forming a select Committee to look into the question of the formation of an Association and it is at this time that the title the *'Cornwall Agricultural Society'* first appears. Seven gentlemen were to *'consider of Plans, and lay the same before a General Meeting, to be held at BODMIN, the second day of the next Summer Assizes'.* These seven gentlemen, consisting of Sir William Molesworth, Bart., Sir Francis Basset, Bart., Revd. H. Hawkins Tremayne, Col. Rodd, Revd. Thomas Trevenen, R. L. Gwatkin, Esq. and John Thomas, Esq. Vice Warden, with the exception of Col. Rodd and the addition of several others were to become the Society's first Vice President's the following summer.

Sir William Molesworth, Bart., of Pencarrow near Bodmin, whose encouragement in 1792 resulted in the formation of the Cornwall Agricultural Society. Painted by Northcott in 1785.

It seems likely that this first meeting was called by Sir William Molesworth, of Pencarrow, near Bodmin as thanks were given to him at the meeting for *'having called the Attention of the County to so useful an Undertaking'*, and it is therefore Sir William who should be regarded as the true instigator of what we know today as the Royal Cornwall Agricultural Association.

From the very start, it was intended that all of the county should be involved not merely those areas adjacent to the then county town. Subscribers were invited and the two Truro Banks, Mr. Wallis' office in Bodmin and Messrs. Glynn, Grylls and Co. at Helston were empowered to accept such subscriptions. Support, even before the Society had been officially formed, was soon forthcoming and by the time the outcome of the December 1792 meeting was published the sum of 47 guineas and 5 shillings had been received. This represented a considerable sum and must have greatly encouraged all those involved at this early stage.

Many names, well known throughout the county and beyond, were represented in this initial subscription list and gentlemen such as Sir William Lemon, Bart., of Carclew and Sir Francis Basset, Bart., soon to be created Baron de Dunstanville of Tehidy for suppressing the food riots in Redruth in 1796, were eager to become involved. Several other families, soon to be important in the development of the Society appear at this stage and include Francis Gregor Esq., the Revd. John Molesworth, William Morshead and Charles Rashleigh of Mennabilly.

As had been agreed, a General Meeting was held at the Assize Hall in Bodmin on July 31st 1793 and with Francis Glanville Esq., the High Sheriff in the Chair it was resolved.

'That an AGRICULTURAL SOCIETY be established, subject to such Rules and Regulations as have been proposed at this Meeting'.

It was further resolved that Lord Viscount Falmouth be elected President of the Society and that Sir William Molesworth, Bart., Sir Francis Bassett, Bart., the Revd Robert Walker, the Revd. H. Hawkins Tremayne, John Thomas Esq. Vice Warden, R. L. Gwatkin Esq., the Revd. Thomas Trevenen, W. R. Gilbert Esq., John Rogers Esq., the Revd. Jeremiah Trist, Edmund John Glynn Esq. and Mr Groffett be Vice Presidents.

The above gentlemen feature prominently in the formative years of the Society and many become President in due course. At this time, Vice President's were obviously regarded as senior officers of the Society, although, of course, now this title is only conferred on those who have actually stood as President, or who have perhaps rendered great services to the Association.

One of the great strengths of the Association has been the very loyal nature of it's supporters and from the very beginning in 1793 to the present day many Cornish families have had an unbroken connection. This can certainly be said of the Molesworth-St Aubyn family and of the Falmouths. It is fitting that the first President and the current Chairman hold the same title of Viscount Falmouth.

John Wallis, an Attorney from Bodmin was elected as the first Secretary and Treasurer to the Society. Wallis, born in Madron in 1759, became a Solicitor in Helston in 1783 before moving to Bodmin

CORNWALL
Agricultural Society.

At a General Meeting,

Held this 31ft Day of JULY, 1793,
At the ASSIZE HALL, in BODMIN,
Purfuant to Public Advertifement,

FRANCIS GLANVILLE, Efq. High Sheriff, in the Chair.

RESOLVED,

That an AGRICULTURAL SOCIETY be eftablifhed, fubject to fuch Rules and Regulations as have been propofed at this Meeting.

That LORD VISCOUNT FALMOUTH be PRESIDENT of this Society.

That Sir William Molefworth, Bart. Sir Francis Baffett, Bart. the Revd. Robert Walker, the Revd. H. Hawkins Tremayne, John Thomas, Efq. Vice-Warden, R. L. Gwatkin, Efq. the Revd. Thomas Trevenen, W. R. Gilbert, Efq. John Rogers, Efq. the Revd. Jeremiah Trift, Edmund John Glynn, Efq. and Groffett, Efq. be VICE-PRESIDENTS.

That JOHN WALLIS, Attorney, BODMIN, be TREASURER and SECRETARY.

That the firft Sub-Divifion Meeting be held at the RED LION, in TRURO, on TUESDAY, the 17th Day of SEPTEMBER next.

That the following PREMIUMS be offered.

	£.	S.	D.
To the beft Ploughman,		3	3
His Boy,		10	6
Second-beft Ploughman,		2	2
His Boy,			5
Third-beft Ploughman,		1	1
His Boy,		2	6

The Experiments to be made, and the Premiums adjudged, on the faid 17th Day of September, in a Field to be provided for that Purpofe, at TRURO.—The Ploughs to be entered at Twelve o'Clock precifely.

SUBSCRIPTIONS to be received at the two Banks in TRURO; Mr. WALLIS's, in BODMIN; and at Meffrs. GLYNN, GRYLLS, and Co. at HELSTON.

By ORDER of the Society,

John Wallis, Secretary.

Printed by JAMES LIDDELL, Printer and Book-binder, BODMIN.

A poster detailing the formation of the Cornwall Agricultural Society on July 31st 1793, and the notice of its first competition, a ploughing match to be staged at Truro, near the Red Lion in the following September.

The first President of the Cornwall Agricultural Society, elected 1793, The Rt. Hon. The Viscount Falmouth of Tregothnan, Truro.

in 1784 and taking over the business of Mr. George Brown. Over the years, John Wallis became a prominent member of the local community and was Mayor of Bodmin no less than seven times. Among the many offices he held was that of the Registrar of the Archdeaconry court (1795-1842) and Town Clerk (1798-1830). His success through his various business dealings enabled him to buy several of the Morshead manors when the estates were sold in 1808.

John Wallis' brother, Christopher, also practised as a Solicitor in Helston, and due in part to his involvement in the sale of Lord John Arundell's lands was able to purchase amongst other propery, Trevarno near Sithney. This link with the Estate has been preserved through the Presidency of the Association of the Bickford-Smith family who currently live at Trevarno. Mr. Peter Bickford-Smith was President in 1990 and his father, Michael Bickford-Smith in 1961.

Better known, perhaps was John Wallis' son, also John, who was Vicar of Bodmin from 1817 until 1866 and who was also a Mayor of Bodmin in 1822. The 'Registers' which he compiled covering the years

1827 to 1838 are important works of reference and a window in the West wall of the parish church commemorates his life as does the more recently named Wallis Road, named after him on the Berryfields estate in 1972.

The plans laid that day in the Assize Hall at Bodmin, would certainly have cheered William Lamport, whose efforts some seven years before had now been rewarded. Now that the Society had been formed, the need to turn good intentions into practical demonstrations of the use of such an undertaking was necessary and it was decided that the first 'Sub-Division' meeting should be held at the Red Lion in Truro, on Tuesday the 17th Day of September 1793. On this day it was decided that premiums (prizes) for ploughing should be offered for the following:

	£.	s.	d.
To the Best Ploughman	3	3	
His Boy		10	6
Second-Best Ploughman	2	2	
His Boy		5	
Third-Best Ploughman	1	1	
His Boy		2	6

It may seem strange to those who know the Show today that the first prizes offered should have been for ploughing and not for stock, but it can be seen that these prizes, which were indeed substantial for the time, created much interest. The winning ploughman, a labourer from the parish of St. Merryn, Matthew Toman and his boy Thomas Toman, were obviously very keen to win the prize offered, particularly when you consider the difficulties in travelling from St. Merryn to Truro in the late eighteenth century.

The second prize was awarded to a yeoman, Thomas Gartley of the parish of Clemence and to his boy Richard Lawrence, with the third position going to a Richard Harding, labourer of the parish of Kenwyn, with his boy Thomas Scobell. Sir William Molesworth, Bart. presided over the meeting that day and such was its success that a second *'sub division'* meeting was at once arranged to be held at the White Hart at Bodmin on Tuesday 17th December. The same premiums were to be offered once again with new regulations introduced as to the ploughing competitions to be held, with *'the experiment to be made in a ley field at Bodmin, where ploughs of any description will be admitted—the ploughing to be performed as for a wheat tillage—each to plough two ridges of 9 feet wide each, and 27 land yards in length'*. Dinner was to be available at the White Hart at 1s. 6d. a head and farmers were requested to attend *'for the purpose of communication and improvement on agricultural subjects'*.

Once again, a successful day was had and thanks given to those who had sent ploughs. The next meeting was scheduled for Liskeard, at the Kings Arms on Tuesday 18th March 1794. This time the *'experiments'* were to be made *'in a turnip field, or a broken arish field, as for the last ploughing of a barley crop'*. Again more rules were introduced, refining the competitions, and *'30 square yards of land, (about the fifth part of an acre) be the quantity assigned to each plough, which is to be ploughed in two or three split ridges, as the ground may admit of; and that one hour and half be allowed for finishing it'*. Entries had to be made at least a week in advance and it was announced that ploughing would begin at eleven o'clock and that *'dinner be on the table exactly at two'*.

It is interesting to see how over a few short months, the ploughing matches, as they would be known today, had developed and the Society appeared to feel that it was time to branch out even further. In February 1794 an announcement was made that the Society intended to offer premiums at a show to be held in either May or June of that year, for the best bull, best stallion, best ram and best boar at a time and place to be decided.

The meeting at Liskeard went ahead as planned and to show that all parts of the county were to be included, another ploughing match to be held at Helston, with a meeting at the Angel Inn was planned for 17th June. At the same time, full details of the very first livestock competitions were confirmed with the announcement reading as follows:

That in order to promote improvement in the Breeding of Live-Stock in this county, the following premiums be offered:

> For the Best Stallion, Five Guineas.
> Best Bull, Five Guineas.
> Best Ram, Three Guineas.
> Best Boar, Three Guineas.

Which are to be shown at Bodmin on Saturday the 24th day of May next, at eleven o'clock precisely.

That it be made appear, to the satisfaction of the inspectors appointed for that purpose, that the owner of each stallion, bull, ram, and boar, produced on this occasion, resides in the County of Cornwall, and has really and truly had it in his possession, and as his actual property, for at least three months previous to the exhibition.

By order of the meeting,
JOHN WALLIS, Treasurer and Secretary

This first show, attracting entries of 11 stallions, 8 bulls, 12 rams and 2 boars and the meeting held on the same day was Chaired by Robert L. Gwatkin Esq. of Killiow. The prizes, which represented

CORNWALL
Agricultural Society.

At a MEETING held at the ANGEL INN, In HELSTON,

On TUESDAY, the 17th Day of JUNE, 1794,

JOHN ROGERS, Esq. in the Chair.

The Premiums for Ploughing, this Day, were adjudged to

	£.	S.	D.
Francis Ferris, Servant to Sir Carew Vyvyan, Bart. best Ploughman.		3	3	
Joseph Martin, his Driver,			10	6
Richard Harding, Servant to R. L. Gwatkin, Esq. Second best Ploughman,		2	2	
Joseph Webber, his Driver,			5	0
Richard Bate, Servant to Mrs. Margery Bryant, of Camborne, Third best Ploughman,		1	1	
Henry Rule, his Driver,			2	6

RESOLVED.

That the Members be requested to attend at the next General Annual Meeting of the Society, to be held at BODMIN, on WEDNESDAY, the 20th Day of August next, (being the 3d Day of the Summer Assizes) when the Premiums for the ensuing Year will be settled, new Officers elected, and other Business of the Society transacted, agreeable to the Rules.

That those who intend to propose Subjects for Premiums, to the Society, are desired to communicate the Particulars to the Secretary, at least ten Days before the General Meeting, that the same may be properly arranged and submitted to the Consideration of the Committee appointed for that Purpose.

By Order of the Meeting,

JOHN WALLIS,
Treasurer and Secretary.

James Liddell, Printer and Book-binder, Bodmin.

A Poster recording the awards made at a ploughing match, staged near the Angel Inn, Helston in June 1794.

a considerable sum to those lucky enough to win, were duly awarded with John Tyeth of Launceston showing the best stallion, John Wevil of Lewannick the best bull, John Slyman of St. Mabyn the best ram and Thomas Hicks of Lanivet the best boar.

Even in these early days, those involved with the Society could see the coming benefits of the development of machinery and at this first livestock show a machine was exhibited by John Luskey of Northill *'for lessening the friction on two-wheel carriages going down hill'*.

The Luskey family, one of whom lived at West Castick, owned a considerable estate on the River Lynher near Trebartha, Launceston, home of Francis Rodd Esq. who was a strong supporter of the Cornwall Agricultural Society and all things connected with the improvement of agriculture. Again the Association's link with Trebartha still exists through Michael Latham, President for 1992 whose family now own this beautiful Cornish estate.

Following the next ploughing match at Helston, the first Annual General Meeting of the Society was called for Wednesday 20th August 1794 at the Kings Arms at Bodmin, to coincide with the Summer Assizes.

A detailed account of this first Annual General Meeting appeared in the Sherborne & Yeovil Mercury on 8th September 1794 with a full list of those who had subscribed in the first year being printed the following week. Lord Viscount Falmouth was once again elected President and a much extended range of premiums published.

It was agreed that again, two ploughing matches, one in Launceston in September and one in Helston in March, would be held with the same prizes being offered with the added condition that the drivers should not exceed the age of sixteen. An additional premium, for the best ploughman and his boy who were parish apprentices was also offered. The ploughing trials to be held in ley fields, as for a crop of wheat.

The livestock classes for a show to be held at Bodmin in June 1795 were to be encouraged by the provision of more premiums, with second prizes being offered for bulls and second and third prizes for rams, in addition to the original first prizes offered the previous year. The prize for best boar was only to be awarded if at least three were forward on the day and the rams were to be shorn on the spot *'under the direction of proper inspectors'*.

A class for sheep shearing was also to be introduced at the 1795 show, with a total of 3 guineas, 10 shillings and sixpence being offered as prizes with *'the sheep to be provided by the society—the shears by the shearers'*.

The last new competition instigated for 1795, which was to be the first of many similar awards was for turnips and was to be awarded *'to the person who grows the greatest quantity of turnips (not less than five* acres) in the autumn of 1795, properly hoed and managed in a husbandlike manner'. The prize for this new class was the magnificent sum of 5 guineas and anyone wishing to compete needed to enter by October 1st so that *'a proper person may be sent to inspect the crops, and make his report thereon at the next meeting of the Society, when the premium will be adjudged'*.

Several new Vice Presidents were elected to the Society and names such as Sir John St. Aubyn, Bart., Philip Rashleigh Esq., Francis Rodd Esq., Weston Helyar Esq., Edward Archer Esq., Revd. Thomas Penwarne and Edward Fox Esq., appear for the first time. A Committee was formed for the consideration of further premiums and it was agreed that the Rules of the Society together with the names of the subscribers should be printed.

The list of subscribers published on September 15th 1794 totalled some 106 people subscribing a total of 143 guineas and 6 shillings, quite a sum, when the average agricultural worker was earning 5 shillings a week.

Over the next months the various ploughing matches and the annual show were held with several names appearing in the prize lists which will be familiar to many. The continuity, which has for so long been a strength of both the county and its institutions can be seen. For instance, when Thomas key of St. Breock won a second prize with his ram at the Show in Bodmin in June 1795, little did he suspect that his descendent, Thomas Key of Bodellick, St. Breock would be a near neighbour of the Association's permanent showground in that same parish nearly two hundred years later. Many such long links exist and a study of the results lists will yield a host of such instances.

By February of the following year, 1796, the Society was already looking at taking an active interest in developments in other parts of the Country. A meeting was called in March to discuss the views of the Board of Agriculture and the petitioning of Parliament for a *'Bill to inclose the Waste Lands of this Kingdon'*. It had been proposed that a committee of Magistrates be elected to correspond with the Board of Agriculture following a suggestion to that effect from the Board's President.

The question of the enclosure of land had for many years created much hard feeling throughout the country, with many traditional lands grazed by those farming on a peasant basis, enclosed by those able to make best use of the legal system. It is said that from the mid eighteenth to the start of the nineteenth century, some 3,000,000 acres were enclosed, with a further 2,500,000 by the 1850's. Such enclosing of land on such a vast scale, was extremely devastating to those who depended on grazing a few stock on what had previously been open land. However, bringing the land into full and profitable cultivation played a vital role in providing food for the ever increasing, town dwelling, population.

Never slow to take positive action, the Society introduced what was probably one of the largest single prizes ever offered, with the sum of 21 guineas allocated to a class for the enclosure of land. This prize offered *'To the person who shall enclose and improve the largest Quantity of Waste Land, in the County of Cornwall, not less than ten Acres, in the best and most effectual Manner, within the course of 7 years'* was to run from July 1796 with the land to be in tillage by Michaelmas of 1798. A further prize of 7 guineas on the same conditions for an area of land not less then two acres, was also offered.

It is encouraging to see the interest taken by the Board of Agriculture in the state of Cornish agriculture, and in a letter to Sir Christopher Hawkins of Trewithen in September 1795, the President of the Board of Agriculture, John Sinclair, refers to the efforts being made by the Society and by Sir Christopher as one of it's Vice Presidents to *'procure, for the Board of Agriculture, a complete account of the Agricultural State, of so interesting a district as that of Cornwall'*. He goes on to say that he hopes to receive the required information in time for the meeting of Parliament and that Sir Christopher should not worry unduly, over the way in which the information is compiled as this can be taken care of at the Board by Mr. Young.

Arthur Young, who in 1793 had been appointed Secretary to the Board, following its formation that year, did much to inform and educate on a host of agricultural topics, particularly through the surveys which he introduced to examine in detail the state of agriculture in eight English counties.

The Society continued to prosper, holding ploughing matches and shows, and bringing those involved in agriculture within the county together to discuss possible improvements. By June 1798, when the annual show was to be held near the Assize Hall, Bodmin, various refinements and additions had been made to the prize list. A prize of three guineas was now to be offered for the best cow having had two or more calves, with two guineas being offered for the *'best made heifer having had one calf'*.

The sheep shearing classes had been split into two sections, with prizes on offer in open and Cornish only classes. A new section for fat sheep was also included with the sum of three guineas *'For the best Fat Sheep, slaughtered, regard being had to the live and dead weight, fed in Cornwall, in the ordinary way, on grass, hay, or turnips, of which a satisfactory certificate (if required) shall be produced'*. A prize of two guineas was also put up, with the same stipulations for the best fat sheep under three years old. The sheep in question were to be slaughtered on the evening prior to the Show ready for judging the following morning.

A printed prize list for 1804 shows the trouble exercised in the judging of such classes, with a detailed account being given of not only the live weight of the sheep but also of the dead weight of the four quarters, the skin, the head, the fleece, the tallow, the entrails and the blood. The winning animal belonging to J. P. Peters Esq. weighed in at 202 lbs., with the four quarters totalling 134 lbs. and 10 lbs. being tallow. The problems encountered in individually weighing the various parts of the carcase and in particular the blood must have been considerable, but the Society obviously felt it an important part of the competition. The knowledge gained from such an exercise was probably of great use when considering future breeding programmes at a time when such detailed information was rarely available.

Also seen at the 1798 Show were wool classes with a prize *'The the person who shall produce the heaviest entire Fleece of Wool, regard being had to the value per pound, the time of growth, and the weight of the animal producing such fleece'* of two guineas and a second prize of one guinea. Another new innovation in 1798 was the introduction of a class for a one horse cart, with the magnificent sum of five guineas for *'the person who shall produce a One Horse Cart, which shall be then most approved of for the strength, lightness and cheapness of construction, and for its general fitness to be used both on the fields and on the roads'*. Such a prize would have provided great encouragement to the improvement of farm vehicles, especially when it is considered that the average agricultural labourer's wage at the time was only five shillings a week.

In June of 1798 a name which is to appear prominently in the early years of the Society appears in an advertisement in the Sherborne Mercury. James Liddell, who described himself as a Printer, Bookbinder and Stationer of Bodmin and *'Seedsman to the Society'* announced to *'the Society and the county at large, that they may now be supplied by him with all kinds of prime NORFOLK TURNIP SEED* he went on to say that *'any other sort of seed, not immediately in the common course of cultivation, may be had on giving timely notice'*.

James Liddell had been born in Stepney in Middlesex in 1765, had been educated in Plymouth and was a printer in Bodmin from 1788. His wife, Mary, was third daughter of John Martyn a farmer of Trevithick, Lower St. Columb. Of their ten children (seven sons and three daughters), many stayed in the Bodmin area, with the third son, Silas Hiscutt Liddell carrying on the family printing business in Bodmin. Another son, Elias Hiscutt Liddell operated a wine, spirit and coal merchants business in Bodmin's Fore Street, whilst Thomas Liddell, farmed at St. Minver near Wadebridge.

By engaging, James Liddell as *'seedsman'*, the Society were making a effort to really alter and improve the state of Cornish cultivation,

Genuine Seeds.

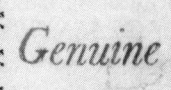

THE CORNWALL
Agricultural Society,

Being defirous of promoting the Cultivation of

GRASSES, and GREEN CROPS,

Which it is thought would be more atttended to,

If GOOD SEEDS could be obtained in an eafy and convenient Manner,

HAVE APPOINTED

J. LIDDELL, PRINTER, BODMIN,
THEIR SEEDSMAN,

For the Purpofe of fupplying Gentlemen, Farmers, and Others, with all Sorts of Agricultural Seeds.

J. LIDDELL, having engaged to fell no other Seeds but what fhall be purchafed by the Direction and Recommendation of the Society, the Public may depend on being fupplied with fuch Sorts only as are GENUINE, *and of the* BEST QUALITY, *and at fuch Prices as the Society may fix.*

LIST of SEEDS which may be had of J. LIDDELL

White Dutch Clover	Lucerne	Beft Norfolk Tranfplanted Turnip — Sugar Loaf and Round
Broad Clover	Burnett	
Trefoil	Buck Wheat	
Marl Grafs, or Perpetual Red Clover	Sainfoin	Round Green Turnip
	Ribwort, or Narrow Plaintain	Round White Ditto
Rye Grafs	Scotch Drum-head Cattle Cabbage	Round Red Ditto
Pacey's newly difcovered Rye Grafs		Ruta Baga, or Swedifh Turnip
	Turnip Cabbage	Wild Endive
Vetches, — Spring and Winter	Turnip Rooted Cabbage	Timothy Grafs

An advertisement for James Liddell, Seedsman to the society, circa 1800

as in the words of William Lamport some years earlier *'a change of seeds'* was of *'great importance for raising superior crops'*. A handbill in the possession of the Cornwall County Records Office details this appointment of James Liddell and states that the Cornwall Agricultural Society *'being desirous of promoting the cultivation of GRASSES, and GREEN CROPS, which it is thought would be more attended to, if GOOD SEEDS could be obtained in an easy and convenient manner'*. The Society went on to say that *'J. Liddell, having engaged to sell no other Seeds but what shall be purchased by the Direction and Recommendation of the Society, the Public may depend on being supplied with such sorts only as are GENUINE, and of the BEST QUALITY, and at such Prices as the Society may fix'*.

Such a move must have been welcomed by many and the seeds available at this stage included White Dutch Clover, which at the end of the eighteenth century was being treated as very much an experimental item. It was of course to be superseded by the indgenous wild white clover nearly a hundred years later. Other legunious crop seeds on offer included Lucerne, Sainfoin, Trefoil and spring and winter Vetches. Lucerne and Sainfoin had been introduced in the seventeenth century as seed crops but never really succeeded well in Cornwall.

Burnett, in the 1790's was being experimented with as a herbage plant but again proved to be of little use.

Ribwort or Narrow Plaintain was also available and together with the various clovers and plants such as trefoil, experiments in the planting of permanent pastures were being made. Until this time, permanent pastures had been an area of the farm regarded as best left alone, but with the availability of such new seed, attempts to improve on the quality of the pasture was something which could be tried and no doubt the talk at many of the Society's meetings was of such new-fangled ideas.

The *'Best Norfolk Transplanted Turnip'* either *'Sugar Loaf'* or *'Round'* and the green, white, red and Swedish or Ruta Baga turnips were much in evidence on this seed list and no doubt the Norfolk four-course rotation consisting of turnips, barley or oats, seeds (clover and rye-grass) and wheat were as popular by this time in Cornwall as in nearly every other part of the land thanks to the promotion of such a system by the Viscount Townshend better known today as *'Turnip Townshend'*. The turnip had soon gained in popularity as both a winter feed for stock and as a plant useful for cleaning the ground prior to other crops.

Further ploughing matches were arranged, this time to be held near the Ship Inn at Stratton and with this announcement in August of 1798 was mentioned that at a General Meeting of the Society it had been resolved that *'it would lend much to the improvement of sheep and wool to establish an* **Annual Fair or Market in this county, for the Sale of Wool**'. A Committee to look at how best to put this plan into operation had been formed and were to report to a meeting at Truro the following March.

An advertisement in February 1799 for a ploughing match at Truro asked all *'Dealers in wool and others interested'* to attend the meeting to be held on the same day as the match. Unfortunately, no further mention of the outcome of this meeting has been found although it will be seen that further attempts to improve the counties wool trade were again tried in future years.

The popularity of the Society's ploughing matches can be seen in a report of a match at Wadebridge in October 1799. Lord De Dunstanville (formerly Sir Francis Bassett of Tehidy) chaired the meeting, as President, at the King's Arms Inn in Wadebridge when the results were announced. Of the 34 ploughs entered, 29 had started including two worked by apprentices, with the top prize of 3 guineas being awarded to Thomas Burt, servant to Mr. William Hick of St. Minver near Wadebridge.

The dawning of a new century saw the classification for 1800 once again reviewed, with the livestock in Bodmin, extended again with a class for the ten best store ewes with a prize of 5 guineas on offer. St. Germans, in October 1800 was the venue for the autumn ploughing match, an area of the county as yet unvisited by one of the Society's events.

The Society was again looking to furthering its encouragement of good farming practices and premiums with this in mind were introduced. 5 guineas was the prize on offer *'to the person who plants, in the best and most husbandlike Manner, the greatest number of Acres of Potatoes, in Proportion to the Size of his Farm, (not less than Five)'*. A further prize of 2 guineas was then offered *'to the labourer who shall raise the best Crop of Potatoes, from a Quantity of Land, not less than one Quarter of an Acre'*.

The interest shown in turnip production was also further developed with the introduction of prizes for turnip hoeing. *'To the Man or Woman who shall, in the Summer of 1800, hoe the greatest Quantity of Turnips, not less than three Acres, in a thorough husbandlike Manner, and at least twice over, the Turnips to be left at proper and equal Distances'*. On offer was the sum of 3 guineas with a prize of 2 guineas for the boy or girl doing likewise.

Great store was set by the competitors judged on the farm rather than at the Show, with several further examples appearing. Water Meadows, which by the end of the eighteenth century were an extremely successful and useful component of the grazing lands of the country, were to be further encouraged through the provision of a 10 guineas prize. A water meadow could be defined as an area of land, which, due to winter flooding, was able to grow grass, often beneath the water, throughout the winter, thus providing early spring grazing for sheep. The Society's prize was therefore *'To the person, being the Occupier, who shall, after July, 1796, and before the year 1800, improve the greatest Quantity of Meadow and Pasture Land, (not usually overflowed in times of Flood) by throwing Water over it, in the most equal and Judicious Manner'*. The area in question had to be of not less than five acres and no doubt a vast number of differing designs of irrigation channels and waterways were constructed with this award in mind. The introduction of these meadows brought into being the person known as a *'drowner'*, skilled in the construction and maintenance of this complicated but effective system.

The growing of *'the best artificial Green Crop, of any Description, for Spring Feed'*, was also now to be included in the prize list with 5 guineas on offer, as were two prizes in connection with apple growing. The first award of again 5 guineas was *'To the Person who shall be the Occupier of an Orchard, (not less than one acre) well stocked with Cyder Fruit, in the best state of Improvement and Cultivation.'* The second prize, which marked a new step for the Society, involved the production of an essay, which would *'shew, from actual Experiment, the best Method of making Cyder'*. The prize for this essay amounted to 5 guineas with a further prize of 10 guineas on offer for an essay concerning wool production.

Obviously still attempting to improve the workings of the wool trade, this prize was designed to find some ideas as how to actually go about making sensible arrangements for the sale of wool between producer and merchant. The wording in the 1800 prize list offers *'To the person, who shall, in an Essay, point out to the Society in a satisfactory Manner, the different species and Qualities of Wool, grown in this County, with the best means to be adopted of promoting the Growth and Sale, of the best Quality'*.

As will be seen later, a vast gulf existed between those growing and those buying wool with little difference in the price offered for the best quality wool as opposed to the lower grades.

Bee keeping also came under the scrutiny of the members of the Society with 2 guineas offered to *'the Laborer in husbandry, who shall be the Owner of the greatest Number of Stocks of Bees'*, however the man's word alone was considered inadequate to gain the prize as *'a Certificate under the hand of the Clergyman of his Parish'* needed to be produced prior to handing over of the award.

The last new award to be introduced for the year 1800 was surely one of the strangest ever given by the Society, although in many ways, probably one that did a great deal of good in the long term. 5 guineas was offered for *'the Person having the greatest Number of Children, under*

21 Years of Age, brought up and maintained without parochial Assistance, and who before the first Day of March next, shall enter into any Friendly Society, and continue therein for at least six months'. This prize was to continue, with minor changes, for a great number of years and presumably proved extremely popular.

The Society's Seedsman, James Liddell, who during the year was actively promoting his service to the county, had extended the list of seeds available with several new turnips joining those previously stocked. Names such as Large Ox, Tankard, Spring Pasture and White Star turnip seed, all from Norfolk, now appeared with the offer of *'orders by post forwarded to any part of Cornwall, by coach, waggon or any other conveyance'*. Also available from James Liddell at this time was a range of bird seed including Hemp, Rape, Maw, Lint and Millet. No doubt the supply of such seed was in response to the popularity at the time of keeping cage birds and this probably proved a profitable side-line.

Over the next couple of years cabbage seed of various types including Cattle, Early York, Sugar Loaf, Early Dwarf and Cornish appears in Liddell's advertisements as do all sorts of garden seeds. Also available are *'steel Turnip hoes of various Forms, Reap-Hooks, Sheep-Shears, Garden Spades, &c.'*. Potatoes also made an appearance, with in 1805 German Kidney's being advertised being *'warranted earlier than any in common use in this county'* and *'very fine Painted Lords and Ladies'* also being available. Strangely enough the German Kidney's were to be sold by the Winchester bushel (approximately 8 gallons) and the Painted Lords and Ladies by the Cornish Bushel (a rather variable amount!).

By the 1801 Show, held in June near the Assize Hall in Bodmin, with new ideas and inventions ever appearing, an award of 10 guineas was made *'For a drawing or specimen of the best Threshing Machine in actual use, with an explicit reference to the person erecting the same, and the expense attending it'*.

This was a real step for Cornwall, as very few such machines were in existence at this time, particularly as far west as Cornwall. Believed to originate in Scotland in the 1780's, the threshing machine, although reasonably well used throughout Scotland by the turn of the Century was not so common in England, although a rapid spread was seen in the next 20 to 30 years. What could well have been the first steam driven threshing machine, was installed at Trewithen Home Farm in 1811. Designed by Richard Trevithick, the machine was to remain in place until 1879, when it was donated to the Science Museum.

By 1802 the efforts of the Society were receiving much attention from the publishers of the Cornwall Gazette, Cornwall's only newspaper at the time with in June, following the annual show a notice being printed to *'congratulate the Gentlemen of this Institution on the Improvements they have already effected in the Agriculture of this County'*.

By March of 1803, when a ploughing match had been arranged to take place near the Star Inn at Marazion, the prizes on offer to the competitors had been changed somewhat. Now instead of purely cash prizes, items of clothing were offered as alternatives to a cash sum. In the case of the Best Ploughman, a prize of either 3 guineas or a coat and waistcoat and for the second best ploughman 2 guineas or a coat. The reason for this change is unknown and it would seem unlikely that most of those competing would prefer clothing to cash!

Amongst the Association's records at Wadebridge is a framed poster, presented to the Royal Cornwall Agricultural Association in 1927 by Sir Hugh Molesworth St. Aubyn. The poster records the resolutions made at a general meeting held at the Society's Room in Bodmin on the 10th August 1803. The meeting was chaired by the Revd. John Molesworth and the resolutions include the introduction of a large number of new competitions, aimed at the active encouragement of improved cultivation. This fascinating document also provides full details of the prize winners at the two previous ploughing matches, at Bodmin and at Marazion and also of those gaining awards at the show in Bodmin of 7th June.

Also listed is a full list of those who had subscribed to the Society in its tenth year with a toal income from subscriptions being recorded of two hundred and one guineas. From this sum a total of one hundred and thirty three guineas and twenty shillings had been provided for prizes with John Rogers of Holwood and J. P. Peters Esq., exhibitors of the best bull and best ram respectively in 1803 receiving the increased sum of 10 guineas each. Mr. Peters of Creegmurrion is described in the *'General View of the Agriculture of the County of Cornwall'* as a man *'to whom the county is indebted for his various and excellent exhibitions of cattle and sheep'* and as someone *'who has not been equalled in either zeal or success by any competitor'*. The above book was drawn up and published by the Board of Agriculture & Internal Improvement in 1811 and provides a fascinating and detailed insight into the state of agriculture in Cornwall in the early nineteenth century and one to which many further references will be made.

The exhibitor of the best stallion, Nicholas Stephens of Bodmin, however, only received the sum of 7 guineas as although a prize of 10 guineas had been on offer, the animal had been *'adjudged as deserving only part of the Premium'*.

Although a prize in the sum of 3 guineas was won by Mr. Thomas Dungey of St. Ewe at Bodmin in June 1803 for the best boar, it would seem that the class was not well supported and such a class does not appear in the classification for 1804.

Presumably the increased wealth of the Society, and the fascination of the time with experiments and innovations led to the vastly increased range of awards on offer for 1804, encompassing many other aspects of agricultural production.

Premium No. 27 for 1804 constituted another new departure for the Society with a class under the heading *'Comparative Culture of Wheat'*. 5 guineas was offered *'For the best set of experiments made on not less than four Acres of Land, two of which to be sown broadcast, and two dibbled or drilled, to ascertain whether it is most advantageous to cultivate Wheat by sowing in the common broadcast way, or by dibbling it'*. One of the stipulations attached to this premium was the requirement *'that an Account of the Expence attending each Mode of Culture, with proper Certificates of the Nature and Condition of the Land on which the Experiments are made, together with an Account of the Produce of the Corn, be produced to the Society'*. Nearly two years were allowed for the *'experiments'* with the final report being required by 1st May 1805.

Another 10 guinea prize for a *'Course of Crops'* showed the long term nature of the thinking of the Society, with the award being made *'For the best Set of Experiments on the Course of cropping lands in Cornwall, to be made in 7 years, from the 25th December, 1803, on not less than six acres; which is to be in Crops of some Description within one Year from Christmas next'*. Again, at the end of the 7 years full details of the succession of crops, expenses and the produce gained were to be provided prior to any cash being handed over!

Also included were three prizes of 3 guineas, 2 guineas and 1 guinea for *'New or Improved Husbandry Implements'* for *'the person who shall produce to the Society, at their next Annual Exhibition of Cattle , any new or improved Husbandry Implement, superior to any in common Use in this County'*.

The 1804 Show was described by the Cornwall Gazette as having *'a very full attendance of gentlemen, farmers and others'* and that *'the shew of cattle was one of the finest ever assembled in this county, and exhibited the most incontestible proof of the advantages the county of Cornwall, and the country at large, have already derived from the laudable exertions of this truly respectable and patriotic society'*.

The evening before the Show, the members of the Society met in what was described as the *'Society's Room'* in Bodmin to discuss the premiums to be offered for the forthcoming year. This reference to a room belonging to the Society presents something of a mystery, as no record appears to exist of such a property although in the accounts of 1804 the sum of 5 guineas appears as *'one year's rent of Society's Room'*. Mention is also made of the fact that any models of new or improved implements etc., which were exhibited at the Society's Shows, became the property of the Society and would be

A Poster giving the 'Premiums and Adjudications' of the annual show at Bodmin, staged by the Cornwall Agricultural Society in 1804. This document was kindly presented to the Association in 1930 by George Johnstone of Trewithen.

'kept at Bodmin, for the Inspection of Farmers, Manufacturers, &c.' This implies that a room was rented by the Society throughout the year and not simply on the occasion of the various meetings.

Only one new class was introduced at this meeting, although it again marked a new venture for the Society. 5 guineas were to be offered *'for the best specimen of cheese made in this county'* with *'the quantity made not to be less than 112 lbs.'* An interesting, but confusing comment made in the Cornwall Gazette to accompany this announcement was as follows *'There is but little cheese eaten in Cornwall, nor will it perhaps come into general use, till beer as a beverage, shall be substituted for grog'*. It is strange that cheese was so unpopular in the County as one of the great cheese producing areas, Somerset was not that far away, and indeed by the late eighteenth century, much cheese and butter were being exported to as far away as America.

This prize, awarded at the Show at Bodmin in May 1805 was won by a Mr. R. Lean of Blisland, near Bodmin.

In October of 1804, one of the regular ploughing matches was held, this time at Padstow, when following the official proceedings *'the Sea Fencibles were drawn up in a field adjoining, and went through the pike exercise with great precision'*. This volunteer force, paid a shilling a day when on duty, numbered some two thousands in the county and were designated to repell invasion.

It was decided that in 1805 a Show should also be held in Helston, as *'the spirit of improvement, which of late years has been diffusing itself throughout this county in a more rapid degree than any other in the kingdom, has met with peculiar attention from the gentlemen and farmers of the neighbourhood of Helston'*. With this in mind and *'in order to encourage so laudable an example'* it was agreed that an annual exhibition *'not interfering however, with the general exhibition at Bodmin'* should be staged in Helston.

August 6th was picked as the date for the first Helston exhibition with a schedule of 8 classes with a total of 22 guineas prize money. The usual livestock classes included a class for 2 year old *'fat wether sheep'* to be slaughtered at Helston on the day before the exhibition, and with *'Dinner to be on the table at two o'clock'* at the Angel Inn all was set for yet another expansion of the Society.

The report of the days events records nine bulls, eleven rams, four boars and two fatted sheep as being produced, and with the attendance of Mr. Rodd, the Society's President together with the 60 people who sat down to dinner, the day *'went off with general satisfaction'*.

Such was the regard with which the Society was held by the Cornwall Gazette that at the end of August in 1805, when details of the competitions and premiums for 1806 were announced, that a very generous offer was then made by the owners of the newspaper. They stated that they *'cannot better promote the laudable views of this excellent institution, than by giving publicity to the premiums it holds out for the promotion of agriculture and industry throughout the county'*. The Gazette went on to say that *'we shall be happy at all times to circulate thro' the medium of this paper, free of any expence whatever, any communications (not chargeable at the Stamp-Office with the duty on advertisements) the knowledge of which may conduce to the improvement of our native county'*. Such an offer must have been gladly received by the members, but how long this assistance existed is unfortunately unknown.

One aspect of the Society's events and indeed those of the Association today, which has not yet been mentioned are the judges, or inspectors, as they were known at the earliest exhibitions. Unfortunately only a few of the inspectors at the first shows are mentioned by name, but it does appear that a system for choosing them, whilst impracticable today, worked well at the time.

Entries for the various classes needed to be with the Secretary, John Wallis, by fourteen days prior to the exhibition. It appears that once these details were submitted, judges were invited to officiate depending on who had entered in the various classes or in the wording of the time *'in order that inspectors as impartial as possible may be procured in time'*. Such a system has many advantages although in 1992 we see judges needing to be invited up to a year in advance, due to the demands made on the time of many of them.

A report of the 1806 exhibition, published, as promised, by what was now known as The Royal Cornwall Gazette & Falmouth Gazette or General Advertiser for the Western Counties on 31st May gives such full description of the scene at Bodmin as to be worth quoting in full;

'The exhibition of livestock at the Cornwall agricultural meeting at Bodmin, last Tuesday, was greater than upon any former occasion. There were 20 bulls and rams entered for the premiums, and many of each sort of a superior breed. Mr. Peters' Devonshire bull, taken in all points is undoubtedly the first animal of the kind ever seen in this county. The rams were, in every respect, deserving of high commendation; and, by the number exhibited from different parts of the county, it appears that the farmer begins to see and heed the advantages to be derived from the improved breed. The slaughtered sheep designed to shew that good breed will, in every instance, claim superior merit; and by comparison of the wether hogs with sheep of two, three, or more years, it is evident, all circumstances considered, that it is more advantageous both to the farmer and the public to kill sheep of one year old than those that are older. Sheep shearing, from the specimens produced at this meeting, seems to have arisen to a degree of perfection here equal to any county in the kingdom, and is an improvement of the highest importance. Several shearers who came from Devon, were prevented from shearing, by their names not being entered in time; but we understand some regulation in that respect is to be adopted for the future.

The husbandry implements produced by Mr. Worgan of Glynn, attracted the attention of a great many agriculturalists, and from their ingenious construction and the easy draught with which they may be worked, appear to be deserving of further consideration.

The secretary exhibited a correct model of a beautiful Devonshire ox in harness, one of the team of the Duke of Bedford at Woburn, in order to shew the advantages of oxen drawing in this manner, over those in yoke; and that neat cattle may, in many instances, supply the place of horses, in drawing carts and other carriages.'

A great number of gentlemen and farmers from all parts of the county attended the meeting, and upwards of 100 dined together in the Assize Hall, where the premiums for the next year were announced. The business of the day concluded much to the satisfaction of all present, and many new members were added to the society.

Such a detailed account gives a vivid picture of the scene in Bodmin in 1806, and whilst the numbers of livestock are not great by the standards of today, when one considers the extreme difficulties which must have been experienced in transporting livestock from various parts of the county, the effort made by many to be present was not inconsiderable.

The fat wether classes had already shown the members the advantages of slaughtering sheep as lambs rather than as older animals, but the suggested use of oxen as carriage animals seems rather less practical!

In the list of premium winners also published, details were given of the implements shown by Mr. Worgan of Glynn. Mentioned are the Great Cultivator, the Lesser Cultivator, a Shifting Double Plough and the united Rollers and Harrow. Little else is known of these implements or indeed if they were fit for the purpose for which they were designed. The early 1800's saw constant changes in the area of farm implements and it appears that Cornwall was providing its fair share of new ideas. The rollers mentioned, may have been of the new cast iron type, which first appeared around 1800, rather than those made of the more traditional stone or wood.

George B. Worgan of Glynn was obviously a man, well known in agricultural circles as when, in 1811 the *'General View of the Agriculture of the County of Cornwall'* was published by The Board of Agriculture and Internal Improvement, it was this same Mr. Worgan who had drawn up the various reports, although it appears that much re-writing was required before it was finally finished.

Quite how Mr. Worgan became involved in agriculture is something of a mystery, as he started out as a surgeon, and in fact accompanied the first convict ships to Australia in this capacity. After apparently experimenting with the growing of vegetables in Australia, and meeting with little success, he returned to England and worked as a school teacher in Liskeard. His next move was into farming, which unfortunately proved to be an unwise change of career, as this venture also apparently failed. His enthusiasm for agriculture and agricultural improvement seems to have outweighed his abilities and he eventually committed suicide, at the age of 80. However, his name lived on in connection with probably the most complete study of the agriculture of our county, although how much of his initial work was published unaltered is open to debate.

This fascinating book, the compilation of which was overseen by the Revd. Robert Walker, the Revd. Jeremiah Trist and Charles Vinicombe Penrose, provides an extremely detailed insight into the state of Cornish agriculture and rural life in general. Although published in 1811, the book obviously took several years to compile, and the foreword, written in November 1808, provides an interesting record of the high regard felt for the agricultural society and also outlines the broad principles of what those involved were aiming to achieve. *'Since the establishment of that excellent institution The Cornwall Society for the Encouragement of Agriculture & Industry in the year 1793, no county in England has, perhaps, advanced more in Agriculture, nor exhibited more striking proofs of the beneficial effects to be derived from gentlemen of property and consequence, and of liberal and enlightened minds uniting with the practical respectable Yeomanry, in promoting rural industry and improvement on the best principles'.*

The book is illustrated with several interesting engravings of livestock, many belonging to those actively involved with the society and provide a picture of the type of stock to be seen at the society's exhibitions.

The Bodmin show of 1807, although not attracting such large entries of stock, did provide the opportunity for the display of a model of a threshing machine, operating it seems on a new principle, although apparently the inventor was unwilling to provide too many details. It had no drum and no extra power was required for its operation when corn was being passed through it and that the idea was equally applicable to hand or horse driven machines. What a shame, no further details were reported! A feature of the exhibition was the shearing demonstration by Mr. John Serle of Trehire in Lanreath, who sheared a standing sheep *'without risk or inconvenience to the animal'*. This method was given much praise, and found to be much easier, particularly when shearing fat and heavy sheep.

The second show of livestock in 1807 was appointed for Stratton in late July, with for some reason Helston, being missed for that year. The show was held near the Ship Inn, with the customary meeting taking place afterwards at the hostelry.

The meeting held during the 1808 show at Bodmin saw representations being made for further study of the improvement of the county's wool trade with the following paper being presented and read; *'We whose names are subscribed, conceiving that great advantages are likely to arise to the GROWERS & DEALERS in WOOL, by establishing a WOOL FAIR in Cornwall, do agree to send our WOOL, or samples of Wool, for sale at St. Lawrence Fair, in August, and that the Wool be sold there at 16oz. in the pound'.*

The above was signed by 43 people, many already involved with the society, and following discussion it was resolved that such a market would be of great benefit to the county and a committee was formed to look into the details of the event. A market at St. Lawrence near Bodmin, was in fact held until relatively recently, and the location was obviously felt to be a reasonably central one.

It was later decided that an August fair was too early in the year and the dates for two fairs were fixed, one to be held at Summercourt in September and the other at St. Lawrence in October. It was further agreed that the sample of wool produced by any one person should be at least four fleeces.

The subsequent reports of this fair at Summercourt vary greatly, depending on who is reporting on the events of the day. First reports record a vast amount of wool on offer, whilst later ones argue this point hotly. Mention is also made of the fact that no wool was actually sold at this first fair, whilst elsewhere is stated that much wool was sold, but with no great difference in price being paid for best quality wool as opposed to the courser.

It appears that the *'wool-staplers'* or dealers were strongly opposed to such a system of dealing in wool, and were particularly against a higher price being paid for wool of a finer quality. One bone of contention was also the society's aim to sell wool by 16 ounces to the pound, as in most other county's, rather than by the traditional Cornish method of 18 ounces to the pound. One interesting point that is mentioned is the fact that in most counties, other than Devon and Cornwall, the wool was washed prior to the sheep being shorn.

The conflict between buyer and seller continued for some time and many would argue until the present day, but the society's fairs, mainly at St. Lawrence, Bodmin continued for some time and presumably provided a much needed service to the farmer. A meeting held by the wool buyers at the Kings Arms Inn in Bodmin in 1810 however, shows the buyers threatening legal action against any grower tampering with his wool by contaminating it with tar, paint, stones etc. in an effort to increase its weight. No doubt the wrangling continued!

The 1809 premium list sees the inclusion of a new award for the person *'who shall ascertain by actual experiment, the cheapest mode of feeding Pigs between Michaelmas and Lady-Day, the weight and value of the Pig when put to feed and when killed, the time of feeding, the nature, quantity and price of the food used, to be satisfactorily certified to the Society'.* Such a competition appears to suggest that the practice of simply producing a pig as fat as possible was starting to be questioned, and that perhaps it was time to look at the cost of feed consumed and the possible advantages of slaughtering at a lesser weight.

During the meeting at the 1809 exhibition, the Secretary, John Wallis, showed those present a flexible tube, which was recommended by the Board of Agriculture as an effective way of *'relieving cattle and sheep when choked by turnips, or blown by eating too voraciously of clover or other succulent food'*. This device had been invented by a Mr. Eagu of Surrey, who had received the sum of 50 guineas from the Society of Arts, for his efforts. It was decided that the instrument should be recommended by the Cornwall Agricultural Society. An advertisement subsequently appeared from W. Corfield of Penryn, stating this recommendation and offering for sale the said instrument together with *'proper directions for the safe and successful use of it'*.

The 1812 exhibition at Bodmin saw the now well established range of livestock and shearing classes, however the wool classes for this year saw a small but interesting change. Instead of simply two classes for ram and ewe fleeces, the 1812 premium list stipulated that the fleeces should be from Spanish sheep. This would almost certainly refer to the Merino breed, which had been seen in Britain as early as the Middle Ages, in small numbers, but which, with a ban on exports from Spain, on pain of death, having been in existence for many years, few were seen in our countryside. The Napoleonic wars had seen the lifting of this export ban, and even in 1811 the formation of a Merino Society. It was to be seen in later years that this breed was not totally suitable for British requirements and therefore numbers dwindled drastically.

However, in 1812, interest in the breed was strong, and at the May meeting at Bodmin, cloth manufactured in Cornwall from Merino wool was exhibited by its makers Messrs. R. E. and T. Pearse of Camelford and it was reported that its *'texture, fineness, and finishing did great credit to the indefatigable exertions of Messrs. Pearse to establish and improve the manufacture of Broad and Narrow Cloth in Cornwall'*.

Such was the enthusiasm shown by those present that it was resolved by those members to appear at the next Annual Meeting in coats made of Cornish wool, manufactured in the county. Unfortunately the account of the next meeting makes no mention of whether or not this resolution was carried out!

1813 saw further premiums offered for crop experiments, with 5 guineas for *'the Best Crop of newly-introduced Spring Wheat, not less than two acres'*. The growing of spring wheat is mainly a twentieth century innovation and these experiments in the early years of the 1800s must have been a great novelty.

The second new prize for a *'person who gets his living as a Rack Renter'* was for *'the best Crop of Turnips after Wheat, or other grain'* of *'not less than three acres'*. The use of turnips was still being greatly encouraged and continued to feature in various of the Society's premiums.

Later that year, in July, Callington was chosen to host the second of the year's exhibitions, with a dinner and meeting to be held at the New Inn after the judging of the stock. The area supported the day well and due to the close proximity to Devon, far more shearers than usual took part in the shearing competitions. Some 100 people partook of the *'ample dinner'* provided, with E. W. Stackhouse, Esq., officiating in the absence of the Society's President Sir Arscott Molesworth. The social side of the Society's gatherings, were always well attended and the extra income generated must have gladdened the heart of many inn keepers. May 1814, for instance, saw Mrs. Jewell of the King's Arms in Bodmin provide dinner *'in the old English stile'* for over 100 people.

By this time, the livestock classes include several premiums for cattle and sheep of varying ages, but the awards for pigs and horses, never particularly well supported, had in the main, been dropped. Occasional prizes for cart horses were offered, but not on an annual basis.

Always keen to promote new ideas, the Royal Cornwall Gazette of June 1814 included an advertisement by an N. Sleeman, Druggist of Truro for *'a sure PREVENTATIVE against the FLY striking SHEEP or LAMB'*. Apparently this new concoction had been endorsed by the principle members of the Society, who had presumably given permission for their endorsement to be used in the press etc. The preparation was said to *'bid defiance to the warmest or wettest season, no Fly ever touching the Sheep after it had been used'*. For such a guarantee to be given, Mr. Sleeman must have had great confidence in this mixture, although one wonders if the 100 5lb. barrels of Prime Catalonia Anchovies, mentioned in the same advertisement, would have sold quite so well!

James Liddell was also still avidly promoting his business as Seedsman to the Society with a range of Norfolk turnip seeds and also *'Winter Tares'*, a type of vetch, presumably used for winter green crop, which he advertised in April of 1815 as being considerably cheaper than the previous year.

The show of 1815 saw probably the largest number of stock yet, but the year is one that appears to have produced several improvements in areas other than stock. Admiral Luke is recorded as presenting a Hainault Scythe to the Society, which *'attracted much attention from its novelty and singular construction.* John Wallis, the Society Secretary also exhibited a rat-trap, which had been made according to the directions given by a Mr. Broad in his *'Disclosure of the Method of attracting and destroying Rats'*. Apparently the vital component was a *'small quantity of the Oil of Carraway applied in a very simple manner'*. It was agreed that descriptions of both these implements, together with details of information from the Bath and West of England Society concerning letters written by a Mr. Craig on the subject of the use of clay ashes as manure, should be published in the Society's next book of premiums.

Also exhibited was a hoe, the property of a Mr. Martyn of St. Columb. The frame was said to be made of wrought iron with the implement being workable by either a horse or a single person. Useful for hoeing and banking turnips, potatoes or other vegetables, it was *'much approved'* by those present.

The exhibition of 1817 also saw several implements being shown, with a Mr. Rennie exhibiting a turnip drill, an iron plough, a horse hoe and various other *'husbandry implements'* said to be of Scottish construction. 1817 also saw the first visit by the Society to Liskeard with one of its exhibitions of livestock in June of that year, with a local man, Samuel Serpell of Liskeard winning the sheep shearing.

The Bodmin show of May 1818, being not so well attended, prompted the Society to alter the way in which future show dates were to be fixed. Previously, the date of meetings had been regulated by the Whitsun Fair, with the show being held on the *'Tuesday next before Whitsunday'*. It was now decided to fix on the first Tuesday in June for the main annual exhibition and meeting.

It was further agreed that the next Bodmin show would include a public sale of *'Live Stock, Husbandry Implements of an improved description, and other articles of Farm Produce'*. These items could either be sold by auction or by private contract and would be advertised to the public at a cost of 1 shilling per lot to the vendor.

The idea of including a sale with the exhibition never really caught on, although at the first sale some 4 2 year old rams, 6 hog rams, 4 rams to be lent for the season, 2 bulls, 8 heifers and several Jersey cows and heifers were entered for sale. Unfortunately no record of the other items sold exists. The show that year also saw a demonstration by a pair of working, six year old, steers, owned by a Mr. R. Retallack of Liskeard. The steers were said to have performed the *'usual work of a farm of 60 acres'* for three years, and then had been fed up since the last harvest. One of them then weighed 12 cwt. and the other nearly 11 cwt.

A proposal for the publication of a *'Cornwall Agricultural Magazine in quarterly numbers'*, receive the *'unanimous approbation of the company'*, but no record is known of such a publication ever having been circulated.

A prize of 10 guineas won by Mr. Rodd of Trebartha, Launceston for his bull at the 1819 exhibition, culminated in the revival of the ploughing matches, which had been discontinued for some years. Mr. Rodd's generous donation of his prize money for use as prizes for ploughing, caused a date to be fixed for a match at Bodmin in October 1819. The resulting competition was said to be one of the best ever staged by the Society with eighteen ploughs out of the twenty four entered starting.

Various interesting combinations were competing with seven of the ploughs being drawn by pairs of horses without drivers. Five *'Iron Scot's Ploughs'* were in use, three of which were provided by Mr. Rennie of Egloshayle, Wadebridge. It was said *'the ease and dispatch with which these Ploughs perform their work, strongly recommend them to the attention of Farmers'*. The umpires for the day were John Permewan, John Lean and William Bicknell with the top premium of 4 guineas for the Best Ploughman without a driver going to William Dunn of Fowey, and the 3 guineas for the Best Ploughman with a driver to Thomas Hawkey of St. Breock.

The next few years saw a progression of exhibitions, mainly in Bodmin, for the most part following the pattern of previous years. Attendances remained good and the interest in stock improvement continued to grow, although the first flush of enthusiasm for all things agricultural wained somewhat. Perhaps many felt that agriculture had reached its peak of improvement and in many ways this could be said to be true. A farm in 1820 would not have looked so different to one in the later part of the nineteenth century, although the picture in the last years of the eighteenth century, when the Society was in its infancy, was of course, very different.

Such was the improved state of things that after the 1825 exhibition, it was decided that sheep shearing classes should be dropped from the premium list, as in the words of the newspaper report sheep shearing *'has arrived to such a state of perfection'* that presumably it was felt unnecessary to further promote its improvement.

The gentlemen guiding the Society through its first thirty years, could have looked back proudly on their achievements and the real developments that, through their encouragement, had occurred in the many facets of Cornwall's agricultural scene in that time.

2. An Amalgamation   1827–1857

The war with France dominated the early year's of the nineteenth century, although whilst war still raged, farmers prospered, with the price of corn, due to the blockades of imports, rising from 43 shillings a quarter prior to war breaking out to 126 shillings a quarter in 1812. When the war was finally over, prices slumped, and the farming population and those who relied on the incomes of the farmers, foresaw very hard times ahead.

To help regulate the situation and provide farmers with a reasonable return, the Government introduced the Corn Laws to artificially support the price of corn at a fairly high level. This move, whilst popular with many living in the country was quite the opposite with town dwellers, and a state of antagonism soon grew up between the two. The Corn Laws were to remain in force until 1846, when they were finally repealed and a free market was to exist.

The main problem facing the agricultural community in the period after the end of war, was the vast amount of imported foodstuffs and other goods appearing in Britain. The main cause for concern were the lack of taxes on imported goods, most of which could be imported free of tax, with the resultant loss of interest in home produced goods.

To help bring the situation to the attention of the Government, an Agricultural Association was formed which met at Henderson's Hotel, Palace-Yard, Westminster in January and February of 1819 under the guidance of one George Webb Hall, Esq. This Association, appearing to relate most closely to the National Farmers' Union of today, encouraged the setting up of similar organisations in each county of the land, with sub branches in the major towns of those counties.

As in 1793, with the formation of the Cornwall Agricultural Society, Cornwall was swift to act and on the 20th March 1819 at a meeting at Bodmin, the Cornwall Agricultural Association was formed.

It was agreed that quarterly meetings, to be held at Jewel's Hotel, Bodmin, should be called to be held on the second Saturday of June, September, December and March. It was further agreed that the formation of similar Associations in the market towns of the county should be encouraged and that an annual meeting with these branch Associations would be beneficial. The first meeting was chaired by John Martyn Bligh, with John Hooper of Penhergard in the Parish of Helland being elected Treasurer, with the power to accept subscriptions to help fund the running of the Association.

Truro was quick to follow the lead and at a meeting at Pearce's Hotel on the 19th May 1819, the Truro branch of the Cornwall Agricultural Association was established under the chairmanship of John Penhallow Peters of Creegmurrion. This meeting further encouraged the formation of more Associations and resolved that *'the formation of such Associations throughout the Kingdom, is necessary in order to obtain extensive co-operation amongst Agriculturalists, and thereby such Constitutional and Legislative protection as is essential to their welfare, and that of the Community at large'.*

Again a fund was begun to receive subscriptions with George Simmons of St. Erme and Matthew Doble of Probus acting as Treasurers.

The June meeting of the Cornwall Association, met as agreed at Bodmin, and heard a petition read which after circulation throughout the county was to be placed before the House of Commons. The petition dealt in the main with the effect the importation of foreign corn and other goods was having on British agriculture, particularly in view of the lack of import taxes. The petition laid out in detail the hardships faced by the farmers of the country and asked for help in protecting the country's agricultural production. The point was made that the country could be virtually self sufficient in most foodstuffs, if the encouragement was given to the producers.

It was agreed that the petition should be circulated by the members of the Association, prior to its being sent to London.

This meeting also saw the formation of a Committee *'appointed for forwarding the views of this Association'* to draw up a set of rules and regulations for submission to the next general meeting. This Committee was to consist of the Chairman, John Martyn Bligh Esq., the Treasurer, John Hooper Esq., John Penhallow Peters Esq., George Simmons, Matthew Doble, R. H. Andrew, W. Norway, Samuel Symons, Mark Guy, Martin West, John Cardell, R. Vincent, J. M. Andrew, Harry Hocken, John Pearse, James Philp, Thomas Bishop, Samuel West, John Grose, Robert Bake and George Martyn. This Committee forms the basis of the current 'Council' of the Association as it stands today and at the next meeting held at Bodmin in September John Penhallow Peters was elected first President and J. M. Bligh, first Secretary of the Cornwall Agricultural Association. It was also agreed that J. M. Bligh Esq. should chair the local meetings at Bodmin.

By October 1819 moves were being made to form a branch of the Association at Launceston, with a meeting being called to be held at the King's Arms Inn.

By the following March, when the quarterly meeting was held at Bodmin, it was found necessary for the Association to affirm their non-political position by disclaiming *'any connection or interference with any Political Party or Parties whatsoever'* and confirming themselves to be a *'loyal and constitutional Association to the King and Constitution of the Realm'*. Those involved in trades other than agriculutre, but who none the less relied on income from agricultural sources, were called upon to add their weight to the petitions placed before Parliament.

Progress was slow, and a meeting held at the Red Lion Hotel in Truro in January 1821 heard that the Select Committee of the House of Commons had been given very limited instructions for the way in which they should examine the whole question of import controls, etc. The meeting called for a further petition to Parliament, whilst thanking the Cornish Members of the House of Commons for their help so far.

By December, the possibility of the abolition of the Corn Laws and the introduction of *'a protecting duty'*, encouraged the Truro Branch of the Association to call a meeting to re-new their efforts in petitioning Parliament, *'while this all-powerful Body is in this happy humour'*. Little was to actually happen but the country gradually adapted and the gradual introduction of method to allow the country's farmers to feed the country's ever-increasing number of town dwellers began. Some of the new ideas, particularly those seen as replacing man with machine, caused ill feeling, but as time went on general acceptance of the new methods was gained.

The Cornwall Agricultural Association appears to have gone into a state of hibernation through much of the 1820's, although by the end of the decade a change of direction was to be seen.

In the meantime, the Cornwall Agricultural Society continued to stage its exhibitions, with it appearing that the last under the banner of the Society was held in Bodmin in 1825. Mention is made of a ploughing match to be held at Bodmin in October 1825, but no record of this event has yet been found and it must therefore be assumed that it was cancelled. The newspaper report on the 1825 show also mentions some proposed changes to the rules of the Society *'in consequence of the improved state of agriculture since the institution of the society'*.

Again, no record appears to exist of any such changes made although various developments can be assumed.

It would appear that the Cornwall Agricultural Society and the Cornwall Agricultural Association merged in 1827, with the main activities of the Association, as it was then known, being centred on Truro. It seems a shame that the older title of 'society' was the one chosen to be dropped, but the more modern term 'association', belonging to the younger, more go-ahead organisation was probably felt to be more appropriate. It is however noticeable, that the activities of the Association from then until the present day, followed the pattern set by the Cornwall Agricultural Society, rather than that of the Association formed in 1819. The petitioning of Parliament, was no longer mentioned, and agricultural promotion of a more local nature was foremost.

The move of activities from Bodmin to Truro is also significant, as although Bodmin was still the county town, Truro was the centre of much of the county's activities. Here were the Assembly Rooms and the fashionable town houses of the great Cornish families could be seen in Lemon Street and Boscawen Street. Truro also had the Royal Institution of Cornwall, founded in 1818 and still flourishing today and also its own Philharmonic Society.

Truro must have seemed the obvious choice as the venue for the meetings and shows of the Association, with so many of those closely involved being often there on business or for pleasure. However, Bodmin was not to be forgotten and in the years to come played host to the show on many occasions.

It should be remembered that the standard of the roads through the county at this time was fairly poor and an important consideration when choosing the time and venue for a meeting or event were the travelling arrangements necessary for those attending. A private coach or carriage was obviously the fastest method of travelling, with the regular coach services also being relatively efficient. It should be remembered though, that even as late as 1855 Kellow's 'Fairy' four-horse omnibus, leaving at 6.30a.m. from Matthew's Hotel in Camborne, did not reach Plymouth until 4.30p.m., after calling at the Red Lion Inn, Redruth, the Red Lion, Truro and Channon's Hotel, Liskeard. The single fare at this time being 14 shillings inside and 11 shillings outside.

The show of 1829, now firmly fixed in Truro, was described as being held in a meadow near the Western Inn, in Boscawen Street. The event was described as a *'shew-fair'* and seems to have coincided with an ordinary market, as prices realised for stock are described as being *'uniformly on the decline'*!

A wider range of prizes for livestock were offered, than in the previous few years, although no mention was made of awards for agricultural experiments, essays, etc. The total prize money offered amounted to £58, with the various prizes now being calculated in pounds rather than guineas. Obviously, these prizes represent quite a decrease in prize money on previous years, and indicate the state of the agricultural economy at the time.

Classes staged that may included awards for; the best-bred bull, the best-bred milch cow or heifer (no buss), (referring to a cow or heifer not rearing a buss calve), the best fed cow or heifer, worth most per cwt., the best pair of fed oxen or steers, worth most per cwt., best pair or working oxen or steers, best-bred ram, best bred hog ram, best-bred 10 ewes that have reared their lambs this season, best-bred 10 ewe hogs, best 10 fat sheep (wethers or ewes), best 10 wether hogs, best-bred boar, best-bred sow, best entire horse, calculated to improve the breed of horses for the saddle in this county and best mare, calculated to improve the breed of horses for the saddle in this county. The last prize offered, given by William Peter Esq. of Chiverton, was similar to many given over the years and constituted a prize for *'a labouring man in husbandry, that has reared the largest family without receiving parochial relief'*. This award, already mentioned as being one of the strangest ever offered, was to be regularly seen over many years to come.

The above classes were to form the basis of the competitions held for several years with small amendments and rule changes occasionally appearing. For instance, sheep being shown at the May show of 1830 were not allowed to have been fed on corn, 'pease' or beans and *'if any doubt shall arise thereon, the Owner will be expected to take an oath'*. Also all exhibitors of sheep, as a condition of receiving their prize money, needed to be in *'the room at Pearce's Hotel'* when the awards were announced.

The size of the shows soon began to increase, with in 1830 some 300 sheep being exhibited and with prices for those sold being at around 6d. per pound. Cattle although not quite so numerous, sold for between 40 and 50 shillings per cwt. Amongst those being exhibited were some *'beautiful cows, of most perfect proportion'* and *'an immense bull of the Durham breed'* from the herd of J. P. Peters of Crigmurrion.

The Secretary and Treasurer of the Association was now one John C. Downe, Esq., having taken over from John Wallis, who guided the original Society through its formative years since 1793.

1831 saw the re-introduction of a class for sheep shearing, with a total of £2 10 shillings on offer. Shearers were only to be allowed to shear one sheep, with the first prize in that year going to one John Andrew. An incident was reported in the press in connection with that year's event, concerning an accident befalling William Peter Esq. of Chiverton. Whilst on his way to the show, his horse fell whilst descending Lanivet Hill, causing Mr. Peter to be rushed to the home of a Mr. Hamley, a surgeon in Bodmin. After being bled and resting for some days, he was reported as being out of danger.

Following the establishment of the Truro cattle market, in which the members of the Association were much involved, it was decided by the Committee, in December 1831 to create a fund to which people could subscribe a maximum of one shilling. This fund was to be used to purchase *'a PIECE of PLATE'* to be presented to a Robert Glasson of the Seven Stars Inn in Truro. It was reported by the Association that Robert Glasson was instrumental in the creation of the market in Truro and that as an appreciation from the farming community, such a presentation would mark this achievement suitably.

Such presentations seemed a popular method of thanking those who had served the community well, as in 1833, it was suggested that a similar subscription, but this time of 2s. 6d. per person be collected to purchase a piece of plate for presentation to Mr. J. P. Peters of Crigmurrion in recognition of his great efforts in the encouragement of the county's agriculture.

The Truro show of 1833 saw the exhibition of a stallion 'Tiger', which had been sent from Glynn, near Bodmin, by Sir Hussey Vivian, Bart. This horse was said to be *'the best horse ever seen in Cornwall'*, and had been purchased to improve the breeding of horses within the county. The stallion was available for the servicing of mares, free of charge, and it was announced that 'Tiger' would be standing in Truro for a week, for the use of the owners of mares in that area.

Such a generous offer would have attracted many takers, and no doubt 'Tiger' spent a busy week in the town!

Following the usual range of livestock awards at the 1834 Show, the now traditional 'ordinary', being a set dinner, was held at Pearce's Hotel in Truro. Some 130 people sat down at 4.00 o'clock with the landlord providing *'with his usual spirit and skill a sumptuous entertainment for his guests'*. Amongst the fare on offer was *'an enormous round of beef weighing between 60 and 70 pounds'*, and if the report in the Royal Cornwall Gazette is accurate much of the rest of the afternoon and evening were spent in the drinking of toasts. The report details some 14 different toasts having been drunk, with *'other appropriate toasts'* then following. Included in this list was the toast *'Corn, Horn, Wool and Yarn'* given by the chairman for the day, the President Mr. Richard Johns of Trewince.

The evening was spent *'in much convivial enjoyment'* and no doubt many plans for future shows and competitions were aired.

Wool, always it seems, a subject of much discussion, was again the cause of great debate during the above dinner with yet another suggestion for a wool fair being promoted. Pearce's Hotel, Truro, was to be the venue, with on the 1st October 1834, the promise of some 60,000 lbs. of wool, *'in the yolk'* to be available for sale. Some 10,000 cwt. of wool *'in excellent order'* was offered for sale by Mr. A. Plummer, one of three *'extensive wool-dealers'* who were in attendance, plus other large amounts from various other sources. The price asked was 1s. 2d. a pound, although unfortunately *'no business whatever was transacted'*.

The steward of the Earl of Falmouth, who was present, is quoted as saying that he expected to receive a commission to purchase as much wool as was available at the above price, within a few days, and it was therefore decided to postpone the proceedings until the first Wednesday in November. No further report appears of this postponed market and it can only be assumed that yet again, the efforts to regularise the sale of wool met with little success.

The premium list for 1835 sees the introduction of an award which is still in operation today, although the competitive aspect of its original form has now been dropped. The award was *'to the Servant or Labourer in husbandry who shall have lived longest period in one continual service'*, with a prize of £2 for the longest serving person and £1 to the second longest serving.

In its first year the prize was awarded to Peter Buddle who had worked for about 50 years at Truthan for Edward Collins Esq. and his late father. The second prize winner was one Richard Dyer who had worked as a labourer for the family of William Phillips of Nance, near Illogan since Christmas of 1784.

1836 saw the introduction of a new set of classes, in addition to those usually staged. In order to encourage the showing of stock by all farmers, and not simply the large landowners, a section was now included *'for stock belonging to tenants, occupying Farms not exceeding £150 per year, getting their livelihood entirely by Farming'*.

Added to the premiums offered by the Association in the above classes, were prizes offered by Mr. G. W. F. Gregor of Trewarthenick, President of that year's show. The prizes already on offer for pigs, horses and sheep shearing were to be awarded for either the open or the tenants section or both, depending on the entries received for the respective classes.

In February of 1839, following the death of John Downe, a general meeting of the subscribers of the Association was held in Truro Town Hall. During this meeting William Floyd Karkeek was elected the new Secretary and those gentlemen who were to form the *'Committee of Management'* for the forthcoming year were chosen. The meeting was chaired by James Hendy of Tretherf, Ladock, Truro who was to be a stalwart supporter of the aims of the Association for many years.

The new Secretary William Karkeek, was a veterinary surgeon, who had taken over the business of his grandfather, William Floyd in 1825, a smith and farrier in Truro who had been in business for some fifty-five years. William Karkeek, based at No. 9, High Cross, Truro, advertised himself as not only a veterinary surgeon, but also a *'Professor of Animal Medicine in General'*; the horse shoeing side of his grandfather's business was continued with *'horses shod on an improved principle'*.

William Karkeek also started a register, for use by those wishing to buy or sell horses. For a registration fee of five shillings and 2½ per cent of the price if sold, he would advertise horses for sale on behalf of the various vendors. One wonders however, how scrupulously this businesss was operated as in an advertisement published in April of 1826 a grey half-bred Arabian mare by Bodkin was *'warranted only to carry colts'*! Such a guarantee must have been very much appreciated by the subsequent purchaser!

Following the appointment of William Karkeek, the executors of John Downe's estate were requested to pass the sum of £137 5s. 5d. to the new Secretary for him to place in a bank account. Then came the problem of choosing which bank to favour with the Association's business, which was settled by the drawing of lots! The bank thus chosen was the *'Cornish Bank'*, which after several take-overs and amalgamations, forms part of what is today Lloyds Bank plc, and with whom the Association still bank.

Little mention has been made so far of the people who played a vital role in all of the Associations events, the judges. The *'umpires'* as they were by then known were chosen by the Committee of Management, with usually three officiating for the cattle classes and three for the sheep. Even in the 1820's and 1830's.

'Umpires' were being invited from quite a distance. For the show of 1839, for instance, six names appear in the Minute Book as having been suggested as suitable candidates for cattle judges for that year. These include John Parsons of Lawhitton, Launceston, John Williams of Trethewey, Ruan Lanyhorne, Benjamin Burn of Gorran, Mevagissey, Mr. Drew of Illminster, Nr. Exeter, Mr. Tresawna of Probus and Mr. Tyack of Helston.

The judges for the sheep classes also included a number from Devon and when it is considered that transport had not yet improved to any great extent, the effort required to travel to Truro was considerable.

Also active within the county at this time was the Royal Horticultural Society of Cornwall, encouraging, again through exhibition, the improvement of horticultural techniques and practices. A communication from their secretary, Dr. Barham, was received in June of 1839 with regard to the possibility of a joint venture with the Association. A sub-committee was formed to meet a sub-committee of the Horticultural Society, and subsequently a general meeting of the Association was called to consider *'certain proposals made on the part of the Royal Cornwall Horticultural Society, for objects common to both Societies'*.

The enthusiasm for this project was obviously not great, as the meeting called for the Town Hall in Truro in October of 1839 only attracted three subscribers of the Association, with one of those being the Chairman!

The meeting was subsequently adjourned until the morning of November 6th, when again only three people appeared. It was agreed to adjourn again, until that afternoon, when thankfully, Sir Charles Lemon arrived, and was able to take the chair and the vast number of seven other members were present!

It was agreed that further meetings should be held with the horticulturalists, and also that Sir Charles Lemon should obtain information with regard to *'opening a communication with the English Agricultural Association'*. This almost certainly refers to the Royal Agricultural Society of England, formed in May of 1838 at the Freemason's Tavern in London. Following their first, most successful *'country meeting'* or show in July of 1839 at Oxford, and the granting of their Royal Charter, news of the Societies activities had spread, and Cornwall, like most other counties, was keen to take advantage of any new ideas that may have been forthcoming.

Sir Charles Lemon, was himself to become closely involved with the Society in the years to come and is indeed pictured in the famous engraving of the Royal Agricultural Society published in 1845.

The meeting, suggested above, was held with it being decided that £20 should be contributed by the Horticultural Society and £10 by the Agricultural Association to some joint projects to be further agreed. The meeting, however, then came to a halt, as Dr. Barham announced that his Society's representatives were not in a position to make any definite arrangements for funding etc., without recourse to a special general meeting of their members.

A letter read at a later meeting of the Association, from Dr. Barham, confirmed that their general meeting had not agreed with any such financial contribution to a joint fund. The whole matter appears to have been dropped and little further communication seems to have taken place.

The Minute Book for 1840 gives a few details of the actual arrangements for the show-yard, with it being agreed that the ground *'shall be staked and roped off firmly'* and the Superintendent of the Police be requested to make various arrangements with regard to the control of the entrances etc.

The costs relating to the staging of the show, bear little relation to the hundreds of thousands of pounds involved today, although up until the later part of the nineteenth century the arrangements were, of course, relatively simple. Over the first fifty to sixty years, costs changed little, with, for instance, the hurdles for the sheep section, costing £2 6s. in 1804 for four years. By 1840, with the numbers of stock involved greatly increased, the Clerk of Truro Market was still only being paid £1 10s. for the hire of hurdles.

The venue for the dinner to follow that year's show was to be the Theatre, if the proprietors were agreeable, with Mr. Pearce of Pearce's Hotel providing the refreshments and Sir Charles Lemon acting as Chairman. It should be remembered that Truro at this time, was very much the social centre of the County, and such places as the theatre and Assembly Rooms, now estate agents offices, between the main Post Office and the Main Doors of the Cathedral, were very popular attractions.

Such facilities and the fact that many of the county families owned town houses in Truro, ensured that the Association and most of its events centred on the area for so long.

Always keen to introduce new competitions, The Right Hon. The Earl of Falmouth, when chairing a meeting in November 1840, suggested that a prize in the form of *'a sufficient number of hurdles' 'for a close sheep fold'* be awarded *'to the farmer that has folded with sheep in the Eastern method, not less than six statute acres of land upon a turnip or green crop as a preparation for wheat or other corn crop'.*

This suggestion found favour with the Committee members present, and the award was included for the coming year with the stipulation that the land in question should be within ten miles of Truro *'for the sake of the convenience of the judges'*.

Such new prizes were constantly being discussed, not all of which appear to be particularly practicable by modern standards. One award, offered by Sir Charles Lemon, Bart., in December 1840, comprised a premium of £3 for the best essay *'on the agricultural effect of the corralline deposits of Falmouth Harbour, with reference to the quantity of each crop produced after its use and the time when certain changes in the vegetation occurred. The essay to comprise also a comparison of its results with the sands which are found in other parts of the County'*.

At a later meeting, the terms of this award were modified somewhat, presumably to encourage competitors, with a statement of the results being required rather than *'the formal character of an essay'*.

One feels, that the amount of time and work needed to complete such an essay, should possibly have been rewarded with more than the prize on offer! However, such premiums, reminiscent of those offered some forty years earlier, offered at least some encouragement to those interested in general agricultural improvement.

This prize was to be followed a year later, by two awards of £2 10s., being offered for each of two sets of experiments, one being concerned with sub soil ploughing and the other with the application of liquid manure. No entries were received for these awards in the first year, resulting in their continuation for 1843. In addition, Sir Charles Lemon, following the success of his earlier prize offered a further prize of £3 *'for the best statement of the result of any experiments, on the application of chemical and artificial manures made in the County'*.

At this time, 1841, the Annual General Meeting of the subscribers of the Association, was normally held in the Town Hall in Truro during December. At a meeting held in January 1841, it was agreed that for the forthcoming AGM, prizes should be offered for a range of classes *'to encourage improvement in the Production of superior agricultural roots, seeds and grain'*.

A schedule of 14 classes was drawn up covering white and red wheat, barley, oats, various types of turnips, mangle wurtzels, carrots, parsnips and cabbage. Three prizes were to be offered in each class, with the first prize ranging from 10 to 5 shillings.

This type of winter competition brings to mind the fatstock or *'primestock'* shows now held in the winter months in many parts of the county. Although not run for long by the Association, this type of competition may well have prompted the later formation of the Truro Christmas Fat Stock Society, now known as the Truro Prime Stock and Produce Society.

Also discussed at the AGM were further refinements to the Association's rules with regard to the showing of livestock. The immergence of true breeds of cattle and sheep and their development, particularly in the early years of the nineteenth century, had led to confusion over what constituted an animal of a particular breed, or more correctly an animal possessing the characteristics of stock originating from a particular area of the country.

For instance the Devon breed of cattle, coming to the attention of those keen on agricultural experiment at the end of the eighteenth century, were soon to be seen in various parts of the country. Thomas Coke's farm near Holkham in Norfolk was the home of a fine herd and the breeds popularity soon grew amongst those who had seen this and other herds.

However, one man, who is said to have done much for the modern Devon we see today, was Francis Quartly of Molland, near South Molton in Devon. When others were selling all their stock, due to the pressures of the Napoleonic Wars, Mr. Quartly took the opportunity to gather a herd of Devon cattle, carefully picking those animals showing the characteristics that he felt to be most sought after. This herd probably provided the blood lines to which most Devon cattle seen today can be traced.

Now that cattle were being more generally classified into distinct breeds, it was felt that this fact should be allowed for in the running of the Association's competitions. Previously cattle classes had, in the main, been simply for *'cattle'* rather than for any particular breed, although a few prizes had already been awarded for cattle of the Devon and *'Durham'* breed. The Durham, not actually being a breed, but more a type, was actually a shorthorn, which had become so named after the huge, 27 cwt. ox, bred by Charles Colling, had been seen on its six year tour around the country.

A new rule was introduced to ensure that the owners of bulls and breeding cows, when they entered for prizes *'declare to the best of his belief, that the animals are of pure blood of the particular breed, under which they shall be respectively entered'*. This type of rule, although differently worded still applies today, guarding the bloodlines of the breeds of livestock for which this country is so famous.

It was further agreed, if the Association's funds would allow, that prizes for breeds other than Devon and Durham should be offered. This was indeed done, with Short Horn Cattle having classes for the Show of 1844.

The AGM of December 1843 saw the unanimous acceptance of a new constitution and set of rules by the subscribers of the Association. Many of the earlier rules were included and the constitution differed little from that drawn up in the 1790's. However, clarification of various points had been required and the writing of a new constitution provided the opportunity for this.

One point to be confirmed, which had been mentioned on previous occasions, was the fact that the Members of Parliament for the County were to be regarded as ex officio members of the Committee of Management of the Association. Further confirmed was the non political nature of the Association, with item 12 stating that *'It is a fundamental rule of the Association that no question shall be discussed, or any tract proposed at any of its meetings of a political tendency, or that refer to any matter to be brought forward or pending in either of the Houses of Parliament'*.

This last point still stands today and forms an important aspect of the Association's current constitution.

The farm competitions, once such a feature of the Association's work, first appeared in 1844 due to the generosity of J. H. Tremayne Esq., by his returning £10 won by him in that year, for use as premiums in these new classes. Although on-farm competitions on specific subjects, such as grain cultivation, water meadows etc., had been a feature of the original Society, a competition to look at the whole management of the farm was a new idea. The prize was to be awarded at the December 1844 meeting of the Association *'for the best cultivated tenant farm, not exceeding 100 statute acres, situated in the parishes of St. Erme, Gorran, Mevagissey, Caerhays, Creed & Cuby'*. The judges for this competition were to be Thomas Julian of Creed, near Grampound, Mr. Slade of Caerhayes and Mr. Trehane of Heligan, who were asked to visit the farms entered during August. To provide guidance for the judges in their deliberations it was agreed that they should *'take into consideration the general state of the land' 'the crops, the state of the fences—of the stock & implements & the value & quantity of the land especially'*.

It was later decided, however, that as the umpires chosen were likely to wish to compete in the competition, new ones should be picked, after the closing date for entries.

To enter for the competition, it was necessary for the competitor to provide the Secretary, by 1st July, with a *'written statement showing the exact extent of his farm, the number and acreage of each field, the system of husbandry he pursued, the quantity of sheep & cattle kept throughout the year & the value & quantity of manure applied by him during the last 12 months'*.

The entry fee for this prize in its first year was five shillings, a fairly large sum for the time and the competitions were to be seen at various times until the late 1950's, when interest in competing dwindled, and the awards were discontinued.

Throughout the 1840's the Association, whilst thriving from the point of view of the interest shown by the community in all that it undertook, had to exercise great economy in all its finances in order that the books should balance at the end of each year.

For the spring show of 1844, it was decided that the six Constables of Police, normally present at the exhibition, should be reduced to five and receive the sum of five shillings each, with the Superintendent of Police receiving the sum of fifteen shillings instead of the £1 previously given.

Advertising costs were also examined, with future advertisements to be placed in only the West Briton & Cornwall Gazette and not the Falmouth Packet.

By 1846, with the financial situation not improving, it was suggested that the various local agricultural societies within the county be approached with a view to their amalgamating with the Cornwall Agricultural Association. It was felt that such a move would strengthen the Association and presumably also encourage interest from all parts of the county. This matter was discussed at some length at the dinner following the Show of June 1846, with many ideas being proposed for how such an amalgamation would work. Many felt that perhaps two or three shows should be held per year, with the venues throughout the County alternating between the major towns.

It should be remembered that at this time, a very large number of small agricultural organisations existed in the County. Many of these groups have long since disappeared or changed name, but the local newspapers of the period contain many lengthy reports of their activities. Such organisations as the Cornwall Roseland Agricultural Association, the East Cornwall Agricultural Society, the Trigg Agricultural Association, the Penwith Agricultural Society, the Kerrier Agricultural Society, the Lostwithiel Agricultural Society, the Stratton Agricultural Society and the East Penwith Agricultural Society, amongst others, were active through the staging of shows and other events.

Mixed feelings greeted the suggestion of amalgamation, with many taking the view that a host of small, local groups, helped in the encouragement of agricultural improvement more efficiently than one large county Association. Much debate ensued, but this was, of course, not the route chosen.

The AGM of December 1847 saw the suggestion made for the further extension of the classes at the future winter meetings with the inclusion of classes for livestock. It was agreed that for 1848, two classes, with three prizes in each, should be staged for the *'best fat cow or heifer'* and *'best fat ox or steer'* at the winter AGM instead of the main summer show. Judges were to be asked to have *'especial* regard in forming their judgement to Quality of Flesh, Lightness of Offal, Age & Early Maturity'. Non subscribers were to be charged the sum of 6d. as admission with a total of 7 guineas on offer as prizes in the livestock section. The resemblance to today's Truro Prime Stock Show is marked although, of course, a far wider range of classes can now be seen. The Show of 1848 is recorded as being held in the Market-House, with those attending almost certainly seeing a scene very much like that seen in the run up to Christmas only a very few years ago, when the City Hall housed all sections of the annual winter event.

It was suggested that for the Christmas show of 1849, the classes for roots and grain should be discontinued, with the introduction of more livestock classes instead, to include pigs fattened by cottagers or labourers and fat wether sheep. It was agreed that the extra stock classes should be allowed but that the classes for roots etc. should be amended rather than cancelled.

Quite strict new rules were introduced to ensure that the roots etc. exhibited were a *'fair specimen of the crop from which they were taken'*. This fact had to be confirmed in the form of a certificate from *'two respectable farmers'* in the immediate neighbourhood of the exhibitors land.

In addition, turnips had to be from a field of at least two statute acres, and mangel wurtzels, carrots and cabbage from fields of no less than half an acre. It was further required that the ground should be *'properly cultivated and clean'*.

The summer show of the Association, held in 1850 on June 5th in a field behind the Ship Inn in Truro, attracted, according to the Royal Cornwall Gazette, a greater attendance of visitors than any previous exhibition. The *'shew-field'* was extremely crowded as were the adjoining streets and the entire centre of the town.

A new introduction that year was a *'sweepstake'* of 5 shillings *'for North Devon Bulls, Short Horn Bulls, Rams, and Boars that shall not be entered for any of the Society's prizes this year. The Sweepstake is open for all England, and to animals which have already carried the Society's best prizes'*. Whether or not the idea found favour is, unfortunately, not mentioned, but no doubt some fun was had on the day of the show!

To further extend the activities of the Association at its winter show, an advertisement in November 1850 encourages *'exhibitors of Implements of Husbandry, and Manufacturers'* to attend the show in the Market-House on 9th December. It is stated that *'excellent accommodation'* will be *'allowed'*, with the request that the selling price of any implements etc. exhibited should be clearly marked.

It appears that the endeavours of the Association met with little support from the manufacturers of implements, with no such items being exhibited at the Market House that December. However, it was agreed at the dinner following the day's proceedings that the idea should be continued for the coming year, if at all possible. Once again, the Association's finances were causing concern, with a balance of only £41 16s. 6d. being held at the end of 1850. It was felt that this represented an insufficient sum to safely run the affairs of the Association and thought was given to reducing some of the following year's premiums, although no actual action was taken.

One matter, however, which did create much interest during the meeting was a letter, from one T. D. Acland Esq. detailing a proposal for a large agricultural society, to cover the six Western counties of the country. It was being suggested that to further agricultural improvement and to aid the disemination of information a large organisation, covering a wide area should be formed to allow agriculturalists from say, Penzance, to meet and compete with those of a similar mind from perhaps Bristol. Whether to extend the activities of the Bath and West of England Society or to form a new society was put up as a topic for discussion. Many of those present supported the general idea, however, it was also felt that the formation of such an organisation may detract, particularly financially, from the Cornish Association, when finances were already strained.

The question of the formation of such a regional agricultural organisation seems to have met with little real support, probably mainly due to the state of the Association's own finances. Savings had to be made, and at the Annual General Meeting of December 1851, it was agreed that the fat cattle classes at the winter show should be dropped for the forthcoming year. However, owing to the considerable saving this was to make, the suggestion to include a show of poultry at the 1852 summer show was met with approval. This proposal, was afterwards changed, with the poultry to be exhibited at the winter show of the following December.

Financially, the situation gradually worsened, although the promise of a prize of £5 from Augustus Smith of Tresco in the Isles of Scilly and a similar amount from that year's President, Richard Davey helped the situation somewhat.

Augustus Smith's prize was to be for *'the best plough or implement' 'for lifting the potatoe crop'*, whilst that offered by the President would be for the re-instatement of the prize *'for the best fatted ox'* or any other award that the Committee felt preferable.

J. Davies Gilbert of Trelissick offered £2 towards the winter exhibition of roots and pigs, with the offer of an extra £3 if a system of judging the roots in the field could be devised.

A meeting the following March decided that the prizes offered that Summer would need to be reduced with *'£14 instead of £18 be offered for North Devon & Short-Horn Bulls, and that £14 instead of £18 for*

Rams & Hog Rams'. Other savings including the suspension of the awards for 3 year old colts were made, totalling a £15 reduction in prizes offered.

Even though funds were being offered by supporters, no winter exhibition was held in 1853, although as the summer show had proved successful and funds had increased somewhat, the proposal to increase prize money back to its 1852 level was accepted. The offer by Augustus Smith had unfortunately not attracted any exhibits, but his increased donation of £10 for the best exhibit of a *'potatoe lifter'* in the coming year was accepted with much thanks by the Committee.

One unusual item which no doubt caused much discussion at that meeting at the Red Lion Hotel in Truro in November of 1853, concerned the showing of rams. On the motion of William James of Merther and seconded by James Tremain of Trevarton it was agreed that all the rams should be shown shorn, *'with the exception of a lock of wool left on the near side of such sufficient breadth to show the character of the staple'*. The sight at the show the next May must have been quite amusing and one wonders how long this practice continued.

Funds for specific purposes appear over the years, to be rather easier to raise than those destined to pay for the everyday running of the Association's affairs. This certainly seems to have been the case when the question of restarting the farm competitions was raised in 1854, some ten years after they had been first mentioned in the premium list.

Some £21 was promised to cover the prizes, a substantial addition to the existing premiums, by four stalwart supporters, Mr. J. Davies Gilbert of Trelissick, Mr. James Tremain of Trevarton, Mr. Richard Davey of Redruth and Mr. Michael Williams MP.

Two classes were to be staged with three prizes of £6, £4 and £2 for farms of 100 acres and over, and three prizes of £5, £3 and £1 for farms of 50 to 100 acres. The entry fee at five shillings represented a substantial sum and in the case of farms entering, the whole of the roots and seeds being grown were to be judged, with only farms with 10 per cent or over of the land in roots being eligible to compete.

The prizes were to be for different districts of the county, changing annually, with the Roseland district being first chosen. The Minute book lists the parishes eligible for the first four years of the competition, with all the areas mentioned being in the locality of Truro. With a few exceptions, most of the members of the Committee of Management of this period resided in the Truro area and although subscribers and exhibitors came from throughout the county, it was felt that such a competition needed to be between farms within a short distance of Truro. There was, however, a note in the minutes, to the effect that if sufficient support were forthcoming from other parts of the county, then the areas listed should be expanded.

1854 still saw no takers for Augustus Smith's potato lifter prize. Following his generous offer for the Committee to make use of the sum for implements in general and the addition of £5 by the Association itself, increased prizes for *'new and improved implements of husbandry'* were included. It was decided that at least £5 should be set aside *'for the best implement for distributing guano and other artificial manures'*. Guano (the solidified excrement of sea birds) had been imported from the islands off Peru since the 1840's and represented one of the few sources of nitrogen apart from farmyard manure etc. available at the time. Other fertilizers such as *'Mr. Lawes's Patent Superphosphate'*, which was available from William Karkeek's Agricultural Stores in Truro, were beginning to catch on, and no doubt Mr. Karkeek, in his position as Secretary to the Association, was able to promote his wares from an advantageous stance.

On the suggestion of, and following a donation from the President for 1856, William Rashleigh of Menabilly, a class for four year old geldings or mares, *'calculated for hunters, or for carriage purposes—bred in the County of Cornwall'* was seen for the first time in the prize list for that year. This type of class marked quite a change from the thinking of previous year's, when the horses exhibited, although possibly for use under the saddle, were mainly to be engaged in agricultural work. It was minuted that the public attending the show would benefit from seeing a collection of quality horses, bred for a specific purpose.

The 1857 show was to be the last of the annual shows held solely in Truro. Times were changing with other parts of the county wishing to play host to the *'county'* show. Many felt the Association to be too Truro orientated and a suggestion as to how to increase the funds of the Association and therefore to enable it to grow and expand met with the support of the agriculturalists of Cornwall.

3. On the Road   1857–1915

Although the general mood of the Committee of Management and of the supporters and subscribers of the Association was one of change, one factor created the climate to cause the annual shows of the Association to be held in various towns of the county. By 1857, finances were once again on a reasonably even keel, although it appears to have been felt that little hope of expansion was likely owing to a fairly static level of income.

In 1846, some ten and a half years after his death, a fund to mark the life and achievements of Francis Hearle Rodd of Trebartha, near Launceston, had been set up. Known as the *'Rodd Testimonial Fund'* and instigated by a meeting of the Launceston and East Cornwall Agricultural Society, the fund's purpose was to remember in a suitable manner a man who was a *'zealous and untiring patron of Agricultural Improvements and the founder of Turnip Husbandry in this County'*.

Subscriptions were sought by public advertisement and it was soon agreed that the setting up of an *'Agricultural School and Model Farm'* should be the aim. The school was to be funded from the testimonial fund, with 4% shares of £25 each being sold to finance the model farm. An estimated £7000 was required to complete the project, a not inconsiderable sum!

Money began to accumulate although by April of 1849 only £774 17s. 9d. had been received by the Trustees. The fund then languished in a bank at Bodmin, accumulating interest, whilst presumably, ideas for its use were considered. It is gratifying to note, that although some 140 years odd have elapsed since the original idea, Cornwall does at last have its own Agricultural College, on the old Duchy Home Farm at Stokeclimsland near Callington.

By late 1856, little hope remained of raising the necessary sum to build the school and it was suggested by John Tremayne of Heligan, a Committee member of the Association, that the show could well benefit from the accumulated money. His suggestion, following a private conversation with two of the Trustees of the fund, was that if the Association were to make itself a truly county organisation by holding exhibitions in various towns of the county, it could apply to the Trustees to receive the income from the fund to use for agricultural purposes.

This idea met with the approval of the Committee, and following the Annual General Meeting of December 8th 1856, a letter was dispatched formally applying to the fund for the grant of an annual sum of money.

The Trustees, presumably feeling that it was time that something was done with the funds in their care, agreed to the request, on certain conditions, and the travelling show, fondly remembered by many when it visited their area, was born. Among the conditions stipulated, for the initial annual grant of £24, was that the money should be used *'as an annual prize or prizes for Agricultural Implements and machinery'* and that the Association should visit in turn the various districts of the county *'never exhibiting in the same locality oftener than once in four years'*. The Trustees further reserved the right to withdraw this financial support, on a years notice, should the original aim of the fund become a possibility.

It was hoped that a scheme to include the funds of the local agricultural societies, in whichever area was to be visited that year could be encouraged, with active involvement by those normally only concerned with their smaller, local event.

The Annual General Meeting of December 1857 saw a host of changes, with the final decision being taken as to the acceptance of an income from the Rodd Testimonial Fund and all that that entailed.

The most significant change at this time was one of title. The *'Royal'* prefix to the name of the Association was added at this time and at last the name which is so well known today came into being. The Association had been receiving the support of the Royal family from at least 1848, when Her Majesty Queen Victoria is recorded as a subscriber to the Association, through the Office of Woods and Forests.

No doubt following the decision to substantially expand the show and become itinerant the Royal approval and recognition was granted to provide a suitable status for such a well respected organisation.

For some years, it was thought that the title of the organisation as it stands today, had been the same from the 1827 amalgamation of Society and Association; in fact this was not the case and 1857/58 marks the year of change.

A new set of rules *'for the management of the Association'* was drawn up and whilst these were largely based on the previous regulations and indeed those originally formulated in 1793, several significant changes were made. For instance, the term *'Committee of Management'* was to be dropped and in its place the *'Council'*, as it is still known today, was to be the governing body.

The Council was to consist of a President, eight Vice Presidents, a Secretary and forty members, chosen from the subscribers of the Association. At the Annual General Meeting, to be held in Truro each December, four of the Vice Presidents and twenty of the ordinary councillors were to retire by rotation, although all would be eligible for re-election. The office of President was to be an annual one, with the same person not being eligible for re-election for four years.

Camborne was chosen as the destination for the first itinerant show and a Committee consisting of William Trethewy of Probus, James Tremain of Newlyn, Samuel Anstey of Menabilly, Sampson Tresawna of Probus and James Paull of Camborne, together with the President, Richard Davey MP and the Secretary, William Karkeek were instructed to meet a select Committee of the East Penwith Society, with regard to the various arrangements necessary for the 1858 show.

The meeting duly took place with £150 in cash, including £10 to be used for local prizes, being offered by the East Penwith Society. Three adjoining fields were inspected and found to offer an excellent site for the show, with one large field to be used for cattle and sheep, one for horses and one for implements.

Members of the Association, subscribing a minimum of 5 shillings were to be allowed free entry to classes, whilst non members needed to pay 5 shillings for an unlimited number of entries. Presumably a good way of encouraging members to join! Members of the East Penwith Society were to also be allowed to enter classes free of charge.

Stewards were to be in attendance by 6 o'clock in the morning, with judging commencing at 8.30, and the public admitted from 10.30 a.m.

An admission charge of 1 shilling was to be charged until 1 o'clock and thereafter sixpence until 3 o'clock, after which admission would be free of charge. Livestock and implements were required to remain in the *'shew-field'* until 4 o'clock, with the stock being issued with green food *'free of expense'*.

The vastly enlarged range of classes for implements were to be a feature of the show and the following extract from the Minute Book, details the awards to be offered for 1858.

THE RODD MEMORIAL PRIZES

Classes *Preparation of the Ground* — Section 1

1	For the Best Plough for general purpose	£2
2	,, ,, ,, Turnwrist Plough, which will efficiently turn the furrow against the hill	£2
3	For the Best Cultivator, Grubber or Scarifier	£2 10s.
4	,, ,, ,, Set of Harrows	,, 10s.
5	,, ,, ,, Set of Seed Harrows	,, 5s.
6	,, ,, ,, Chain Harrow	,, 10s.
7	,, ,, ,, Clod Cruncher or Pulveriser	£1 10s.

Cultivation of Crops — Section 2

8	For the Best Corn Drill for small occupations for or Hilly districts	£3
9	For the Best Turnip & Mangold-Wurzel Drill for Ridge or Flat, depositing manure with Seed	£2
10	For the Best general Manure Distributor	£2 10s.
11	For the Best Horse Hoe for Green Crops on the Ridge or Flat	10s.

Harvesting Crops & Preparing for the Market — Section 3

12	For the Best Hay Working Machine	£1 10s.
13	For the Best Horse Rake, for collecting Hay or Corn having a mode of delivery more or less self acting	15s.
14	For the most practically useful Portable Threshing Machine with Riddle & Straw Shaker, not less than 3 Horse Power	£3
15	For the best Winnowing Machine, which shall also be convertible into a simple blower	£1
16	For the best one horse Cart for general purpose	10s.
17	For the best two horse Harvest Waggon	£1

IMPLEMENT PRIZES GIVEN BY THE SOCIETY

Preparation of Food for Stock — Section 4

18	For the best Chaff Cutter worked by horse or steam or water power	£1
19	For the best machine for Grating or Pulping Roots	10s.
20	For the best Turnip Cutter	10s.
21	For the best Corn & Pulse Bruiser worked by horse, water or steam power	£2
22	For the best Oil Cake Crusher	5s.

Miscellaneous — Section 5

23	For the best Sheep Feeding Hurdle Rack	10s.
24	For the best Cookery Apparatus for Farm Kitchens	15s.
25	For the best Field Gate, not less than 9 feet in length fitted with Hangings and Fastenings	10s.

In addition to this comprehensive range of awards a further £5 was to be available to the judges for presentation to any implements, not falling within the categorys listed, but which they felt to be worthy of merit.

This list of premiums, not only shows the importance felt at the time that should be given to the encouragement of improvement in agricultural machinery, but also gives a fascinating insight into the range of equipment which could have been found on the larger farms of the mid nineteenth century.

Various changes and additions were also to be made to the list of livestock premiums to be offered at Camborne with two classes to be staged for Devon bulls, a class each for Devon cows and heifers, with a similar set of classes being offered for cattle of the Short Horn breed.

Sheep classes were to move away from the mixed classes of previous years, with for 1858, five classes being staged purely for Leicesters, as well as two ram classes, one for Cotswold sheep and the other for South Downs.

Of all species, sheep had probably seen more changes due to breeding programmes than any other. The traditional type of Cornish sheep, described by Carew in his Survey of Cornwall, published in 1602 'had generally little bodies, and coarse fleeces, so as their wool bare no better name than Cornish hair'. He went on to say that they had *'grey faces and legs, coarse short thick necks, stand lower before than behind, narrow backs, flattish sides, a fleece of coarse wool, weighing about two or three pounds, of eighteen ounces; and their mutton seldom fat—from eight to ten pounds per quarter'.*

By 1811, with the publication of Worgan's General View of the Agriculture of the County of Cornwall, it seems that the type of sheep described above had already almost completely disappeared, although in Worgan's view this fact was not one to be lamented! By this time much crossing had been attempted, with rams from Exmoor, Dartmoor, North and South Devon, Dorset, Gloucester and Leicester being used. Leicester flocks were to be seen at the home of the Revd. Robert Walker at St. Winnow and at Killiow, the seat of R. L. Gwatkin Esq., whilst South Downs could be seen at Tregothnan, the home of Lord Falmouth and at Tehidy Park, the home of Lord De Dunstanville.

Detailed descriptions of the weights of wool and meat produced can be found in Worgan's book and show the great efforts being made to improve the type of sheep to be found in the County. No doubt all those concerned would have been gratified to have seen the improvements already made by the 1850s.

In the Horse Section, four classes for agricultural horses and four classes for Hacks or Hunters were to be included, although as yet no mention was made of individual breeds of agricultural horse such as the Shire.

Pigs were, as yet, to see no particular breed classification, with the four classes being split into simply large breed and small breed. Sheep shearing continued to be included as did the awards for farm labourers long service.

Two special prizes, to be offered by the East Penwith Society, took the form of two classes with six prizes in each for cows and pigs belonging to labourers.

In the words of a report read at the Annual General Meeting in December 1858 the move to a *'migratorial existence'* and *'the experiment of co-operative action with local societies' 'has proved more successful than even the most sanguine promoters of the scheme had dared to expect'*.

The move to an itinerant show was obviously a great success with a very large number of stock, implements and visitors being seen at the show. Subscribers, from every part of the County had now become members of the Association and two invitations had already been received from towns wishing to play host to the 1859 exhibition.

The assistance of the East Penwith Society had proved invaluable as had the support of the Directors of the West Cornwall Railway who had most generously transported the stock and implements to and from the exhibition free of charge.

Receipts for the year had totalled some £557, with a profit of nearly £90 remaining after all bills had been settled. Already the idea of extending the show to two days was being mentioned. One black spot however, which must have marred the year's proceedings, was the death of William Karkeek, the Secretary of the Association, who since his election in 1839 had seen the Association through many difficult times.

The man elected to take over as Secretary and who was to be in office for fifty years, was Henry Tresawna of Probus. The Tresawna family had been Committee members, judges and great supporters of the Association for a great number of years and Henry Tresawna was to see the Show through one of its periods of great expansion.

Henry Tresawna of Probus, Secretary of the Royal Cornwall Agricultural Association from 1859 to 1909.

Invitations for 1859 had been received from Trigg Agricultural Society representing Bodmin and the Farmers Club of St. Austell and after much deliberation the town of St. Austell was picked as the venue for the second itinerant show. The move to a two day event was agreed, with the first day to be devoted to the *'trial of implements'*.

Hereford cattle classes were also to be seen for the first time that year, although sheep shearing and the agricultural long service awards were for 1859 to be dropped. Amongst the additional awards in the implement section was to be one of £5 for reaping machines and £10 *'For the best combined steam threshing machine'*.

£14 was to be paid to a Mr. Dunn for the use of three of his fields for the show, with the Committee making the various arrangements employing Mr. Dunn to fence the lower hedges of two of the fields and the lane hedge for the sum of ten shillings! Two further fields, belonging to one Mr. Kellow were to be used for the trial of implements.

The larger the event, the more complicated the necessary arrangements and 1859 saw for the first time tenders being accepted from the Innkeepers of St. Austell *'for the privilege of erecting a booth for the sale of refreshments'* and for the provision of a *'cold dinner'* in the Market House. Two tenders were received with that from Mr. Job of £16 being accepted. The quotation of Mr. Walter Hicks, Honorary Secretary to the Farmers Club, was accepted to provide a meal at 3/- per head on the last evening of the show.

The Show began to reach out to an ever wider audience and advertisements were ordered to be placed in the *'Mark Lane Express'*, the *'Plymouth Journal'* and the *'Exeter Western Times'* in addition to the usual local newspapers. Following a suggestion from the Committee formed in St. Austell for the occasion, the question of whether or not to include a large poultry show was raised. Most Council members felt that the suggestion had merit, but as to whether it should be included as an integral part of the show; this was another matter! Finally, following a vote by the Council, it was agreed that such a section should be included and a grant of £15 was made towards the costs on the understanding that anyone visiting the poultry show, must pass through the main show entrances rather than a separate one. Thus was borne the first of the *'section shows'*, to be organised by local committees wherever the show travelled. These shows within a show, have formed an important part of the Royal Cornwall for a great number of years and now cover areas such as horticulture, dogs, pigeons, bees and honey, rabbits, cage birds and goats as well as poultry.

Once again the show proved a great success, with the venue for 1859, St. Austell playing a very gracious and welcoming host to all who attended. Some 100 cattle, 182 sheep, 70 horses and 23 pigs were on display, representing the largest yet display of livestock at a show in the County. The railway had again proved invaluable and no doubt was expressed as to the benefits of extending the show to two days.

An invitation had now been received from the Penwith Agricultural Society for the 1860 exhibition to visit Penzance with again £150 being contributed by the local society. This invitation was gladly accepted with thought already being given to the destination for 1861.

A communication from the Bath & West of England Society and from a Truro Committee indicated that in 1861 the Bath & West of England Show would be visiting Truro and the request was made for the Royal Cornwall Agricultural Association's show for that year to merge with that of the Bath & West. This suggestion met with the approval of the Council, although a special meeting of subscribers, called to discuss the matter, had to first agree to staging two successive shows in the western division of the county, as that year's show should have moved to the east.

A sum of *'not less than £150'* was voted by the Association towards the costs of this joint venture and planning progressed, although the preparations for the forthcoming Penzance event were of course foremost in the Council minds.

A new innovation, for 1860, which does not appear to have been included in previous years, is the appearance of a silver cup, valued at 10 guineas, given by the President, John St. Aubyn, for the best four or five year old mare or gelding calculated for a carriage. The award of silver cups and trophies now such a part of the livestock section of the Royal Cornwall Show, was not a regular feature until the later part of the nineteenth century and of course at that time trophies would have been won outright, rather than represented each year.

The choice of site at Penzance, always a matter of great concern, found the Council considering three possible pieces of land suggested by the Penwith Agricultural Society. After much deliberation the offer of part of Colonel Scobell's lawns at Nancealverne amounting to approximately 8 acres plus an adjacent field of 4 acres was chosen. Colonel Scobell's kind offer to grow an acre of *'rye corn'* in preparation for the *'trial of reaping machines'* was also gratefully accepted. A piece of land of about 4 acres was also offered by a Mr. Matthews of Trengwainton for the trial of implements, situated about a mile from the main showyard.

Some 144 entries, from 17 different exhibitors, were to be seen in the implement classes at that year's show with a total of some 467 head of livestock. Covered accommodation was provided for the bulls and the arrangements generally held as the best to date.

Once again the railways in the form of the South Devon, Tavistock, Cornwall & West Cornwall Railways had provided much service to the Association, by conveying stock, etc. at *'very liberal terms'*.

The actual activities of the Association for 1861 obviously centered around the visit of the Bath & West to Truro during June and the merger of the two shows for that year. The event was heralded as a great success and felt to be one of the most successful ever staged by the Bath & West of England Society. At the Association's Annual General Meeting, local exhibitors were congratulated on the vast number of prizes achieved whilst in competition with those bringing stock from other counties of the South West. The visit to Truro was the first by the Bath & West to the County and it was to be not until 1868 that a return visit by that Society would be made, at that time to Falmouth.

The merged show of the Bath & West and Royal Cornwall at Truro in 1861, taken from an illustration in the London Illustrated News in the Authors collection.

The year passed with the invitation of the East Cornwall Agricultural Society for the show to visit Liskeard in 1862 being accepted. Some £350 was offered in prize money, the most yet, plus a silver challenge cup valued at £40, offered by *'six Gentlemen of the County'*.

Truro was to once again see the show in 1863, following the inability of the East Penwith Society to offer more than £100 as their contribution as part of their invitation for the show to visit Redruth. The sum was felt insufficient for the show to be efficiently run and therefore the offer by the people of Truro was accepted.

Long woolled rams, of breeds other than Leicester, steam threshing machines and thorough bred stallions were all recipients of new prizes at that Truro show.

Breaking from tradition, a new Committee was then formed for the choosing of judges, a role previously in the hands of the full Council. Three judges were at that time invited to judge each section of the show, those being Implements, Devons/Herefords, Short Horns/Pigs, Sheep and Horses. For each section two judges were to be invited from within the County and one from further afield. This Committee *'the Judges*

Selection Committee' still exists today, although of course, the required number of judges per year is rather greater than in the 1860s!

Following a suggestion from the Truro local Committee, a meeting of the Council discussed the possibility of providing funds for the running of a *'floricultural show'* to form part of the main exhibition for 1863. The sum of £15 was granted in that first year, on the condition that no additional charge be made to those visiting that part of the show.

The AGM later that year heard that this new addition had proved extremely popular and *'was the means of attracting a large number of ladies to the showyard'*. Thus the Flower Show as it is known today was born, although the costs of staging it now somewhat exceed the £15 originally spent!

The most important event of the year, however, and one which marked the start of a long tradition was the grantings of a request made by the Association in January 1863. The request, through the Duchy of Cornwall Office in Buckingham Gate, invited His Royal Highness the Prince of Wales to offer his *'patronage and support'* to the Royal Cornwall Agricultural Association.

A reply in June 1863 confirmed that His Royal Highness *'has graciously consented to become the Patron of the Association and has sanctioned an annual subscription to its funds of £10 from the Duchy revenues'*.

Thus began a tradition of Royal Patronage and interest spanning the years to this day.

The request from Saltash to play host in 1864 met with the ready approval of the Council, with the close proximity of Plymouth being viewed by many as an opportunity for an extremely large show attendance.

For the first time, the judging of the livestock was to be open to the public who in previous years had only been admitted once the awards had been made. An admission charge of 2/6 each was to be charged for the privilege!

The Saltash show, attracting record crowds and a huge display of implements numbering some 300, proved most successful and from a financial viewpoint a great boost to the Association. A profit on the show of just over £415, swelled considerably the Association's reserves, and allowed for the possibility of further expansion. A *'Horticultural'* rather than *'Floricultural'* show had also been staged and once again had proved popular.

Extra sheep classes, with classifications for the Shropshire Down breed were now introduced with further awards for sheep of *'Other Downs'* breeds. Falmouth and Penzance competed for the presence of the show in 1865, although as Penzance had already been visited and as stock etc. could be *'conveyed'* by rail to Falmouth *'without break of gauge'*, Falmouth was subsequently chosen.

The event was to see for the first time the inclusion of a dog show, with some 240 dogs being entered, including approximately 50 pointers and setters and 20 to 30 harriers, all housed in a marquee of 150 × 30 feet.

The Falmouth show, although again a great success, was somewhat overshadowed by developments in other areas. The *'cattle plague'* was sweeping the country with some 233,000 animals to die before the epidemic was over. The actual disease in question was Rinderpest and a special meeting of the Council was held in August of 1865 to consider what steps could be taken.

The Secretary, Henry Tresawna, was instructed to write to the Secretaries of the local agricultural societies to seek any information they may have as to the situation in their own localities. He was also to print and circulate, through the Board of Guardians in each part of the county a letter outlining the symptoms and treatment of the disease.

Mr. Tresawna was empowered to call a meeting of the Council when new information came to his attention. He was to call such a meeting in November, when a letter from a Mr. Lamb, a Veterinary Surgeon of Liskeard had been received. Mr. Lamb described the situation, after a visit to Truro as *'the most frightful scourge that perhaps ever visited the British Farmer'*. Calls were made for the Board of Trade to be requested to *'prevent the landing of imported foreign cattle'*, and the meeting agreed to meet a few days later to take further action.

The subsequent meeting heard of the ravages caused amongst the cattle of the county and a motion was passed calling upon the Board of Trade and through them the Privy Council to take action to curtail the importation of live cattle. This point was further endorsed at the Association's AGM and a request forwarded to the High Sheriff to call a county meeting to discuss possible ways of combating the disease.

An invitation had been received and accepted from the Launceston Agricultural Society for the 1866 exhibition to take place in that area. However, it was felt by the Council that no show should take place in that year as the movement of livestock around the County could only add to the problems being created by the rinderpest. Subscriptions were instead to be used for the purchase of *'plant'* for the permanent use of the Association and the invitation to Launceston was to stand for 1867 if a show were to be possible for that year.

By the AGM in December of 1866 the situation had improved somewhat with no cases of the disease having appeared in the county for some months. It was therefore felt that the 1867 event, at Launceston, should go ahead, and plans were put in hand. Dartmoor sheep classes were to be staged and the decision for all aspects of the show to be held over the two days was made. Previously, the first day had been devoted to the trial of agricultural implements with the livestock being displayed purely on day two. A poultry and dog show were to again form a part of the show, and on application from the Launceston Agricultural Society the sum of £15 was voted as a grant towards the expenses. This grant would enable a total of £80 to be offered as prizes in these sections, no doubt encouraging exhibitors from far afield.

However, as late as April, with the show due to be held in late May, an order was issued by the Privy Council *'that no exhibition of horned cattle should be held this year'*. Immediate discussions ensued, with it being agreed that the show should go ahead, minus cattle classes, but with the addition of butter classes, with a minimum of 3 lbs. being exhibited.

The local Society's contribution, in view of the loss of the cattle, was to be cut from £150 to £75, to presumably offset a possibly reduced income.

In fact, no such reduction in income resulted, as the show turned out to be the best attended to date. A surplus of nearly £1100 was made, with the decision for the show to be fully operating on two days judged fully justified.

1868 was to again see the Bath & West of England Society visiting the county at Falmouth and once again no actual show organised by the Association was to take place, with instead a merger of the two shows. £150 was voted as a contribution to the proposed Falmouth exhibition with a further £50 to be added *'in case the Local Committee of Falmouth were compelled to construct a railway siding'*. If only the costs of such a venture were as modest today!

The show, one of the largest staged by the Bath & West Society, saw the added benefits of a larger number of counties involved, through their amalgamation with the Southern Counties Association. This fact also relieved those of the Council who disliked the interference of the visiting Bath & West, although of great agricultural use, with the regular running of the Association's own exhibitions. The addition of more counties to the area covered by the Bath Society meant less frequent visits to Cornwall and therefore fewer years without a show run by the Association.

At about this time, it appears that the Association was instrumental in the formation in the County of a Chamber of Agriculture to work in conjunction with the Central Chamber in London. Although never actually involved with the work of the Chamber, due to its involvement in matters political, great efforts were made by the Officers and Council of the Association to see its formation and thereafter to stand back from its actual operation. This Chamber, appearing to be a lobbying body for legislative reform, possibly forms an early move towards the formation of the National Farmers Union in 1908.

Penzance, the destination for 1869, was to see the first classes for cows and heifers *'of any other pure breed.'* For the years previously, the only pure bred, female cattle, were of the Devon, Shorthorn or Hereford breed. Some classes for cross-bred females had also been introduced, but now a need was to be seen for classes for cattle of the purely dairy breeds.

'NARCISSUS – a Devon bull bred by and the property of the Rt. Hon. Evelyn Viscount Falmouth, Tregothnan, Cornwall. Calved Sept. 30 1867, sire Sunflower (937) gd sire Duke of Chester (404), dam Picture the 4th (2225) by Davy's Napoleon (464) gd dam Picture (337). In 1869 he gained the first prize at the Royal Cornwall Society's show at Penzance, as the best bull not exceeding 2 years old, and in the same year, the 2nd prize at the Royal Agricultural Society's show at Manchester, for the best yearling bull above 1, and not exceeding 2 years old. In 1870 he carried off the 1st prize of the Royal Cornwall Society's show at Launceston, for the best bull above 2 and not exceeding 3 years old, the 1st prize in class 1, at the Bath & West of England Society's show at Taunton, for the best bull not exceeding 4 years old, and also 1st prize at the Royal Agricultural Society's show at Oxford for the best bull not exceeding 3 years old. In 1871 he won the 1st prize of the Royal Cornwall Society's show at Truro, in Class 1, for the best bull above 3 years old, at this meeting he was also awarded the special prize for the best bull of any age in the yard; and he gained the 2nd prize at the Royal Agricultural Society's show at Wolverhampton, for the best bull above 3 years old.'

Painted by Whitford and still hanging at Tregothnan, this painting and its fascinating inscription details the marvellous success enjoyed by 'Narcissus' in the 1860s and 1870s.

Also mentioned for the first time is a payment to be made to a J. Olver of four guineas as a fee for acting as Veterinary Surgeon to the Association during the period of the show. No doubt the range of ailments encountered during the show were rather simpler than the current Veterinary team have to sometimes deal with!

After the show at Penzance, again a most welcoming host town, Launceston and Bodmin were to vie to act as venue for 1870. Bodmin, unfortunately not at that time directly linked by rail to the rest of the county, lost to Launceston, partly on the grounds that the last Launceston show in 1867 was *'imperfect'* owing to the absence of cattle classes.

Launceston was to see the introduction of championship prizes, an innovation to be sponsored by Viscount Falmouth with £10 *'for the best animal in the Cattle Classes'*, £10 *'for the best sheep or pen of sheep in the Sheep Classes'* and £5 *'for the best pig or pen of pigs in the Pig Classes'*. Never before had the idea of class winners competing against each other been introduced, and no doubt the ensuing competition led to a great deal of debate around the judging rings.

One aspect of that show in 1870, deemed worthy of mention in the report given at the December AGM, was a working demonstration of a steam plough in a field near the Showyard. The plough had been lent by Sidney Davey, a farmer from near Redruth and created great interest amongst those attending. Some 13,000 people visited the show over its two days, and the resulting surplus caused the Association to agree to a sum *'not exceeding £1000'* being invested in railway debentures.

Following the year's success, the Truro show of the following year was to benefit with an extra £50 being added to the prizes on offer. The *'any other pure breed'* prizes for cows and heifers were to now change to classes for breeds of the *'Channell Islands'*. Jersey and Guernsey cattle had been well known in England for their milk producing abilities for many years, but numbers were relatively small and it was probably not until such moves as the establishment of a herd book for Jersey's, for instance, in 1838, that their popularity and numbers began to significantly increase on the mainland.

A matter which created much discussion and which marked a distinct move on the part of the Association for the 1871 show, was whether or not to employ the Marine Band to entertain visitors to that year's event. No such entertainment of a non-agricultural nature seems to have been a part of the show prior to this year, and this introduction marks a distinct change of approach to the event as a whole. The sum of £22 was voted by Council with a further £5 being added by the Truro Local Committee to cover the price quoted by the band of £27 for one day's appearance. Although band fees are today extremely high, this sum in 1871 represented more than was granted to the running of the Poultry, Dog and Horticultural shows in total! Compared to the salary of the Secretary, albeit part-time, which was raised in 1871 from £35 to £50, the band do not appear to have done too badly!

A report on the preparations for the show in the West Briton, gives an idea of the work put in by a town hosting the show. *'On Tuesday evening most of the decorations of Truro were completed. On either side of Lemon-Street graceful specimens of coniferous trees were planted, and the view from the top of the street was very pleasing to the eye, and led many to ask why our streets are not permanently adorned and their monotony relieved with rows of trees. At the bridge at the foot of Lemon-Street there was a triumphal arch, bearing the word "Welcome". This arch, like the others in the town, was made with poles covered with evergreen and surmounted with handsome banners. Another arch was placed at the entrance to the show, with the words "Success to the Royal Cornwall Agricultural Association". In front of the Town-Hall, the Royal Hotel, and Red Lion Hotel trees were also placed. At the entrance to Prince's-Street there was another arch, bearing the words "God speed the plough", over which was a shield with the Prince of Wales's feathers. On the other side of the bank and at the entrance to Duke-Street was a similar arch, bearing the words, "Fish, Tin and Copper"; the produce for which Cornwall is noted. At the junction of Victoria-Square and St. Nicholas-Street there is an arch with a fine span, and it bore the words "Agriculture, Fish, Tin and Copper", in large and conspicuous letters, suspended beneath were a plough and some spades. The arch is also ornamented with sheaves, some handsome banners, &c. The steeple of St. Mary's Church is hung from top to bottom with flags, borne by projecting poles and the Lander Column, too, is gay with bunting; lines of flags cross the principal streets, and the town is altogether en fete'.*

The show is believed to have been held in a field at the top of Lemon Street, above the Lander Monument, and the tradition of arches, described above, was to continue wherever the show went for many years.

1872 saw the show return to Bodmin, following an absence of 47 years, since the last show organised by the original Cornwall Agricultural Society had been held in that town. Throughout the 1860s and early 1870s the horse classes had seen a steady increase in the number and complexity of awards offered, with an ever larger number of classifications by age etc. appearing. 1872 was to see the first classes for ponies introduced, together with two classes of different heights for cobs. Also new for that year were horse shoeing classes with three prizes of 3 guineas, 2 guineas and 1 guinea on offer. These classes were to continue into the 1950s, when due to a lack of competitors the awards were dropped and the necessary equipment, anvils etc., owned by the Association were sold for scrap.

The earliest known photograph of the Royal Cornwall Show, taken it is thought at the show in Truro in 1871. The President, Edward Archer of Trelaske, Launceston, can be seen standing in the grey top hat, with his wife in a white dress sitting beside him. The Secretary, Henry Tresawna, can be seen kneeling.

Following the Penryn show of 1873, held in the grounds of Tremough the home of Mr. Shilson, which unfortunately had been hit by bad weather, St. Austell was chosen as the venue for 1874. Held on Mr. Coode's lawns, the show was the best attended staged to that date. Over 17,000 people paid for admission over the two days, not including the subscribers and exhibitors. No prizes for implements were offered, but instead the £25 10s. donated by the Trustees of the Rodd Testimonial Fund was used for the trial of implements in nearby fields.

Truro's successful bid for the show for 1875, was to be marked with the Presidency of His Royal Highness the Prince of Wales, Patron of the Association. Unfortunately, His Royal Highness was eventually unable to actually attend the show and the Earl of Mount Edgcumbe officiated during the show in his place. A request by the Truro Committee to extend the show to three days nearly introduced the change that was not to be seen until 1974, almost 100 years early. The question was debated by Council, but following a vote the proposition was defeated by 16 votes to 10. In fact, the Liskeard show of 1876 did in fact extend to three days, although the added costs were greater than the extra income generated and the third day was not to be seen again until 1974.

Significant sums were, at this time, being offered for various aspects of the arrangements for the show, the public catering rights in 1875 attracted four tenders, with a bid of £55 from Mrs. Jenkins of Falmouth being successful. For the first time mention is made of the tenders for the copyright of the show catalogue, possibly meaning 1875 to be the first year that a complete catalogue was published.

Once again, the records of previous years were broken with over 24,000 people paying to attend the show, hence the possibility of extending the show to three days was raised, with it suggested that the third day should be at the cheaper admission price, as on the second afternoon of the show.

The Liskeard show saw for the first time the erection of a grandstand, with £90 being paid by visitors for the use of it and although the cheaper third day had proved a useful experiment, the general feeling of the AGM was against continuing with such an extension.

Camborne in 1877 was followed by another visit to Saltash in 1878 when for the first time driving classes, as they would be known today, were introduced. Two prizes of £10 and £5 were to be offered for mares or geldings, over 14.3 hands and under 5 years of age *'to be driven and tested on the show ground in single harness, which shall be deemed the cleverest single harness horse'*. The harness and carriage was to be provided by the exhibitor of the horse for *'testing his horse'*. At various times, classes for carriage horses had been seen, but this *'testing'* was something rather different.

1878 also sees the arrangements for the show gradually becoming more complex, with, for instance, the normal stewards of livestock being augmented with extra stewards for forage and finance. Detailed instructions, outlining the way in which stewards should operate at the show were to be forwarded to all those chosen to officiate and arrangements made for the various parades of cattle and horses. Presumably, the parade of livestock, today such a memorable feature of the show, was a new idea for the time, with one parade of cattle taking place during the first afternoon and two on the second day. Similar parades for horses were also to be arranged, although no mention of the sheep being included is made.

Large entries of stock made their way to the Saltash showground of 1878, with the Hereford cattle being particularly mentioned for the improvement in numbers and quality on previous years. Of the three main breeds exhibited, Devons saw 44 entries, Shorthorns 39 and Herefords 37, and this relative equality between the breeds was felt to show the great increase in the numbers of Herefords in the county generally.

Unfortunately that years show did not receive the attendance by the public that had been expected, with only 13,394 paying for admission over the two days. The absence that year of a horticultural exhibition was felt to be possibly to blame, as no fault could be found with the siting of the showyard, in a position convenient for the railway station.

The railway, always playing such an important role in the success or otherwise of the show, had for many years further supported the Association, by offering free tickets to Council members if journeying to or from a meeting of the Association. This arrangement was, however, to come to an end in 1879, when a letter was received by the Secretary from Mr. Frierson, Secretary to the Great Western Railway Company stating that the Directors of the Company had decided to discontinue the practice, although they would be prepared, for one year, to offer Council members return tickets at the single fare price.

No doubt such privileges were gratefully received by those having to attend meetings from various parts of the county, and any help was better than none. The support of the different railway companies in the transportation of stock and implements, does however, seem to have lasted for many years with arrangements for free or low cost transport being made well into the twentieth century.

Falmouth in 1879, welcomed a large number of visitors, only beaten by the attendance at the Truro show of 1875. Stock numbers were slightly down, with for some reason the entries in the Devon cattle classes, moving the breed from the best represented to the least, with only 14 cattle forward. A covered grandstand, which had previously been open to the elements, met with great approval as did the introduction of a show of cage-birds and hurdle jumping in the horse section.

The show catalogue for 1879 gives a wonderful picture of agricultural development by the later part of the last century, which apart from the lists of livestock exhibitors, gives details of the forty trade stands exhibiting.

Many names, some still in existence, can be recognised and the following extract lists a few of those who were present:

William Brenton, East Cornwall Implement Works, Polbathic, St. Germans
Davey, Sleep and Co., Crafthole, St. Germans
John Glasson Clark, St. Austell
Anthony Sauto, Castle, Bodmin
James Buckingham, Bathpool, Northill, Launceston
Stephenson Brothers, Plymouth
John Huxtable, Brayford, Southmolton, Devon

J. G. Mitchinson, Penzance
W. M. Nicholson & Son, Newark-on-Trent
John Prout, Kelly's House, Lewanick, Launceston
N. Sara, Penryn
William McLean, Bawdoe, St. Winnow
James Pugsley, North Street, Ashburton
T. Bradford & Co., Manchester
Henry Hodge, St. Austell
T. B. Burns, Camelford
Joseph Thomas Skelton, Lostwithiel
John Curtis & Co., Merchants, St. Columb
Oatey and Martyn, Wadebridge
John Draydon, Helland Bridge
J. H. Day & Sons, Crewe, Cheshire
William Harris, Coach Builder, Lostwithiel

Amongst the items for sale were every possible type of farm implement and the inclusion in the catalogue of the prices completes the picture of the requirements of a Victorian farm. However, the items on sale covered a far wider spectrum than purely agriculture with, for instance, a quite large range of garden equipment also being available.

Three coach builders, Crabb Brothers of Bodmin, William Harris of Lostwithiel and John Beale of Truro exhibited a range of carriages etc. with, for example, Waggonettes at £25, Two-wheel Dog Carts ranging from £14 to £21, and Spring Carts from £5 10s. to £9 10s.

J. D. Williams & Son of West Street, Tavistock, showed a large range of saddlery and harness, with an *'elegantly handworked Quilted Lady's Saddle, with leaping head and patent safety Stirrup'* on offer at £8 10s., and *'William's Excelsior Saddle and Brown Leather Polish, where no labour is required'* in 1s. and 1s. 6d. bottles.

J. H. Day & Sons of Crewe, with a range of the latest animal medicines, could no doubt offer a cure for just about any complaint, with *'The Black Drink'* being sold at 19s. per doz., or if you preferred *'purified Driffield Oil'* at 2s. 6d. per bottle.

James Hearle & Sons of Gwarder offered the newest types of mowing machines and lawn mowers with a *'New patent Villa Lawn Mower, to cut 9 inches wide, 'suitable for tall or short people, easily managed by a lady'* on sale at £2.

Moving to the larger items on show, Wallis and Steevens of Basingstoke could supply you with a *'one 8-Horse power Traction Steam Engine, with all fittings complete'* for the magnificent sum of £400. A differential gear would have cost an extra £20, with an enlarged fire box or water lifter each being an extra £10.

One hopes that business was brisk, although, agriculture and trade in general, were facing hard times with the local Committees finding it increasingly difficult to raise the funds to cover their portion of the costs. The number of members, however, subscribing to the Association, continued to climb although the wish for more farmer members as opposed to simply landowners was often expressed.

Plans were laid for the 1880 visit to Lostwithiel, with a few small reductions being made in the prize money on offer in the livestock classes, no doubt a reflection of the state of the local economy. The regulations regarding the exhibition of livestock annually came under close scrutiny and by this time constituted a fairly complex set of rules. One new addition for 1880 was the introduction of a regulation allowing the Council to fine an exhibitor a sum *'not exceeding 10s.'* if a case of *'disobedience of orders or want of punctuality'* was found amongst his *'servants in charge of the animals'!*

The lengthy meetings that must have been required to actually impose such fines, must have made many Councillors question the sense of including the rules in the first place!

The choosing of the Council of the Association had, by this time, become a matter for an annual meeting of members which was held in a marquee in the showyard during each year's exhibition. A similar system to that which exists today was in use by the 1880s, whereby the Council was split into those elected from either the Eastern or Western Division of the County or *'Irrespective of Locality'*. A certain number of members of Council had to annually retire, these people ineligible to stand again for one year. Nominations at the show meeting to decide the successful candidates.

As the system has lasted for well over 100 years, with few changes, something must be said in its favour, although, even today, it does sometimes cause a certain amount of confusion amongst those retiring and being nominated.

Mention of the general state of the county's economy and in particular the *'depressed state'* of agriculture and mining, was again made in the report to Council at the Annual General Meeting of 1880. Although the show at Lostwithiel, held on the beautiful lawns of Lanwithan, the home of Mr. Foster, had been little hit by the depressed economy, with sales of implements reported as *'very satisfactory'*, the year ahead worried some of those present. The subscribers of the Association, at that time providing some £360 a year, were only covering a portion of the amount offered in prizes annually (for 1880 some £800). The need to encourage more members to subscribe or to reduce the prize list was discussed at length. The dependence by the Association on a well attended show, with large numbers paying for admission, was paramount in making ends meet. The Redruth

Committee, had already requested that their contribution to costs be reduced to £100 from £150, and the necessity to look carefully at the Association's finances was stressed.

Some savings were in fact made that following year, with a general decrease in the prizes offered being made throughout the cattle section. Entry fees for horses were also increased to 5s. per horse *'in order to make up a portion of the costs incurred in building the horse boxes'*.

At this time, the costs of actually staging the show were increasing annually, with more livestock entries dramatically adding to the costs of the provision of covered cattle accommodation and horse boxes. These high annual costs, were of course to be one of the main reasons the Association moved to a permanent site at Wadebridge some eighty years later. Since the show had begun to grow and with the need each year to build facilities for stock, large amounts of timber and other materials were needed annually. After each show it was the practice to dispose of the bulk of this material by auction sale, rather than to incur haulage and storage costs by keeping this type of equipment for the next year's show.

In 1878, for instance, over £128 had been raised through the sale of timber used for that year. This practice was to continue for many years and was a well known event, after the actual show had been broken down. Once or twice, a fee of perhaps 25% of the timber value was paid to a merchant for the hire of the wood, but this system seems to have proved less successful.

By the end of 1881, a significant increase in the number of members was to be seen and the prizes for the 1882 show at Launceston were therefore increased slightly, with, for instance, the first prizes in the bull classes rising by £2 to £10. The prizes in the Guernsey and Jersey cattle classes were also increased somewhat to equal those offered in the Devon, Shorthorn and Hereford sections; not something which had previously been the case.

Thought was also given to whether or not the system for collecting the sums subscribed by the members should be altered. At this time, a collector, Mr. Martin Magor of Truro was employed to collect subscriptions throughout the county, receiving the sum of 5% of the subscriptions collected as a fee. His retirement in 1881 after many years of service and his subsequent replacement by Samuel Harvey, also of Truro, caused discussion as to the future system to be employed. No firm decision was taken however and no changes made for the following year.

By the 1880s, with rail links operating throughout most of the country, the distance from which visitors came to the show was increasing all the time. The Committee appointed to invite judges, for instance, for the Launceston show of 1882, were far reaching in areas chosen to provide the officials. In the Shorthorn classes, the two judges for that year, Mr. J. Wood and Mr. J. Thompson came from Ripon and Chippenham respectively. No doubt the true Cornish hospitality they received on their arrival in the county, matched that offered to those judging at the show today, when an invitation to judge at the Royal Cornwall is highly sought after. Those officiating at Launceston that year were accommodated at either the White Hart or the Kings Arms, as scribbled notes against their names in the Minute Book indicate.

An added attraction for the Truro show of 1883 was an auction sale of bulls, to be held on the second afternoon of the show. An auctioneer, Mr. Wesley Stephens, was appointed and the necessary rules drawn up. Animals entered in the young bull classes were to have a reserve price of 20 guineas, whilst the older bulls would have a reserve of 30 guineas. From the selling price 2½ per cent commission was to be deducted, with half being paid to the auctioneer and half being added to the Association's funds. Cash sales only to be allowed! It was later decided to also hold a sale of rams, after the bull sale had been concluded, with reserves of £10 to apply.

The site found for the show was described as one of the best so far and was offered, free of charge, by Lady Smith at her estate of Tremorvah, which was within easy reach of the railway and the town.

The practice of requiring the host town to provide a sum of money, normally £150 towards the costs of staging the show, met with much debate during 1883. A decision to change this request was made and in return for dropping this financial request, it was to be instead required, that towns provided a show yard, *'with proper approaches and fences and a sufficient supply of water'*. The site was to be inspected by a Committee of the Council, prior to any final agreement being reached, to ensure its suitability for the purpose.

The work of the Council of the Association had, by now, fallen into a fairly set routine for the organisation of each years show. Applications were examined at a meeting, usually in July, from towns hoping to host the show in the following year. A meeting in early October, examined the year's accounts, ready for the AGM later in the month, and also selected a Committee to make any alterations to the Prize List that were deemed necessary, and to then report back to Council. A November meeting, confirmed or negated, the recommendations made by this Committee and decided on such matters as tenders received for printing costs etc. Further meetings in the early spring chose judges and dealt with any other details of the show's organisation.

Following a decision to offer no prizes for implements at the Bodmin show of 1884, the question of what to do with that year's grant from the Rodd Testimonial Fund was raised. Obviously, any use of the sums offered had to be of an agricultural nature and suggestions were requested. The possibility of once again introducing farm competitions met with a great deal of support, and it was agreed that two prizes of £15 and £10 should be offered for the best managed farms, exceeding 150 acres, in the Eastern Division of the County. An additional prize of £10 was then offered by that year's President the Rev. Sir Vyell Vyvyan, Bart., and a second prize of £5 from Colonel Goldsworthy, on the same basis, but for farms exceeding 50 acres but under 150 acres.

Arch erected for the Royal Cornwall Show of 1884, outside the George and Dragon Public House in St. Nicholas Street, Bodmin.

The competition was to be open to tenant farmers, with a 10s. entrance fee for members and a fee of 20s. for non-members.

Judges were to be asked to especially consider the following points:
1. General Management with a view to Profit.
2. Productiveness of Crops.
3. Goodness and suitability of Live Stock.
4. Management of Grass-Land.
5. State of Gates, Fences, Roads, and General Neatness.
6. Management of the Dairy and Dairy Produce, if Dairying is pursued.

Medals or Certificates were also to be available for award *'to any really deserving persons employed on any of the competing farms for distinguished merit in the discharge of their duties'*.

All farms entered for the competition were to be inspected twice, once at around the middle of June and again in early October, with the judges results being forwarded to the Secretary, Henry Tresawna by the end of October. A written report on the competing farms in question was also to be provided by the judges, to be with the Secretary by 1st December.

The judging of such a competition was no small task to undetake and the men in question would have needed to allow a substantial length of time to complete the inspections and paper work.

Only two entries were actually received for the 150 acre and above setion, with the first prize being awarded to Mr. Richard Olver of Trescowe in the parish of St. Mabyn, a farm on the Pencarrow estate of the Molesworth family, (today of course the Molesworth-St. Aubyn's). Second prize went to a farm occupied by Mr. James Holman of Treberrick in the parish of St. Michael Caerhayes.

Another change, made that year, was in relation to the procedure to encourage members to subscribe to the Association. Following a suggestion by Richard Olver of Trescowe, a person was appointed for each of the various districts of the county, in the hope that that person could endeavour to persuade farmers etc., from that district to subscribe. This system of districts was to be based on the Poor Law Unions of the county and subscribers to the Association can be seen listed by union, well into the twentieth century. This system seemed to successfully create an atmosphere of competition between the districts and did in fact, prove a worthwhile means of encouraging members.

By the AGM in October 1885, the membership of the Association was reported as being 645, with a steady increase having been seen during the year. Penzance, the venue that year, had again proved most successful with a good entry of stock. No doubt the prize for thoroughbred stallions had seen hot competition with the premium being raised from £20 to £60, although the *'roadster'* stallion class had for that year been abolished.

The premium list for St. Austell in 1886 was again to see several changes with a new class for Jersey bulls *'with two heifers of his progeny'* being suggested and sponsored to the tune of £5, by the Jersey judge of the previous year, Mr. Francis Le Brocq of St. Peters, Jersey. This prize was to be augmented by a second prize of £2 given by the Association. The Royal Cornwall Show, always well supported by exhibitors of Channel Island cattle was even at this time seeing judges from the islands officiating, and this gesture by an islander was much appreciated at the time.

A further change to the cattle section was the introduction of a special prize of £5 given by Mr. Arthur Tremayne, *'for the best milch cow in the yard of any breed, the bona fide property of a Cornish tenant farmer, to be tested for quantity and quality of milk'*.

A prize based on milk production figures was a new venture for the Association, with today, of course, such awards being quite commonplace. The prizes for pairs of agricultural horses were destined to be dropped at that time, whilst pony breeding was to be encouraged with a new class for *'the best pony mare and foal, or in foal, not exceeding 13½ h.h.'* Great changes were annually seen in the prizes offered for horses, complicating dramatically attempts today to see a clear picture of the development of showing of horses and indeed the use of horses generally, either for work or pleasure.

Protests against prize winners, always a feature of any show, were very much in evidence at the end of the last century. Written protests, lodged with the Secretary, often caused lengthy discussions at the Association's meeting, with in many cases the same matter having to be referred back to several meetings, before a satisfactory decision was arrived at. One instance in 1885, concerned a protest made by a Mr. Cardell over a mare exhibited at the Penzance show by Messrs. Yeo. The complaint stated that the mare was in fact a noted prize winner called 'Elegance' owned by a Captain Moreton Thomas of Coity Mawr in Breconshire. The evidence of one William Powell, groom to Captain Thomas was put forward to state that he had had charge of the mare 'Elegance' since his master had owned her and that he was positive that she had never been at Penzance. A very lengthy discussion ensued, with eventually, Mr. Cardell's protest being upheld *'the weight of evidence being in his favour'*.

Thankfully, protests only happen rarely today, with a host of safeguards in the form of rules and regulations being in use to guard against the need for protests on the show day itself. Also of course, the existence of a host of Breed Societies, representing virtually every breed of cattle, sheep, pigs and horses, play an important role in regulating the activities of those breeding and raising the stock in question.

At this period, that is, the last years of the nineteenth century, there appears to have been no fixed days of the week on which the annual show should be held. Weekends were avoided, but local conditions dictated the actual days on which the show was to take place. For instance, after fixing the date of the 1886 show for St. Austell on a Thursday and Friday in late June, the problem of clashing with St. Austell market day on the Friday was raised. The show was subsequently altered to the Wednesday and Thursday, although by the middle of this century, prior to becoming a three day event, it was always fixed as being on these two days. The dates of the show were also a fairly moveable item, with many considerations affecting the final decision. The fact that the shows of the Bath & West and those of the Royal Agricultural Society of England moved around to various locations and therefore dates, was always borne in mind when fixing the Royal Cornwall Show date. If either of the other shows were anywhere in the South West, pains were taken to avoid a clash, which would particularly affect the numbers of livestock being exhibited.

Nowadays, an unofficial national calendar of major agricultural shows exists in theory, to hopefully avoid the problems of clashing. Obviously, some conflict does arise, from time to time, although if shows are in very different parts of the country such problems are not too serious. For instance, The South of England Show virtually always directly clashes with the Royal Cornwall although of course, exhibitors should always choose to head for Cornwall!

Camborne in 1887, saw the show taking place in a summer *'exceptionally favourable'* *'for outdoor gatherings'* with a correspondingly large attendance and entry of stock. A new venture for this year, suggested and organised by a special Camborne committee, was the construction, in a building provided by the Association, of a working dairy. This experiment was apparently much appreciated by those visiting the show and provided a talking point for that year's event.

The question of the efforts being made to establish dairy factories within the county was raised at that year's AGM and the possibility of assistance from the Association mentioned. The suggestion of a series of lectures to take place in various parts of Cornwall, on matters connected with dairy farming, met with the approval of those gathered and further discussion of the subject ensued.

The Earl of St. Germans had presided as President that year, to be followed by Mr. John Charles Williams of Caerhayes Castle. Mr. F. Julian Williams, also of Caerhayes Castle is, of course, Vice Chairman of the Association today and again the strong family link is still very much in existence, as with a great many families up and down the county.

The question of promoting improvements in dairying was again raised some weeks later with three Council members being elected to form a Committee to examine a Danish pamphlet on dairy management, and to report back to the full Council with an estimate as to the likely costs of its reproduction.

The series of dairying lectures was also further discussed with a Canon Bagot being suggested as a well regarded authority on the subject, with the added possibility of staging *'exhibitions of practical butter making in the principal towns of Cornwall'* being discussed. The costs of such an exercise were then to be investigated, with a further report to be made to the Council.

The report by the Committee on Dairy Management was in due course heard, with unfortunately little progress having been made on the question of the publication of the pamphlet written by Mr. Boggild on Danish dairying. Those forming the Committee, namely J. C. Williams, W. Trethewey and J. Tremayne had studied the translation into English and found it to be well written and informative. Unfortunately, the author was disinclined to allow its publication in England. The likely costs of a series of lectures were thought to be quite high, with the danger of not attracting the small farmers who could most benefit, being of concern. However, the suggestion of butter making demonstrations, perhaps during the market days of various towns, was thought very worthwhile.

A sum of £40 was requested to enable such a scheme to be put into action, with perhaps eight or ten different venues to be visited. The project met with the Council's approval and the towns of Launceston, Liskeard, Bodmin, St. Columb, St. Austell, Truro, Redruth, Falmouth, Helston and Penzance were visited.

J. C. Williams Esq., President for the Newquay show of 1888, was to offer special prizes in the cattle classes, for both bulls and cows or heifers, for cattle of breeds not having their own classes. This therefore was to offer the opportunity for a wider range of cattle to be presented at the show and only in the last few years a similar move has again been made with the introduction of the *'any other pure breed'* classes in both the cattle and sheep sections.

The extra breeds of cattle seen at this time were to be Aberdeen Angus and South Devons, both of which were to see their own classes before long.

An added problem to concern the show at Newquay was the fact that the date chosen was found to clash with that fixed for the Devon County Show. The Council, with typical Cornish diplomacy, resolved not to change the date of their show, but at a joint meeting of members of both the Cornish and Devon Committee a compromise was reached.

If either show were to change date, a sum of £10 would be paid by the other Association to help defray the extra printing costs etc. incurred by the change of date. In the end, Cornwall kept to its date of the 20th and 21st June!

It is at about this time that the shows begin to be referred to as the *'Royal Cornwall Show'*, rather than simply the annual exhibition of the Royal Cornwall Agricultural Association. Why this had not happened at an earlier date, is rather odd, but perhaps the formality of the Victorian period was to blame.

Helston fought off a bid from Falmouth and succeeded in playing host town in 1889. Again, the horse section saw a multitude of small changes, mainly in the Hunter and Hack classes, with in the jumping section a stipulation being made that the course to be jumped should include hurdles and water and *'be taken in hunting form'*. Ponies were to also see a new class instigated for ponies *'to be tested in harness'* and for the first time in the pig section, breed classes were introduced. Previously pigs had purely been shown as either of large or small breeds. The Berkshire was to be the first breed receiving specialist awards, with two classes being staged, one for boars and one for sows.

Pigs, unfortunately never a very large section of the show, were to have a chequered career at the Royal Cornwall. Today, classes are not staged at all, owing to the lack of suitable pedigree herds within the county willing to exhibit their stock.

By 1889, the Rodd Testimonial Fund prizes were again taking the form of awards for *'new or improved implements or machines'* rather than the field trials of a few years earlier. Medals were awarded to the winning exhibitors, and mention of these awards can be seen in many of the implement advertisements of the period.

Much debate accompanied the arrangements for the show of 1890, owing to the added complication of the annual show of the Royal Agricultural Society of England being that year to be staged in Plymouth. An approach from John Winter, Secretary of the Devon County Agricultural Association in November 1888, was made to the Council, suggesting that for 1890 no county show be held in either Cornwall or Devon, but that arrangements be made to involve the two Associations in the exhibition of the national Society. It was further suggested that the members of each organisation should obtain free admission to the Plymouth event and that perhaps, special, local prizes could be offered.

This approach met with a mixed response by the Council of the Royal Cornwall Agricultural Association. Many felt that the Association should continue to run its own show for that year, whilst a wish to join forces with the RASE was expressed by others. After a long series of meetings it was eventually agreed that the Royal Cornwall should be held, but that the sum of £100 would be offered to the RASE for use at its show at Plymouth. This sum was to be raised through the sale of stock held in the Lancashire & Yorkshire Railway.

Next came the problem of where to hold the 1890 Royal Cornwall so as not to conflict too greatly with the Plymouth event. The rules of the Association stated that the show should be held in alternate divisions of the county, with the eastern division's turn being in the forthcoming year. Obviously, to stage the county show anywhere near Plymouth, could have spelt disaster and therefore a special meeting rescinded the problematical rule in question for one year, allowing the western division to play host two years running.

Truro's application, with the usual offer of a suitable site and most importantly, an adequate water supply, met with the Council's approval and yet again Truro was to be the destination.

The prize list was again examined with a few changes made, with for the first time a class to be offered to exhibitors of donkeys, with entry fees of 2/6 for members and 5/- for non members to be charged.

One proposition to come from the Truro local committee concerned a possible grant from the Association to help cover the cost of introducing *'lady riders and other attractions'* for the 1890 show. The Council decided that they *'could not entertain the proposal'*, and so what exactly the inhabitants of Truro had in mind is today rather unclear! Obviously something not thought to be in keeping with the traditions of the show at the time!

Perhaps such an additional attraction ought to have been allowed, as the Truro event, hampered by bad weather and the show of the Royal Agricultural Society of England at Plymouth, was not well attended, although the number and quality of stock was well up to previous years. For the first time, a loss on the running of the show was made, with £327 12s. 1d. having to be found from reserves. Some more of the Lancashire & Yorkshire Railway stock was sold, to cover the amount, although, at least, a reasonable profit was made on the stock when sold! Applications were also received from the caterer, Mr. Mark Bull, and the publishers of the show catalogue, Messrs. Hartnoll of Newquay, for a reduction in the fees originally offered by them, owing to losses made at the show. No such reduction was considered possible, with it felt that the risk should be taken by those submitting the tender.

Some £1700 was at this time held in reserve, and whilst this provided a cushion for the Association to fall back on if necessary, caution was advised by the Auditors, when considering the 1891 prize list.

An arch erected outside The New Inn at Tywardreath, for the Royal Cornwall Show at Par in 1891.

Par's invitation for 1891, after reassurances with regard to a query over the amount of hotel accommodation available in the area, was accepted and was to constitute a visit to a totally new area. Fears as to the somewhat scattered nature of the population of the area proved groundless and the show proved a great success.

One event, to mark that show at Par, was the presence for the first time of a member of the Royal family. His Royal Highness the Duke of Edinburgh, visited the show and given the fine weather conditions, Par was declared to be a very worthy host to the Association. The local Committees support, was to also be remembered by their generous presentation to the Association of a magnificent silver cup at the AGM later in the year. The cup, to be known as the *'Par Challenge Cup'*, was valued at 50 guineas and was to be for presentation to the best three year old bull. The exhibitor winning the cup either three years in succession or any five times, would win it outright.

Also presented to the Association during the year, were 6 volumes of the Herd Book of the Devon Herd Book Society, accompanied by a letter from the Society's Secretary, Mr. Risden, stating that the books were to be for the use of the Association's members.

The subject of agricultural education was one too often raised during 1891, particularly following the suggestion by Mr. C. D. Acland MP, that the Association should branch out from merely holding the annual exhibition, into other areas of agricultural advancement. His letter outlined the idea of a series of experimental fields, in various parts of the county *'wherein experiments might be tried upon subjects of importance in the farming of the county, under the management of one or more Committees of practical farmers, guided by advice and assistance of a competent scientific man'*. Mr. Acland further suggested that application should be made to the President of the Agricultural Department for a grant to carry out the above project.

A Committee of the Cornwall County Council, looking at the question of *'technical instruction in agriculture and in fishery'*, also suggested that the Association might like to take a more active role in such educational matters.

The Committee had suggested that it would probably be prepared to forego part of its main grant in favour of the Royal Cornwall Agricultural Association and the County Fisheries Committee, if those two bodies were to take up the challenge.

Much discussion then ensued, with a Committee of the Association being formed to consider the various suggestions and to draw up proposals to be made to the General Purposes Committee of Cornwall County Council.

The proposals made were not to be accepted by the County Council, however, they did decide on three items to be taken up by the various county organisations. The first concerned the employing by the County Council of a *'migratory Dairy School'*. Secondly a grant of £200 was to be made to the agricultural societies of the county *'for instruction in sheep shearing, thatching, hedging and other handicrafts connected with agriculture'*. The third and last item agreed by the County Council was to be the provision of a grant of £100 towards the type of field experiments suggested by Mr. Acland.

The President, C. Davies Gilbert Esq. and five other members of the Council, in October 1891, called a general meeting of the members of the Association to look into the possibility of staging a stallion show in the spring of each year. Those gathered for the meeting, approved the idea, on condition that the show be held in the same town as the main Royal Cornwall Show. A further meeting confirmed the premiums to be offered, with three classes to be staged, to each have a first prize of £15 and a second prize of £10. Class one was to be for thoroughbred stallions, class two for *'roadster'* stallions and class three for stallions for *'cart or agricultural purposes'*.

Presumably, the term *'roadster'* refers to an animal to be used for drawing a carriage or any vehicle to be used on the road. The show was staged at Redruth on 29th March, with the prize winners being also exhibited at the Royal Cornwall Show, also in Redruth, in June. Two judges officiated, one, Mr. Calmady of Tetcott, near Holsworthy, to judge the thoroughbreds, and another, Mr. Wills of Berry Barton, near Totnes, to judge the agricultural stallions. Both judges jointly judged the *'roadster stallion'* class.

Unfortunately the show of stallions met with little success and at the AGM at the end of 1892 a decision not to hold a similar show the next year was made. The Royal Cornwall Show at Redruth, was, however, extremely well attended with over 20,000 people passing through the gates.

Four towns, Wadebridge, Liskeard, Launceston and Lostwithiel, sent deputations to try to secure a visit by the show in 1893, with after two ballots, Liskeard, being chosen.

South Devon cattle, now such a part of the Royal Cornwall Show, were to see two classes being staged for the first time in 1892. 1893 saw £50 being allocated as prize money, with four classes being staged:

For the Best Bull calved before 1892	£7	£4	£2	
For the Best Bull calved in 1892	£6	£4	£2	£1
For the Best Cow or Heifer calved before 1892	£6	£4	£2	
For the Best Heifer calved in 1892	£6	£4	£2	

Leicester sheep, with numbers on the decline within the county, were to be dropped from the prize list for 1892, which left awards being offered for South Hams, Long Wools and Down sheep, although classifications for Leicesters were to be seen on and off in the years to come.

Dairying was again to be the subject of much discussion, with the Dairy Committee being asked to communicate with the Technical Instruction Committee of the County Council with regard to the possibility of making arrangements for a *'competitive exhibition of proficiency in butter making, milking and other operations in connection with the dairying interest'*. A set of classes were agreed upon, with the sum of £50 being voted by the Association to this purpose. Nine classes were staged in a marquee in the showyard, under the supervision of Mr. and Mrs. Warren, as follows:

1. Dairy Work viz: churning, butter making, salting and packing for students who have passed through a course of training and won certificates.
   Six prizes of £5 £3 £2 £1 10/- and 5/-
2. Dairy Work, as above, for non students.
   Five prizes £3 £2 £20/- 10/- and 5/-
3. Fancy Butter, not more than 3 lbs.
   Three prizes 40/- 20/- and 10/-
4. Fresh Butter, to be delivered 10 days before the show, in 1 lb. plain rolls to test its keeping quality.
   Three prizes 30/- 20/- and 10/-
5. Best method of packing butter in lots of not less than 2 lbs. or more than 11 lbs.
   Two prizes 20/- and 10/-
6. Best method of packing cream.
   Two prizes 20/- and 10/-
7. Milking Test   Adults £2 £1 and 10/-
   Juniors under 21 years £2 £1 and 10/-
8. Milking Machines.
   Two prizes £1 and 10/-
9. Test Appliance for testing amount of butter fat in milk.
   Two prizes £1 and 10/-

Lectures on dairying also formed a part of this section of the show that year at Liskeard, and were apparently well attended, with the money expended by the Association being thought well spent.

Another feature that year, following the provision of £40 by the Technical Education Committee, were demonstrations and lectures on the shoeing of horses. The displays *'excited a great deal of interest and drew together a large number of shoeing smiths to witness the mode adopted by the Demonstration in making and fitting shoes to horses'*.

Shoeing competitions had been a part of the show some years earlier, but this new idea of demonstrating improved techniques drew a large, audience of those actually involved in the trade in the county.

The following year at Falmouth, the show was to see both the dairy and shoeing demonstrations and lectures continued, with, in addition, competitive classes for shoeing held. Of the two classes staged, that on the first day dealt with the shoeing of hacks and that on the second with the shoeing of cart horses. Prizes of £3, £2 and £1 were offered, with again, £40 being granted towards the costs of the lectures and the provision of a suitable temporary building, by the Technical Committee. This Committee were to also generously offer £65 towards the costs of the Dairying Demonstrations, with £50 being granted by the Association itself.

The Dairy show was much enlarged in this its second year *'and the very liveliest interest was shown in the Butter Making Competitions'*. In the shoeing section, Professor Penberthy, a member of the Council of the Royal Veterinary College in London, lectured on both days of the show to large audiences.

The Wadebridge show of 1895, saw the second highest number of livestock shown in the history of the Association to that date. At 495, the total entered were only one less than at that great show in Truro in 1875, and with a very large number of people attending the decision to visit Wadebridge was seen to have paid off. Only a few years earlier, Liskeard had been chosen in preference to Wadebridge, as no direct rail link at that time existed. However, by 1895, the North Cornwall Railway had reached Wadebridge, and the town was no longer felt to be cut off from the bulk of the population of the county.

December 1895 saw the Association approached over the subject of the banning of the importation of live cattle, sheep and pigs in an effort to control the possible spread of disease. A letter had been received from the Central Chamber of Agriculture suggesting that stock should either be slaughtered prior to shipment or at the port of arrival in this county, immediately they were disembarked.

A deputation was to be received by the Board of Agriculture in London during December, to consist of representatives of the Royal Agricultural Society of England, the Smithfield Club, the Shorthorn Society and various other societies. The Royal Cornwall Agricultural Association were asked to support this deputation and Mr. Hart Key of Wadebridge, who was to be in London on that day offered to represent the Association.

A resolution, on the motion of Colonel Tremayne, was passed at the meeting for communication to the Board, *'that this meeting agrees with the suggestions of the Cattle Diseases Committee, and hopes that they will receive the immediate attention of the President of the Board of Agriculture'*. No doubt, following the ravages of the *'cattle plague'*, only a few years earlier, those involved with stock in the county, felt very strongly about this possible source of disease.

Prize winning Shorthorn cattle, photographed at Trenethick Farm, Wendron, Helston. Owned by Thomas Francis Roskruge, 'Roan Ruth' gained a 2nd at the Royal Cornwall and a second at the Devon County in 1896 and a 1st at Helston Show in 1897. 'Tolgullow Jim' gained a 1st at Helston Show in 1892 and a 1st at the Royal Cornwall Show of 1893.

St. Ives, also a town not previously visited by the Association was to receive the show in 1896, and whilst the entries of stock were very large and the number of visitors well up to previous years, a loss was sustained. A deficit of some £170 was to be seen in the accounts and grave concern given to the possible ways in which extra income could be received or costs cut. It was agreed that the prize list for the following year's show, to be held at Lostwithiel, be carefully inspected with a view to making reductions where possible.

Such reductions were in fact made, with for instance, no classes being staged that year for riding ponies. Several of the cattle classes saw reduced prize money being offered and it was even decided that the music to entertain those visiting the show, should not cost more than £30.

The Annual Meeting of Members, held in the showyard, during the show at Lostwithiel, passed a resolution, with acclamation, which had been proposed by the Chairman of the meeting, Mr. Richard Coode. Mr. Coode's proposal was to state that as *'they were a Royal Society and that HRH The Prince of Wales was their Patron'* he hoped that the following message would meet with the approval of the members of the Association:

'We the President, Council and members of the Royal Cornwall Agricultural Association desire to offer to Her most gracious Majesty, our most respectful and cordial congratulations upon the attainment of the 60th year of Her most illustrious reign and to place on record our gratitude for the great material, intellectual and moral progress that has so eminently distinguished the period during which she has ruled the nation, and to assure her of our most heartfelt wishes for her future welfare'.

The meeting actually went on to elect Richard Coode, of Polapit Tamar, near Launceston to the office of President for the forthcoming year. That year's show at Lostwithiel, was unfortunately, not the success that had been hoped for with bad weather, and the many festivities planned for HM The Queen's Diamond Jubilee celebrations clashing somewhat. Again a loss was made, but the decision was made, not to draw on the Association's reserves, but to encourage new members to subscribe and reduce costs were possible.

The Penzance show of 1898, thankfully helped to put the Association back into the black, with a most successful and well attended event. The show was visited by several members of the British Dairy Association whose annual conference was that year staged in Cornwall, at a time to coincide with the dates of the Royal Cornwall Show. The members were said to be particularly impressed with the dairy exhibition and demonstrations, still at this time being supported by grants from the county's Technical Instruction Committee.

Yet more breeds were to see their own classes in the following year at Launceston. Dartmoor sheep and Large Black and Large White pigs were each to receive two classes, with all the other pig breeds being placed in mixed classes for pigs 'of any other breed'.

Launceston, although proving a success in many ways, with excellent entries of stock and implements, did however, sustain a small loss to a poor attendance of visitors. The show seems, at this time, to have reached a peak as far as the attendance of the show was concerned, which was to only vary depending on the area of the county visited. However, success in many other areas had led to a vast array of prizes being offered and expensive covered facilities etc. needing to be provided for the various exhibits. Great care needed to be exercised in order for the Association to balance its books each year, with the subscriptions of still more members being the only real answer.

Truro was picked to play host the following year, with it being seen at the time as being appropriate that the first show of the new century, should be staged in the town which gave birth to the Association as we know it today, although of course Bodmin, rather than Truro, should be regarded as its real starting point.

One suggestion for prizes which was however, not, at this time taken up, came from the Devon and Cornwall Branch of the National

Association for the Prevention of Consumption or other forms of Tuberculosis. This Association, with its rather lengthy title, was keen to see the Association give annual prizes for animals free from tuberculosis. Much discussion followed, but no prizes were to be offered on such a basis at this time.

The show of 1900, as was usual when visiting Truro, was a great success, although wet weather on the second day, dampened the hopes of a record number of people attending. The first day did however create a record, with more attending on that day than at any time previously.

A surprise visit to the show by Her Royal Highness the Princess Henry of Battenburg, must have created a few headaches for the officials. Quite how this unexpected appearance came about is unknown, but it is certain that a true Cornish welcome met this, the second member of the Royal family to visit the show.

For the first time that year, the Dairying Exhibition was organised by the Association itself, rather than by a local committee in the area the show was visiting. It was felt necessary for such an important section to be directly in the hands of the Council of the Association, and the decision to make this change appears to have worked well. Once again, the dairy section proved most popular and at the meeting of members during the show the hope had been raised that the County Council would see fit to establish a Dairy School, as had been done in various other counties.

Erected for the Show of 1901, this arch in Fore Street, Bodmin was outside the old Market House, as can be seen from the bull's head, which is still today part of the Market House facade, just visible in the top right hand corner.

'Romping Girl', a Shorthorn heifer owned by Thomas Francis Roskruge of Trenethick Farm, Wendron, Helston, winner of a 3rd Prize at the Royal Cornwall Show of 1902.

The Grand Parade of Cattle, circa 1900, with, as today, photographers in action.

Bodmin, the shows real home town, again played host in 1901, in a year which saw the death of one of the shows greatest supporters. Arthur Tremayne of Heligan had played a vital role in the success and expansion of the Association, proposing its migratorial type existence and then again proposing the addition of the show's second day. He had been known to have been involved in all that went on from at least the 1850s and was to be sadly missed by those involved with the Association.

Camborne in 1902, unfortunately, again hit bad weather on the second day, with the resultant loss having to be added to those made in the previous two years. This year was also to see no classes being staged for Hereford cattle, owing to no entries being received for the classes in 1901 and the awards therefore being dropped for the show at Camborne.

However, membership was steadily increasing and the lower attendances were only as a result of poor weather, rather than a lack of interest in the activities of the Association.

By this time, the role of the Secretary, Henry Tresawna, had gradually increased and it was felt that as the post was of course only part time, an assistant should be found. The post of assistant to the Secretary drew an amazing 38 applications, with that of Mr. J. F. Crewes proving successful. A salary of £20 per annum was agreed and the appointment confirmed. It would of course, be J. F. Crewes that was to take over from Henry Tresawna on his retirement as Secretary in 1910.

Moves were made to obtain the attendance of HRH The Prince of Wales at the St. Austell show of 1903, owing to the fact that he would already be in the county around the period of the show. Unfortunately, the reply from his Private Secretary, Sir Arthur Bigge, pointed out the difficulties of such a visit in that year. A further approach was made to seek the Presidency of HRH The Prince of Wales in 1904, and therefore his attendance at the show in that year.

The answer received pointed out that as His Royal Highness would be visiting the county in 1903, he would be unable to visit again in 1904! What a difference from today, when various members of the Royal family are often seen in Cornwall at many times throughout the year!

A VIP visit in 1903, which did, however, come about was that made by Mr. Hanbury, President of the Board of Agriculture. The invitation was proposed by a Mr. James Thomas, and then left in the hands of the Agricultural Sub Committee of the County Council. During the next year, a request was received from the Board of Agriculture, for the Association to appoint correspondents, to keep the Board informed of agricultural developments within the county. The first correspondents consisted of the President for 1904, Mr. John Williams of Scorrier and the Secretary, Henry Tresawna.

Up until this time, the judging of livestock and horse classes had all taken place on the first day of the show, but a request from the Local Committee saw 6 horse classes being included for the second day in 1903. The idea was to fill the judging rings for more of the show, and of course, today judging is seen throughout the show period.

Also begun in 1904, and carried on today, was the practice of measuring horses on the showground, if entered in classes with height restrictions. This task, always time consuming and sometimes unpopular, is however, a necessary practice although the introduction of life certificates has obviously eased the situation.

1904 also saw the appointment of an Assistant Veterinary Surgeon at a salary of 4 guineas per annum. No doubt, the work of the main show vet had also increased over the years, leading to the need for extra help. The gentleman appointed to the position was a Mr. E. R. Smythe of Falmouth, who was initially appointed for a period of one year, as assistant to Mr. Thomas Olver.

It is interesting that in 1905, for the show at Newquay, when the various tenders were considered for the copyright of the show catalogue, the name of Oscar Blackford of Truro appears for the first time with the winning bid of £40 5 shillings. Blackford's were to produce the catalogue for a great number of years, with Netherton and Worth, also of Truro, often winning the contract as well. Blackford's, although now in different hands and moved to St. Austell, have printed this history, and along with the other items they produce each year continue their long association with the show.

The stand of C. H. Harris, engineers of Wadebridge at the Royal Cornwall Show, circa 1905.

A busy showground scene; the Royal Cornwall Show, circa 1905.

Newquay in 1905, proved an extremely welcoming host town, with the arrangements made by the local committee finding great favour amongst those visiting. The dairy section continued to grow, with for that year a silver cup on offer for butter making, given by the Hon. Mrs. Tremayne. The Judge, Mr. Lorham, a well respected authority, reported that some of the exhibits were *'the best he had ever examined'*, although he went on to stress *'the necessity for greater attention being given to the drawing off of the butter-milk, and thoroughness in washing'*.

In 1906, Redruth was again to see a visit of the show, and with a wonderful attendance, probably due to the good weather, a substantial profit was made, amounting to some £496, a very considerable sum for the period. The Band of the Grenadier Guards performed on the band stand, and the permanent grandstand belonging to the Redruth Exhibition Society must have proved very useful and saved the necessity of erecting temporary seating.

The Trustees of the Rodd Testimonial Fund were still providing annual prizes for award to agricultural implements, with often the prizes taking the form of silver and bronze medals. For 1907 at Liskeard, the normal prizes *'for new and improved machinery and implements'* were offered and were supplemented by a silver and bronze medal to *'be offered for the Best Hay Turner—Exhibitors to provide horses—the field and grass to be provided by the Society'*. Such practical demonstrations of new machinery must have created much interest and probably considerably helped the sales of the firms concerned!

December of 1907 saw an approach from the Royal Agricultural Society of England, requesting the cooperation of the Association with

A packed showground at Redruth in 1906. Notice the stand of Mallett & Son of Truro, still exhibiting at the show in 1993.

Part of the machinery section of the Redruth Show of 1906.

89

A view of the Main Ring at the Redruth Show of 1906.

The Band of HM Grenadier Guards on the bandstand at the Redruth Show of 1906.

Ready for any emergency, the Ambulance Station at the Royal Cornwall Show at Redruth, probably 1906.

regard to tuberculosis in cattle. A meeting had been called in London for later in the month, and Mr. William Hoskin was chosen to attend as a representative of the show.

This same meeting also heard, through a letter from Mr. Cressy Treffry, then the Director of the Showyard, that prizes for horses were available, through a rather novel source. It was learned that the Government, through the War Office would give prizes for 'remounts' to the tune of £9. The term 'remounts' refers in this instance to three types of horses: viz,

Cavalry Horses from 15½ to 15.2½ hands, from 4 to 6 years old

Draught Horses from 15.3 hands

Cobs from 14.3 to 15½ hands.

It was agreed that the above three classes should be staged, with a further £9 from the Association being added to the prize money offered by the Government; the classes to be held on the second day of the show.

June 1908 saw the Association celebrating fifty years of migratorial shows, with a visit to Helston under the Presidency of His Grace The Duke of Leeds, through his lands at Godolphin, was a keen supporter of the Association and chaired that year's meeting of members during the show. At the Annual General Meeting in October, much was made of the progress of the Association since the introduction of its itinerant status. Some 700 people were reported as then being members, and with an ever growing number of classes the future looked very secure.

The stand of T. F. Hoskin & Co. of Marazion and Penzance, at the Royal Cornwall Show of 1908 at Helston.

The dairy section of the show had grown significantly by this time, with for 1908, 162 entries being seen in the buttermaking classes, more than double the number of 4 or 5 years before. Mention was also made in that year's annual report of the important work being undertaken by the County Council Dairy School, who with *'Miss Nicholas' capable instruction'* was of *'great value'* to the dairy farmers of the county.

By this time, the cost of staging the show, not including the local shows run by the local committee, had risen to over £2,000, with just over £800 being distributed as prizes for livestock and implements.

St. Columb, a town never before visited by the Royal Cornwall Show was chosen as the destination for 1909. This was to be Henry Tresawna's 50th and last show as Secretary, although as with the last show of the immediate past Secretary, Albert Riddle, the occasion was to be graced with a Royal visit.

Both their Royal Highnesses, The Prince and Princess of Wales visited the show at St. Columb that year. The Prince of Wales, as Patron of the Association, was to later write to Henry Tresawna, expressing his *'sincere regret'* at his resignation as Secretary, and fully acknowledging the *'great debt of gratitude'* owed by the Association for all that Henry Tresawna had done whilst in office.

A testimonial fund was set up to solicit subscriptions from members, for sums not exceeding £1, to make a suitable presentation to the retiring Secretary. Just over £200 was raised through the subscriptions given by some 353 members and a *'cabinet of silver'* was eventually presented. Unfortunately, Henry Tresawna was unable to

To help commemorate the visit of Their Royal Highnesses, The Prince and Princess of Wales to the Royal Cornwall Show at St. Columb in 1909, this arch was erected at Black Cross, between St. Columb and St. Columb Road.

enjoy his retirement, as he died shortly after giving up his position within the Association.

The post of Secretary was advertised and a Committee designated to examine the applications received. A salary of £55 was offered for the post, which was, of course, part time, and after deliberation the Assistant Secretary, Mr. Crewes was appointed.

Whilst at the show in 1909, The Prince and Princess of Wales were presented with a *'beautiful casket'* by the tenants of Duchy property who also offered an address to the Royal couple, who of course, were soon to become King George V and Queen Mary.

Her Royal Highness The Princess of Wales visiting the Royal Cornwall Show at St. Columb in 1909. Her Royal Highness is accompanied by Major Charles Edward Whitford, Chairman of the Local Shows Committee.

Following the death of His Majesty King Edward VII, a telegram was forwarded to King George V offering the *'most loyal and respectful sympathy in the great grief which the whole nation is sharing'* of the Association of which His Majesty had been Patron for nine years. A reply was subsequently received from the King thanking the Association for *'their kind message'*.

September of 1910 saw for the first time a reigning monarch as Patron of the Association. His Majesty King George V had graciously consented to continue as Patron as King, following his previous support as Prince of Wales. His Majesty also consented to His Royal Highness The Prince of Wales acting as Vice Patron.

On a more mundane note, catering, always a subject of much discussion at any show, was the cause of a great deal of debate, even in 1910. Annual tenders were received for the provision of catering facilities within the showyard as well as for the sale of fresh fruit.

The arrival of Their Royal Highnesses, The Prince and Princess of Wales, at the St. Columb Royal Cornwall Show of 1909.

However, standards at the various shows were at this time somewhat variable and a method of better regulating the position was sought.

One Council member even suggested the Association take on the catering themselves, although thankfully this suggestion was turned down.

A system of prices was agreed for various sized refreshment tents with luncheon and tea tents of 20 ft. wide being charged at 1/- per foot run and tents 'used specially for drinks' charged at 2/- per foot run. Applicants for luncheon tents also needed to provide information on the menus they were to offer at 2/6 and 1/6 per head.

The problem of the transport of the 'show plant', the equipment owned by the Association for staging each year's show was also creating a problem at this time. It was suggested that the various equipment should be moved direct from the previous year's site to the showyard to be used in the following year, rather than having to move everything twice. For this plan to succeed, the exact location of the next show needed to be chosen much further in advance and the possibilities of this change of practice led to much discussion.

With no permanent showground, much work was needed annually to prepare for the arrival of the show at it's chosen site. A presentation to mark the assistance given by one such man was made during 1911. Mr. W. H. Joery of Probus, had, in that year completed 50 years as Show Yard Contractor and to mark the event, a cheque in the sum of £30 together with a suitably inscribed silver cup, given by various members of the Association was presented. Once again, the longevity of people's association with the show is shown.

95

The Official Opening of the Nature Study and Handicrafts Exhibit, probably circa 1910.

The Exhibit of the Cornwall County Farm and Dairy Co-Operative Society, probably circa 1910. On sale was Soda and Milk at 3d., Milk only at 2d. and 1d., Sponge Cake at 1d., Fruit at popular prices, butter and cream.

During the latter part of 1911, discussions were held with the Mayors and other representatives of the towns of the county, with regard to the possibility of their issuing an invitation to the Bath & West & Southern Counties Society to visit the county during 1913. An invitation was issued on behalf of Truro and subsequently accepted by the Bath & West. It was agreed that no Royal Cornwall Show should be held in that year, but that £100 should be granted for the provision of local prizes at the Bath & West show, plus the sum of £150 that would be made available to the local committee in Truro.

In the meantime, the town of Penzance, who had been one of the contenders in the issuing of an invitation to the Bath & West, were to welcome the Royal Cornwall instead in 1912. This year saw the name *'South Ham'* sheep giving way to the term *'South Devon'* sheep, and further second day horse classes being staged for both *'jumpers'* and *'harness'* horses, who had competed on the first day but not won a prize.

Thought to have been taken in 1912, a view of Alverton in Penzance, welcoming the Show.

For the first time, cattle from the Royal Farm at Windsor were to be seen at the show, with a great deal of interest being shown by those viewing them.

Plans for the Bath & West progressed with a large county committee being formed, including all members of the Council of the R.C.A.A. and also of the County Council together with the Mayors and representatives of many of the towns of the county. A total of £1,700 had to be raised within the county to assist with the costs of staging the show, with for instance some £200 having to be paid to

The Truro Water Company for the supply of water to the show yard. It was hoped that perhaps half of this sum could be recovered through the sale of piping etc. after the show.

The Main Ring, thought to be the Penzance Show of the Association in 1912.

The transport of stock by rail, played an important role in the staging of the Show. Seen here is a pig in a crate being unloaded from a Great Western Railway cart.

A decoration committee considered how to spend the £100 allocated to them from the county fund for the adornment of the city with street decorations planned:

From the station, Richmond Hill, Ferris Town, River Street, St. Nicholas Street, Boscawen Street, Lemon Street, to the Show Yard.
From Princes Street to Boscawen Bridge.
From King Street to the Post Office.
Plus the erection of *'one or two good arches'* at various points.

In addition it was suggested that some fir trees could be obtained to decorate either side of Lemon Street.

The show was to be held on Higher Newham Farm, now occupied by the Burley family, who for many years have been ardent supporters of the Royal Cornwall. Stock was to be transported by train to Newham and then up the private lane to the farm.

The Bath & West at Truro in 1913 proved, once again, Cornwall's ability to play host to the farming counties of the South West. Over 60,000 attended the show over its five day period, with a meeting of the Royal Cornwall Agricultural Association's Council being held in the Council Pavilion during the show. Many local competitors were to be seen amongst the prize winners and the whole event was described as a great success. At the meeting during the show, three towns applied to invite the Royal Cornwall to visit them in 1914, these being Launceston, Bodmin and Fowey. After much debate, Fowey won the vote, with J. de Cressy Treffry being unanimously elected President for the ensuing year.

A new class was again to be introduced for the second day of the show, this time for ladies' hacks, to be a mare or gelding, any height or age, to be ridden by a lady. The President offered to give a silver cup, valued at 5 guineas, to be offered as a first prize for the new class, with the balance of the prize money being made up by the Association.

The efforts by the War Office in relation to the provision of horses for the army were again to be seen at the show in 1914. Four infantry and two cavalry horses were to be sent to Fowey by the War Office for parade in the Show Yard, to show the horse breeders of the county the type of animal required for the army.

For the first time, the show was to see covered shedding being provided for the exhibitors of implements which met with general approval, as did the exhibits displayed that year by the Australian Commonwealth and the Canadian Government. What with the presence of The Royal Garrison Artillery Band of the Citadel of Plymouth and the great support of *'the Squire of Place House'* as Mayor of the newly formed Town Council a memorable visit to Fowey was made.

Also, a *'new permanent facade'* was said to have greatly improved the look of the main entrance to the show, although the cost, at £95, appeared at the time to be quite high.

Prize winning bull, circa 1915.

Sheep judging, a very serious affair! Probably circa 1915.

Shortly after that show, War was declared and at the next meeting of the Council in October 1914, the sum of £50 was voted to be contributed to the National War Relief Fund. A meeting in February of 1915 considered in great detail whether the show at Camborne in that year, should in fact be held. The Local Committee had expressed grave doubts as to whether or not they would be able to raise the necessary sum, owing to the amounts being subscribed to funds set up to assist the war effort.

The Council felt it important for the show to proceed if at all possible and agreed to pay the cost of the rent of the show yard to help the Local Committee. On this basis the plans for the show were to proceed.

Also at this time, the Association's help was requested by His Grace The Duke of Portland, President of the Royal Agricultural Society of England. The co-operation of the Association was needed in the carrying out of the objects of the Agricultural Relief of Allies Committee and Mr. M. H. Holman was elected to serve as the R.C.A.A.'s representative on the Central General Committee. At a later meeting the whole of the Council of the Association together with others nominated by the Cornwall Farmers Union formed a County Committee to assist with this relief work.

The show at Camborne in 1915 was surprisingly successful, with good entries of livestock and one of the best shows of horses seen for

Side saddle jumping at the Royal Cornwall Show of 1915, held at Camborne.

some years. During the show *'detachments of the Military'* were admitted to the show on both days and resulted in 40 people being recruited for the army. Red Cross nurses made a collection at the show and raised the magificent sum of £50 towards the work of their Cornwall Branch.

The Annual Report for the year reported, amongst other deaths, the loss *'on the field of battle'* of Capt. the Hon. T. C. Agar-Robartes, M.P., of Lanhydrock who was one of the younger of the Association's Vice Presidents and who had stood as President of the Show in 1911 at St. Austell.

Plans had been instigated for the show to be held at Callington in 1916, following an invitation having been received from that town. Doubt had been expressed as to the facilities available, and a delegation sent to view the proposed sites, as well as the hotel facilities and railway yard etc. Those who had visited the area were impressed by the enthusiasm of the local people for a visit by the show and apart from a request to construct extra sidings at the railway station, Callington was expected to make an excellent venue.

The impetus given to agriculture in the area by the establishment of the Prince of Wales' *'Royal Farm'* at Stoke Climsland was also thought to have been a factor in Callington's enthusiasm to host the show.

However, by July 1915, the Council had decided that no further preparations should be made for the next show, until *'peace had been declared'*.

Unfortunately the declaration of peace was to be rather farther into the future than had been envisaged and although the meetings of the Council continued, no show was to be seen until 1919.

4. In War and Peace   1919 – 1939

Throughout the hostilities, support of the war effort was continued, with it being decided that subscriptions should still be paid, to enable the Association to make donations to the various war funds. £25 was granted in October of 1915 to the Cornwall Branch of the British Farmers Red Cross Fund, with the sum of 250 guineas being granted to the Agricultural Relief of Allies Fund in April of 1916. To assist with the fund raising and in view of the lesser amount of work entailed through the war years, the Secretary, Mr. Crewes, readily agreed to reducing his salary by half, to £50. Throughout the war the Association, through its Secretary strived to raise money for the Allies Fund with the co-operation of the Farmers' Union Branch Secretaries and Collectors. By October of 1916 £1,000 had already been promised, and by the time the fund closed in 1921 some £3,760 had been raised by the county under the guidance of the Association's Secretary.

A letter of thanks from His Grace The Duke of Portland and a Diploma from the Ministry of Agriculture of the Government of Belgium was received in recognition of this help, and in total £251,814 was distributed amongst the farms of our allies.

A meeting held on the 1st January 1919, heard that at last it looked possible to stage a show, as the Ministry of Munitions had withdrawn restrictions and the Great Western Railway had offered no objection to the proposal. It was thought sensible to choose a main line town as the venue, with Truro being picked as most suitable, for what was envisaged as being a very large event. Callington, was asked to forego their claim to the show that year, and in fact were to act as host in 1920.

Plans for Truro in 1919 progressed with several changes being seen to the arrangements made for the pre-war shows. For instance, for the first time, arrangements for motor cars are mentioned, with cars being allowed to enter the showyard at a charge of 4/- each *'exclusive of chaffeur'*!

The members of the Press were to be catered for with two Press tickets being issued each to The Western Morning News, the Western Daily Mercury, the West Briton and the Royal Cornwall Gazette. The Secretary also negotiated for the *'cinematograph'* rights for the show, with the sum of £7 being paid with a further *'acknowledgement fee'* of 10/6 being offered for other photographic rights.

The visit by His Royal Highness The Prince of Wales, Duke of Cornwall, was a great highlight in what was to be a record breaking show, with a paid attendance of 29,768, which was not to be beaten until the first show after World War II in 1947.

Permission was given for 120 members of the Women's Land Army and a similar number of Boy Scouts and members of the Cadet Corps to form a Guard of Honour near the grandstand, where tickets were to be on sale for reserved seats at 5/- each. Admission, to the show was to be 3/- on the first day until 2p.m., and thereafter 2/-, and on the second day, 2/- and then 1/6 as a reduced rate.

The Band of H.M. Royal Marines was booked to entertain in the Show Yard, with further concerts being given on each evening of the show in The Market House.

In the catalogue of that year, a host of names, familiar to many who have regularly attended the show, can be seen. Cattle included Devons from the herd of HRH The Prince of Wales, K.G., at the Home Farm, Stoke Climsland, with the judge, W. Brent of Clampit, Callington, being from a family still very much involved today. Others mentioned include the Eustices' of Bezurrell, Gwinear, the Blights' of Tregonning, Breage, the Bolithos' of Trengwainton, Penzance, the Williams' of Scorrier House, the Falmouths' of Tregothnan, the Hoskings' of Fentongollan, Probus, the Warnes' of Tregonhayne, and the Johnstones' of Trewithen. All of these families still have members on the Council of the Association today, and in many cases also show regularly.

Some 45 firms exhibited at the show that year, with those from Cornwall including: The Bodmin Granite Co. Ltd., J. Carne & Sons, St. Keyne, A. D. Brewer & Sons, Grampound Road, Wm. Brenton, Ltd., St. Germans, M. L. Blamey & Son Ltd., Truro, Veryan and Penryn, John Edwards & Sons, Truro, Hosken, Trevithick & Polkinhorn & Co. Ltd., Truro, Royal Cornwall Gazette & Cornwall County News, Truro, West Briton, Truro, John Tonkin, Truro, Taylor's Garage Ltd., Penzance, W. Penrose & Son, Truro, and Bullen Bros. Ltd., Truro.

Many of these firms are still in being today, if not perhaps under the same name and were supplemented that year by many others from all parts of the country.

Callington, so keen to see a visit of the show in 1916, was at last to receive that long awaited influx in 1920. The town and surrounding area gave a hearty welcome to the Association and with the Presidency of His Royal Highness The Prince of Wales, the two days of the show were a success.

Unfortunately, The Prince of Wales was unable to visit the show owing to being overseas on a tour of some of the *'Dominions of the Empire'*. During a general meeting of members on the second day of the show it was decided to ask Mr. Peacock of the Duchy of Cornwall Office *'to cable to Australia loyal congratulations to our President on the success of the Callington show—great stock and implement entries, splendid attendance'*.

This message together with the news of the many prizes won by the Prince's livestock, no doubt came as a welcome reminder of home.

That year's show saw the introduction, for the first time, of prizes offered for Gloucester Old Spot pigs and also the return of a class for wool. Many new types of agricultural machinery could be seen with tractors being heavily promoted by such firms as Hosken, Trevithick, Polkinhorn & Co. Ltd. In their case *'A Tractor for Every Farm'* was the slogan with such makes as Austin, International Junior, Titans, Saundersons and Cleveland Catarpillar tractors on offer.

The next year, when the show was to visit Falmouth, the prize money in several classes was to see significant increases, with many increasing by 25% to 50%. From about this time, the Association began to affiliate to several of the main breed societies, with the Shire Horse Society being first in 1921 at an annual subscription of 2 guineas. This move gradually led to a more regulated approach to the provision of competitive classes, with approved judges needing to be chosen and strict breed society rules adhered to.

By 1922, the Association was also affiliated to the English Guernsey Cattle Society, at a cost of 21/- per annum, in order to obtain that Society's Special Prizes. This was also the first year that British Friesian cattle were to be granted classes at the show, although in at least the first and second years, half of the prize money had to be provided by the British Friesian Society as the entries were so few in number.

For the following year the milking trial classes, growing in popularity since the war, were to also be extended with three prizes of £7, £4 and £2 for the quality of milk and three similar prizes for the best butter test from the milk produced from the animals entered. The regulations for these trials were to be submitted by Capt. G. H. Johnstone of Trewithen and the, now famous, Miss Nicholas the dairy instructress. It was also hoped that assistance could be obtained from the Council of the Cornwall Milk Recording Society.

The two cheese classes which had been run for a few years, were, for 1923 to be amalgamated into one class for any variety of hard cheese, with four prizes of £3, £2, £1 and 10/- also being offered by Mrs. Williams for milking competitions for girls and boys under 16 years of age.

An application was received for the introduction of classes for 'Long White Lop Eared' pigs, but it was felt unnecessary to stage special classes and that those who wished to show should enter in the 'Any Other Breed' classes.

November 1922 saw the resignation through ill health of the Association's Secretary, Mr. J. F. Crewes, who was to unfortunately die within the next few weeks. It was resolved that the post should be advertised at a salary of £100 per annum, plus out of pocket expenses and 5% of the subscriptions collected and the following particulars forwarded to applicants:

'The duties of the Secretary are to take and keep the minutes of the transactions of the Association, conduct the correspondence, keep the accounts and collect and receive the Subscriptions which he shall pay into Lloyds Bank.

He shall issue Prize Schedules and receive entries, and shall also generally arrange for the laying out of the show-yard for the annual exhibition in conjuction with the Director and as Executive Officer shall be responsible for the routine work of the Association.

The salary is £100 per annum and out of pocket expenses plus 5% on subscriptions received for which he will be required to provide office accommodation and clerical assistance.

Applications in applicants own handwriting should be sent in a sealed envelope endorsed "Secretary Royal Cornwall Agricultural Association" to the President, P. D. Williams Esq., Lanarth, St. Keverne R.S.O. not later than Friday 15th, third class railway fare within the county will be allowed to selected candidates—

Canvassing any Member of the Council will be deemed undesirable'.

Following a meeting of the Council, on December 22nd 1922, the post was offered to W. Courtney Hocking Esq. of Midland Bank Chambers, Truro, who was to serve the Association until up to the second World War. Many applications were received for the post and owing to the high standard of those applying, six people instead of the suggested four were put forward for final selection.

Pictured at the Camborne Royal Cornwall of 1923, is 'Chilcampton Day Dream' and her foal, winner of the Agricultural Mare and Foal Class for Mr. L. B. Beauchamp of Norton Hall, near Bath.

One problem which faced the Council when making the necessary arrangements for the Camborne show of 1923, was the lack of plans and specifications relating to the erection of the show yard. Presumably, owing to the death of the previous Secretary, such information had gone astray! The decision was made to employ a Mr. A. J. Cornelius as Surveryor to the Association at a salary of 20 guineas with the assistance of Mr. Doble Joery, who had previous experience of building for the show, in a capacity of Clerk of Works.

The services of the Cornish Tent Company were engaged *'at a less price than last year'*, and the R.A.C. employed to oversea the parking of cars, with the Association receiving 75% of the charges made to drivers. The possibility of issuing members with badges, rather than the customary admission cards, was discussed, but a change not recommended for that year.

The show was held at Lower Rosewarne Farm, Camborne, and the efforts of the new Secretary were rewarded with a most successful event, with the new Milk and Butter Trials and Milking Competitions and the expanded jumping competitions, now affiliated to the Showjumping Association, much praised.

Wadebridge was again to see a visit by the show in 1924, with Colonel C. R. Prideaux Brune of Prideaux Place, Padstow, acting as President. An introduction for that show at Wadebridge, was to be a class known as the *'Premium Bull (Ministry of Agriculture Live Stock Improvement) Class'*. This scheme designed to encourage the breeding of better quality stock, lasted for many years and was a popular aspect of the cattle section of the show.

Another innovation that year was to be the inclusion of classes for young farmers for the judging of livestock by points. For 1924 two classes to cover the judging of a dairy cow and a long wool sheep were included, with for 1925 the addition of a third class for the judging of a bacon pig. Obviously, the formation of the actual Young Farmers' Clubs was still some years ahead, but this move to encourage youngsters to practice their livestock skills marked the type of atmosphere which was to lead to the creation of the Clubs later on.

Each year, in preparation for the show, a host of small details were annually debated, which today, in the main, stand from one year to the next. The tenders received for the various types of catering and other rights, which represented an important and much needed income to the show, as indeed they do today, were a cause for annual concern. With the gradual growth and therefore complexity of the show such things as the provision of toilets etc. were of prime importance! For instance, for the show of 1924 not only did Messrs. Terrell & Sweet of Tavistock win the tender for the sole rights for the sale of tobacco, chocolates and sweets, they also won the rights for the provision of cloak rooms. This right was granted in return for a fee of £17, however, a condition was imposed that the cloak room charges should not *'exceed 3d. per person admitted'*!

Other tenders at the time included the rights for the sale of *'green fruit and ices'*, won in that year by Messrs. Eathorne & Son of Helston for the sum of £8 15/-. The rights for ices had only been seen for a couple of years by this time, thus reflecting the difficulties that must have been experienced by those keeping the ice cream at the correct temperature.

F. Rodda of Camborne, a well known family of the area, won the rights for the sale of drinks at £12 10/-, with Mr. A. W. Pascoe of Helston acting as the main caterer and paying a royalty of 2d. per meal sold, to the Association.

To encourage non agricultural visitors to visit the show and to educate those who came, special parades of the various breeds of cattle were planned for that year, for both winners and non prize winners, to give an opportunity to display the breed characteristics of the beasts on show. From a livestock point of view, one matter which was causing concern during this time was the possibility of foot and mouth disease. Regulations had been brought into force to help combat the disease, although concern was expressed at a Council meeting of the Association in April of 1924 of the loophole created locally by Plymouth. Both Cornwall and Devon had adopted 'protective measures' but it seemed that Plymouth, with its imports and exports of cattle was a possible area for concern. A request for the branding of cattle imported through Plymouth for slaughter was made, following a resolution passed by the meeting.

That year's show at Wadebridge met with bad weather conditions and a loss was subsequently made. The Annual General Meeting, held in the Will's Memorial School, Egloshayle Road, Wadebridge, heard that just over £250 had been lost on that year's event, although, owing to the inclement weather, it was felt lucky that the deficit had not been greater.

Throughout the 1920's gradual changes and improvements were to be seen throughout many areas of the show, with, for instance, the introduction of such things as the provision of accommodation for herdsmen and grooms, in areas separate to the actual cattle sheds and horse boxes. A separate *'canteen'* was also to be provided for those looking after livestock and the officials of the show. Much thought was also being given to the provision of a *'garage'*, the term then used for car parks, at each year's show. Whether to farm this item out or to hire in labour was discussed, but the need to find a suitable site to allow for the parking each year, was, by now, an accepted fact.

The costs of running the show also met with much detailed scrutiny with many refinements being suggested. The improved accommodation being requested by those exhibiting in both the livestock and trade sections, had to be paid for, and a revised scale of charges for both entries and trade stand space was agreed.

These charges, for the Helston show of 1925 were fixed as follows:

	Members	Non Members	
Horse Entries	20/-	30/-	
Cattle ,,	10/-	20/-	
Sheep ,,	7/6	15/-	
Pig ,,	10/-	20/-	
Shoeing		5/-	
Dairy		4/-	
Trade Stand Fees:			
Ordinary Shedding		5/-	per foot
'Machinery in Motion' Shedding		6/-	,,
Open Ground		2/-	,,

In comparison with 1992, some of these charges appear quite high, with for instance the average horse entry costing £5.50 for members and £8.00 for non members and open trade space being sold at £5.05 per foot run. Many would say however, that the value of the prizes offered in the 1920s was greater in real terms than today, which is probably true, but of course the costs faced by the show today, bare little relation to those experienced some seventy years ago.

An important change for that show of 1925 was the employment of a main showyard contractor, with much of the equipment necessary for the staging of the show being hired and then taken away, rather than purchased and then sold at auction as had been the practice for a great number of years. The contractor who was to be employed in this manner, and who is still the main contractor to the show today was the firm of L. H. Woodhouse & Co. of Nottingham.

That first contract, for a period of two years, proved most successful and marked an important development in the expansion of the show and a happy and mutually beneficial relationship lasting to this day.

The Judges and Secretaries of the Poultry Section, Royal Cornwall Show, Helston, 1925.
From left, C. H. Horne, Judge; J. Pryor, Secretary; W. Powell-Owen, Judge; R. Fletcher-Hearnshaw, Judge; E. J. Davis, Secretary and John Roskrow, Judge.

The show at Lower Windmill at Launceston in 1926, met with a serious problem in a form of a very restricted train service, forcing many who would normally travel in that way to take to cars instead. The site, covering some 20 acres, met with the approval of the Council and a successful two days were had.

A meeting of members during the show heard the news that His Royal Highness The Prince of Wales had accepted the Presidency of the Show for the forthcoming year. 1927 was to be celebrated as the Centenary of the formation of the Association, which was of course the case, although the true roots of the show go back to Bodmin in 1793, with the formation of the Cornwall Agricultural Society. The change of title and area of activity, with the move to Truro in 1827, was thought to be the focus of a host of special activities connected with the show in that centenary year.

One major aspect of 1927, was the provision of 57 special centenary cups to be won outright by the exhibitors winning them. These cups were made to a standard design by Messrs. Page, Keen and Page Ltd. of Plymouth, with each one being made of solid silver and weighing 13 ounces. At a cost of £5 each, all of the cups were donated by various supporters of the Association and offered for competition throughout the livestock and dairy classes of the show.

In 1993 when we are celebrating the Bi-Centenary of the true formation date of the show, the cost of such an exercise would unfortunately be prohibitive, with each cup probably costing some £500 plus!

New also for the 1927 show were classes for Aberdeen Angus cattle. Now no longer seen at the show with their own section, the breed were that year granted four classes, with two Centenary cups being offered by The Rt. Hon. the Viscount Falmouth and Admiral Sir Charles Graves Sawle for the best male and female exhibits. In addition, a Silver Medal was given by the English Aberdeen Angus Cattle Association for the Champion Aberdeen Angus Beast.

Amongst the prize winners that year, the President, His Royal Highness The Prince of Wales, was to win no less than three of the special centenary cups, with one being won in the Devon cattle and two in the Shorthorn cattle classes.

Amongst other prizes offered, His Royal Highness also gave a cup valued at £25 for the best premium bull of either a dairy or beef breed, which along with a cup for the best bull bred in Cornwall, was won by A. Jeffery & Son of Mullion. This success marked a long and fascinating involvement for a local family who were to win many awards with cattle from their very small herd of shorthorns of the Scottish type. The herd started with cattle purchased from the Royal farm at Stoke Climsland, then under the management of Mr. Annen, saw great success, with many major trophies won outright. One magnificent cup, given by HRH Prince Edward, Duke of Cornwall, was to be re-presented by the family for 1933, after being won outright after three successive wins, only to be won outright by W. Jeffery again in 1935.

The show of 1927 was held at a site at Treliske, on the outskirts of Truro, on the estate of the late Sir George Smith. For the first time, one admission fee covered the entrance to all sections of the show, including the *'local shows'* of poultry, dogs etc., which previously had charged separate fees. This change was to be continued, with today, only the Flower Show charging an additional fee.

The show was declared open by His Royal Highness The Prince of Wales, whilst standing on the bandstand, and his speech was heard throughout the showyard through specially positioned loud speakers.

During the year, an important presentation to the Association was made, bringing into its possession the oldest document, relating to the early days of the show in its keeping. A framed *'Premium List'* detailing the prizes to be offered for the year of 1803 was presented by Sir Hugh Molesworth-St. Aubyn of Pencarrow, and in 1930, a further similar document, detailing the awards made at the show of 1804 was presented by George Johnstone of Trewithen. Both of these documents now hang in the offices of the Association at Wadebridge, whilst various other items, relating to the very first years of the original society are in the keeping of the County Record Office at Truro.

Following the excitement of the so-called centenary celebrations, the thanks and congratulations of many organisations involved with the show that year were to be received. The dairy section again met with great enthusiasm and the Dairy Superintendent, George Fortescue and Miss Nicholas, the Dairy Instructress and her staff were thanked for all their hard work.

The Police, who always play an extremely important role in the smooth running of the show, were also congratulated on the efficient manner they had looked after their duties. The Deputy Chief Constable of Cornwall, Inspector Hosking and Sergeants Wherry and Prout received a special vote of thanks and the congratulations of the Association were passed to the Chief Constable of the county, Col. Sir Hugh Protheroe Smith, on the honour recently conferred on him.

A note, added to the Minutes of the AGM for that year, records that some 30,000 gallons of water were used in conjuction with the 1927 show and that this fact should help in the planning of the water supply for future shows!

Among those thanked, was Eric Westby, the first of the contractor's foremen, who, on behalf of L. H. Woodhouse & Co. of Nottingham, was to oversee the construction of much of the showyard up until the early 1960's. In appreciation of this hard work, the Association , in 1927, granted him the sum of ten guineas as a gratuity for his help over the previous three years.

1928 saw the show returning to Bodmin for the first time since 1901 and the efforts made by the local committee, under the presidency of Mr. Gerald Harrison MP, the chairmanship of the Mayor, Mr. Browning Lyne and the secretaryship of Mr. S. T. Hore, saw the sum of over £800 being raised to cover the expenses levied on the host town.

The show was held on a site at Pryors Barn, about half a mile from the town on the Bodmin to Liskeard road and was to see one of the largest attendances ever, although of course not competing with that of either the first show after the war or of that of 1927.

A new feature for that year was *'the Avenue of Progress'*, where exhibits included the Educational Exhibits of the Cornwall County Council and Seale-Hayne College and the Ministry of Agriculture with their Pig Marketing Demonstrations.

'Henfors Goalkeeper', exhibited by Mr. C. Tregoning of Gulval, Penzance, at the Penzance Royal Cornwall Show of 1929.

The following year, in Penzance, the show was to occupy what was described as one of the loveliest sites ever chosen. Following the election to the Presidency of Lt. Col. E. H. W. Bolitho, his offer of a site at his home, Trengwainton, was quickly accepted and the annual report describes the site as being *'on high ground, beautifully wooded and affording a magnificent view of Mount's Bay and the surrounding country'*.

The pleasant outlook of the site, seems to have particularly appealed to the exhibitors of livestock, with the second highest number of entries ever on show. Over 240 horses, and over 300 cattle were to be listed in that year's catalogue, along with a host of other competitive entries, ranging from acetylene welding, first introduced in 1927, to the judging of butter by points.

Seen for the first time, in the Main Ring that year, were to be demonstrations by *'Champion Sheep Dogs'*, which apparently proved very popular.

The show at Tremabe, Liskeard in the following year, 1930, went well, although a dispute concerning the cost of reinstating demolished hedges and fences lasted for several months after the show. The local committee had failed to take note of their obligation to cover the cost of any such work, and on receiving an invoice from the Association's contractors, refused payment. Although efforts to reconcile the situation were made, several meetings passed before a satisfactory compromise was reached. This type of problem, is typical of the minor hiccups experienced whilst the show was travelling and as the show grew so did the range of difficulties to be overcome!

One more major hurdle concerned the venue for 1931. Redruth, the likely host town for that year, had unfortunately had to decline the honour owing to the *'acute distress prevailing in the mining division'* and the likely problems that would have been experienced in trying to raise locally the necessary funds.

St. Ives were next to offer a home for the show that year, but after a detailed survey, various problems were encountered. The sites around St. Ives had traditionally created difficulties owing to the small nature of the fields and the narrow roads. The Chief Constable advised that St. Erth would be the nearest railhead, for the transportation of stock and machinery, etc. and that virtually all of the show traffic would then need to pass through the town of St. Ives, on their way to the showyard. After much debate, the Council of the Association came to the decision that such disruption of the arrangements for the show could only cause much dissatisfaction and that another venue should be found.

St. Columb, thankfully, came to the rescue, with the offer to bring their invitation forward from 1932 to 1931, and following a special resolution, to allow the show to be held in the Eastern division of the county in two consecutive years, plans proceeded to be put into action.

The schedule of prizes for that year's show, saw the introduction, in the place of the *'Handy Hunter Competition'* of a *'Hunter Selling Class'*, for animals not exceeding four years of age which had been sired by a Premium stallion. The maximum selling price was agreed at eighty guineas, with the relevant prices to be stated in the show catalogue.

Winner of the class in that first year, with a selling price of the maximum eighty guineas, was Mr. E. P. Adams, of Clifton Gardens, Truro, with 'Aboukir'.

By this time, the number of trade stands at the show had risen to 105, with many of those exhibiting, still being well known regulars at the show today. The *'local shows'* section had grown dramatically with in 1931 some 87 classes being staged in the dog section alone. The poultry section, which today has grown to be one of the largest summer shows of poultry in the country, had by that time reached 115 classes, with a host of special prizes being offered by many of the national breeding clubs and societies.

The pigeon section, again still thriving today, saw 27 classes, covering such wonderfully named breeds as Norwich Croppers, Dragoons and Flying Tipplers, with in the working homer section, most of the classes being staged for the *'likeliest flyer's'*. To add encouragement to those entering, amongst the prizes on offer was a *'genuine Wedgwood vase, given by F. Capern Esq., Bristol, for best Working Homer conditioned on All-a-Fire'*!

The fur section of the show, which at that time, not only catered for rabbits but also cavies and even cats, was again supported by a vast number of clubs and societies. Although 1931 saw the cancellation of the two cat classes, due to a lack of entries, 27 other classes were scheduled, with the judge officiating, being a Mr. F. B. Roberts, who had travelled from West Croydon for the occasion.

Amongst the prizes in the Cage Birds show were some rather strange offers from sponsors from various walks of life. The best parrot or parrakeet (single or pair) had the opportunity to win a tin of cigarettes, give by C. Roskruge Esq. of St. Columb, whilst the best canary in the show would receive a 1lb. tin of *'Pulvex Vermin Powder'* given by Messrs. Cooper, McDougal & Robertson Ltd., of Berkhamstead. Somewhat strange prizes, although certainly not quite as unusual as the *'Volume of Sunny Mag'*, given by Messrs. George Newnes Ltd. (tit bits) of the Strand, London to the exhibitor of the best bird in the local classes!

21 classes were staged, although entries in that year were not strong and the same situation has dogged the section for several years.

Within the Horticultural Show, classes were during the 1930's staged in three sections, one being for open classes, the second for *'home gardeners and cottagers'*, whilst the third section was for commercial growers. Within the various sections, classes for *'plants in pots'*, cut flowers, vegetables and fruit were scheduled, and although no classes for floral art, as we know it today, existed, there were classes for, for instance, *'best arranged vase or bowl of garden flowers'*. In addition, for 1931 a class for the *'best decorated dining table, 4ft. × 3ft.'* was included; the start perhaps of the themed floral art exhibits we see today.

The last of the section shows, run on slightly different lines than those previously mentioned, was the honey section. Run by the

Cornwall Beekeepers' Association Show, in conjunction with the Royal Cornwall Show, the section included similar classes to those seen in the 1990s, with both Cornish and Open classes on offer, although of course, the home made wine section, so popular today, was yet to be introduced.

Dairying, was to play an important role at the Show for a great number of years. Pictured in 1932, this class of buttermakers, seem to be enjoying the competitive spirit.

The visit of the show to Penryn in 1932, proved very worthwhile, although of course, farming and the economy in general, were, as today, suffering hard times. The 23 acre site at Roskrow, on the Truro to Helston road, about one mile from the town, suited the needs of the show admirably, as did the one-way traffic systems, put into force by the Chief Constable and Supt. Basher!

The President for that year, Capt. C. H. Tremayne of Carclew, Perran-ar-Worthal, was the third member of that family to officiate in that capacity, with Capt. Tremayne's grandfather, acting as President in 1869 and his father in 1909.

An interesting special feature at that Penryn show was an exhibition of Penryn Corporation's *'antiques'*. This display featured many of the items owned by the ancient borough, including the famous Jane Killigrew Cup, valued in 1932 at £3,000, which had been given 299 years earlier by Lady Jane Killigrew for help and protection given to her by the borough. A 200 year old fire engine and the *'old-time'* stocks, were amongst other items on display.

The competition, which had been run on and off, and on varying lines, for a number of years, for the farm labourer who had served the longest period on any one farm in Cornwall, was to be won that year by a Mr. H. Magor of Carlean, Breage. Having completed 64 years on one farm was quite a record, although it is probable that an award will be made at the Royal Cornwall Show in 1993 to a gentleman who has completed 65 years of service.

New for 1932, although of course, a traditional part of the show today, was the issuing of badges to members, rather than simply tickets as in previous years. This new idea, had been seen to work well at shows such as the Royal, and the original design for the badge was in fact based on those issued to members of the Royal Agricultural Society of England. It was thought that being seen to have a badge, could also help with the recruitment of members!

From about 1930, the possibility of again inviting the Bath & West to Cornwall had been discussed on various occasions. However, by 1932, when the likely cost to the host town was estimated to be in the region of £800, the idea was dropped owing to the economies needing to be made by both the Association and the towns it visited. In fact, Cornwall was not to see a visit by the Bath & West until 1955, when Launceston would be the venue.

1933, when the show visited St. Austell, was again an occasion on which the Association was to be honoured by a visit from a member of the Royal family. His Royal Highness The Prince of Wales, Duke of Cornwall, the Vice Patron of the Association, visited the show on its second day and after touring the various exhibits and sections of the show, decided to give an impromptu address to those gathered.

Colonel Lord Vivian, of Glynn near Bodmin, invited those present to enter the Main Ring and the Prince proceeded to give an address from the grandstand. It was reported that those hearing the speech were much impressed at the Prince's knowledge of agriculture and his stessing of the need to specialise in certain products. The Prince of Wales also referred to his recent visit to Denmark, and his thoughts on the importance of *'concentrating on the right type of pig for bacon purposes'*.

New classes were also introduced that year, for ornamental and agricultural ironwork, which it was hoped would encourage the blacksmiths of the county to diversify, in response to the ever decreasing number of horses to be seen in the county.

Instead of the usual military band, the local committee decided to make a change and St. Hilda's Band, world champions in 1912, 1920, 1921, 1924 and 1926, were engaged to give morning, afternoon and evening concerts on each day of the show.

1934 saw much debate entered into within the Association as to the state of the finances of several of the small local agricultural societies. It was felt vital that these organisations should be kept going as they formed the basis of much of the support of the county show.

However, although the finance committee felt that some financial assistance should be offered, the Council were to disagree and decide to take no action at that time. One local society that was however successful, and which in 1934 celebrated its centenary was the Stithians Agricultural Society. The sum of five guineas was paid as a donation to their funds from the Association, to help mark their historic year.

Always keen to promote improvement in all fields of agriculture, the 1934 show, staged at Treswithian, Camborne, saw the erection of a model cowshed, which as part of a *'clean milk department'* endeavoured to show the best ways of making sure of the cleanliness of the milk produced.

Another feature that year, which was described as outstanding, was an exhibition of handicrafts and other work done by children attending schools in the Camborne/Redruth area. Staged by the Cornwall Education Committee the varied display included cookery, botany, book-binding, hand-loom weaving, biology, handwork and needlework.

The Ministry of Agriculture's display of *'National Mark Produce'*, encouraging the increased production and consumption of foods of reliable quality, depicted the ways in which produce made its way through the various marketing boards and *'National Mark packing stations'* to the urban consumer.

The total entries for the year, numbering 1368, were a record only beaten by the centenary show at Truro.

Towards the end of the year, the Council heard that the firm of Messrs. Climas Bros. of Redruth, who had previously hired to the show the equipment for the shoeing competitions, were retiring from business. The four sets of equipment, including forges, anvils and vices were on offer for sale for £26, although the Council tried to knock down the price somewhat! The full price of £26 was finally paid; a bargain by today's prices!

The Newquay show, staged at Trethellan, on the Pentire side of the town, in 1935, saw the horticultural section nearly wiped out by strong winds. On the evening prior to the opening of the show, the large marquee containing all the exhibits was blown down, but the strenuous efforts of the local committee allowed the section to open as normal the next morning.

A new facility provided for that year's show, was the provision, for the first time of a member's pavilion. This move was met with great approval by the Association's members and plans for a much enlarged facility for the following year's show were set in motion.

During the meeting of members held in the showyard during the show, an event, the first in the already long history of the show took place. For the first time a lady President was elected, with Mrs.

Charles Williams of Trewidden, Penzance, being invited to officiate for the ensuing year. The annual report mentioned the hope of the members that this move may encourage more ladies to take an active interest in the activities of the Association.

Of course, since this time, many lady President's have held office, with the first Royal lady President, Her Royal Highness The Princess Royal, standing for 1993.

The report for the year 1936, contains, amongst other items, a tribute to the late King George V, who for many years had acted as Patron of the Association, although never able to visit the show, his interest in the activities in the county were well known. The news that King Edward VIII, who had previously as Prince of Wales, acted as Vice Patron of the Association, had graciously consented to become its Patron was met with acclamation.

Following the gradual need to provide more all round entertainment, *'trick rides'* by the Inniskilling Dragoon Guards, were for 1936 a major new feature of the Main Ring timetable. At a cost of over £325, this represented a very substantial cost, but the extra appeal of the show, to a perhaps wider audience was thought to make it worthwhile. Those involved in the display, were, however, expected to earn their keep, with displays not only being featured during the day, but also, for a separate admission fee in the evenings as well!

Also new for the year, was a *'milk cocktail bar'*, following a suggestion for such an innovation to be included from the Finance Committee. Quite what form this facility took is unknown, but images of an agricultural version of a scene from a Noel Coward play are conjured up! The idea must have proved successful, as it was repeated for the following year at Wadebridge, with the caterer being instructed that all milk sold, should have been produced by the working dairy on the showground.

Unlike the shows last visit to Wadebridge in 1924, good weather accompanied the event and a substantial attendance was attained. The show, held on a site covering some 30 acres, was on a part of Bodieve Farm, on the outskirts of Wadebridge, which is today the home of the Phillips', the parents of Thelma Riddle, wife of the Show Secretary and author of this book.

A local committee, presided over by Eng. Capt. T. W. Cleave, and chaired by Mr. S. T. Button, took the organisation of the local side of the show in hand with all arrangements being reported as working well.

Unfortunately, owing to the unavailability of military personnel, no grand military display could be arranged, as in the previous year, although great efforts in this direction were made. Instead a parade of tractors and farm implements *'demonstrating modern mechanical possibilities on the farm'* was staged, and with sheep dog demonstrations and the Band of H.M. Black Watch, together with *'pipers and dancers'*, no doubt the programme was suitably filled.

One problem which did cause some concern were the restrictions placed on parts of Devon in connection with foot and mouth disease. However, apart from the absence of some of the usual exhibitors from that area, no real difficulties transpired.

The membership of the Association had by this time gradually risen to 1,328, showing an increase of 39 on the year, with the total income approaching £6,500 for 1937. The accounts also show a donation by the Association during the year to the Royal Cornwall Infirmary Extension Fund to the tune of £50.

1938, when the show was to visit a site at Nansloe, Helston, already well known to the Association, was the first year at which specialist classes for the newly formed Young Farmers' Clubs were to be staged. In this first year, three classes were staged, all relating to livestock, as follows:

1. Young Farmers' Club exhibiting the best cared for exhibits—Cattle only confined to Cornwall—Award £5
2. For members of any Cornish Young Farmers' Clubs competing in the Students Judging By Points Contest—First £2, second £1 (pigs excepted)
3. For members of any Cornish Young Farmers' Clubs, competing in the Students Judging of Bacon Pigs Contest—First £2, second £1, third 10/-.

Also seen for the first time, was to be a parade of fox hounds in the Main Ring on both days of the show. The Four Burrow hounds, led by their master Percival Williams made a *'gay and enthusiastic scene'*, and started a long tradition of such parades.

Unfortunately, by this time, foot and mouth had actually been discovered in Devon, and therefore entries of certain types of exhibits from that county were to be missing that year. One highpoint, however, was the success in the Devon cattle classes, of His Majesty The King, with his cattle sent from the Duchy Home Farm at Stoke Climsland.

The stand of the Women's Institutes, expanded for 1938 saw demonstrations of the preservation of food given by members of various branches, in response to the *'international situation'*.

Still relying on and much involved with the activities of the railway companies great concern was being expressed in October 1938, over the plans by the Great Western Railway to electrify the line from Taunton to Penzance. The idea was thought by many in the Association to be extremely dangerous both to human and animal life, and the possibility of making an official protest was discussed. A resolution was finally passed to make representation to the company with regard to the possibility of installing an overhead live line rather than a live track. Quite what response this met with is unfortunately unknown!

Plans progressed for the visit to Bude in 1939, with much discussion taking place over the proposed visit by the Royal Agricultural Society of England to Plymouth in 1941. The possibility of a visit by the Bath & West to Cornwall in the same year had been raised, but owing to the visit of the Royal, such plans were thought better left until perhaps 1945. Actually all thoughts of agricultural shows, were to be overruled by rather more serious events to come.

However, Bude played a welcoming host to the show, with any doubts as to the suitability of the town as a venue, owing to its somewhat isolated position being quickly dispelled. Held on land belonging to the President, George Thynne of Trelanna, Bude, the site at Broadclose Farm, which was within about ten minutes walk of the town suited the purpose admirably.

The President, on declaring the show open, from the bandstand, spoke of the importance of agriculture to the economy of the country, and of the difference of opinion being shown by the people of Great Britain, towards agriculturalists in view of the current events in the *'international sphere'*. Their vital role was being recognised and the fact that their own *'salvation may depend, in times of crisis, upon its (agricultures) well being'*.

Goats were to make an appearance for the first time in 1939, and with 35 entries proved a popular addition. Milking tests, similar to those for cows were performed with, it is said *'surprising results'* being recorded. The classes for Hampshire sheep, introduced for the show of 1938 were expanded to four and a children's jumping class made its debut.

As late as August 1939, all plans were being prepared for a show at Falmouth in 1940, however, following the outbreak of War, a meeting of the Advisory Committee in October, recommended *'that no show be held in 1940 and that all activities of the Association be suspended for the duration of the War'*.

It was agreed that the Officers of the Association as well as the Council and the various Committees would remain in office for the duration, with Lord Seaton of Bosahan, Manaccan, remaining as President until the end of the next show *'whenever that may be'*.

Prior to the closing of the meeting the sum of £100 was voted to be donated for the use of the Cornwall County Red Cross and St. John Ambulance within the county.

The meeting was then closed by the Chairman, George Johnstone, who could not know that it would be June of 1954 before the Council would again meet.

5. A New Beginning 1945 – 1956

Following a meeting of the Joint Advisory and Finance Committee in May of 1945, the Council met during June to look at plans for the staging of a show in 1946, following the receipt of an invitation from Truro. It was reported that during the war, the Royal Cornwall Agricultural Association had contributed £1,000 to the Red Cross Agriculture Fund, an amount equalled only by the Royal Agricultural Society of England, from amongst donations given by county and other agricultural shows.

The sad news was also related, that the Assistant Secretary, Anthony Courtney Hocking had died on active service and the appreciation of the services he had rendered to the Association was asked to be minuted.

It was also reported that the Royal Agricultural Society of England were proposing to only allow cattle to be exhibited that were from herds that were TT Attested or from herds with other special qualifications. This plan was not to the liking of the Association of that time, although they hoped that breeders would make moves to become attested prior to the need for the introduction of such a regulation.

However, by October of 1945, the possibility of staging a 1946 show had diminished, with the likely shortages of materials and equipment making it unlikely that a show anything like on the scale of pre-war Royal Cornwall's could be held. Rather than hold an unsuccessful event, plans were postponed for a further year.

The year passed and unfortunately this period saw the resignation of the Secretary, W. Courtney Hocking, through ill health. In appreciation of his 24 years as Secretary, it was unanimously resolved that he should be elected a Vice President of the Association.

The question of a successor then arose, and advertisements, offering the post at an inclusive salary of £300 per annum were placed. 14 applications were received and after various meetings and ballots, Anthony Williams of Elm House, Ponsanooth, Truro, was elected the new Secretary.

The estimated costs of staging a post-war as opposed to pre-war show were soon to become apparent. The estimate from the contractors alone, showed a 100 per cent increase on the show staged in 1939. The local Truro committee faced the same problems, with materials etc. not only being hard to come by but also very costly. To bring the shows finances back into shape, several fairly drastic measures were felt to be necessary.

The charges for trade exhibits were doubled, with significant increases being made both to the price of admissions and entry fees. The Association itself, also took over many of the expenses, previously met by the local committee, such as the rent for the showyard, the removal of hedges and the installation of a suitable water supply.

Prize money was however, also to be increased with £2,000, instead of the prewar £1,766 being offered in the horse, cattle, sheep and pig sections. The classes previously staged for Aberdeen Angus cattle were to be dropped due to lack of interest, with Ayrshire cattle being seen for the first time at the show.

Those who had expressed doubt as to the sense of reinstating the show so soon after the war saw all fears of failure dispelled even before the show opened. Held in July, rather than the usual June, due to the need to fit in with the contractor and austerity restrictions, the show attracted a record 2,000 subscribing members before the start of the first day.

Entries were well up to previous years, with the horse section in particular showing a large increase on other shows. The sheep and pig sections were not quite up to their usual numbers and unfortunately the butter-making contests, so popular before the war, were unable to be held for that year.

Very large entries, kept the County Organiser of the Young Farmers' Clubs busy and the interest in the milking competitions was so great that barriers had to be erected at the last moment to keep the crowds back.

Over 44,000 attended that show at Truro, helping to dispel the gloom of the period of austerity that was still very much in evidence. The Association made a massive profit of over £5,000 allowing plans for the following year to be put into action without too much fear of financial difficulties.

At the end of 1947, and following the great service done to the dairy section of the show over a great number of years, the Association willingly subscribed the sum of £20 to a subscription list formed to mark the retirement of Miss Nicholas, the County Dairy Instructress. Miss Nicholas had seen the butter making and other dairy classes grow at the show from nothing to a magnificent display of the finest products the county could produce, and no doubt her retirement marked, in many ways, the end of an era.

Bodmin in the following year, was again to benefit from the enthusiasm created in Truro. With the membership swollen to 2,500, over 500 up on the previous year, all looked well. The entries exceeded all expectations at a total of 2,794, breaking the previous record by almost 1,200. This huge increase, although spread through many areas of the show, reflects the great interest in the YFC classes, with some 1,660 entries being received for this section alone.

Towards the end of 1948, moves were made to commemorate the great services done to the Association over a long period, by George Johnstone of Trewithen. It was finally agreed that a portrait should be commissioned of Mr. Johnstone and an artist, Mr. Thompson, RA., was subsequently engaged at a cost of £250, and the painting presented.

Erected for the Show of 1948 in Priory Road, Bodmin, this arch stretched between the church wall and what was then Robartes Gardens, now removed owing to the widening of the road.

As early as 1949, mention of the advisability of the Association purchasing a permanent site was being made. In fact, in April 1949, a special committee to examine the proposals made by the Advisory Committee for such a purchase was formed. For some time, the possibility of purchasing Polwhele Army Camp, near Truro, was carefully examined, and in fact an offer for the site of £3,000 for approximately 39 acres was made. However, following further negotiations and the debate as to whether or not the site would in the future prove to be large enough, no action was actually taken. It was however, still thought that a permanent site should always be borne in mind for the future.

Following a successful visit to Falmouth in 1949, which amongst other things saw the re-introduction of goat classes, plans progressed for the show at Callington in 1950.

Never before had such plans been laid, that preceded that show at Callington. For the first time in its long history, the show was to be graciously honoured by a visit from the reigning monarch, who was of course, also Patron of the Association.

George Johnstone of Trewithen, Grampound Road, Chairman of the Royal Cornwall Agricultural Association from 1950 to 1957.

Not only His Majesty King George VI, but also Her Majesty The Queen and Her Royal Highness Princess Margaret were to make their way to Callington, where one of the largest one day attendances ever, were ready to greet them. A full day was to be spent at the show, with the Royal party timed to arrive at 11.00 a.m. and not to leave until approximately 4.30 p.m. A specially built Royal Pavilion was erected in the showyard, where the President and Vice Presidents entertained their Royal guests to luncheon.

To comprise of a drawing room, dining room and kitchens etc., the pavilion erected, was beautifully decorated and furnished, and stood within its own garden. The interest it created was so great, that a decision was made to allow those that wished to look inside, the opportunity to view the interior from the verandah, through the open windows, after the Royal Party had left!

The Royal guests were met on their arrival at the Showground, by the Director of the Showyard, Edwin Baker of Treniffle, Launceston, on a white horse. After being conducted to the Pavilion, a tour of the show was made including the judging rings of the Cornwall Federation of Young Farmers' Clubs, the cattle judging rings, the exhibit of the Cornwall Agricultural Executive Committee and the stand of S.W.E.B. Her Majesty The Queen, following for part of the morning a separate route, visited the exhibit of the Women's Institute, the local shows, including the Flower Show, and the creche staffed by the Women's Voluntary Service.

After luncheon, and again preceded by the mounted Director of the Showyard, the party visited the Main Ring, where the presentation of three of the Association's Long Service Awards, including one to Mr. Isaac Mallett took place. Also presented were two Long Service Medals awarded by the Royal Agricultural Society of England, after which the Grand Parade of prize winning cattle and heavy horses entered the ring.

During the parade, the Queen presented *'The Antony Challenge Cup'* which had been given by that year's President, Sir John G. Carew Pole, Bart., D.S.O., for the Champion Hunter. The cup which in that first year, was presented to Reg Lobb of Norton, Bodmin, has recently come back into the possession of the Association, after being won outright many years ago.

After the presentation, a class of jumping was then watched from the Royal Box before a visit to the Duchy Tent to meet tenants.

Such a visit to the county caused a great deal of excitement at the time, with many people never before having seen Their Majesties and HRH Princess Margaret. One event which added a special touch to the day, was the presentation to both the Queen and Princess Margaret of bouquets of flowers by Katherine and Mary Eustice, the two young daughters of one of the forage stewards, George Eustice of Tregotha, Gwinear, who was in later years to be Director of the Showyard.

When Her Majesty again visited the show in 1985, as Queen Mother, after looking at an album recording that visit of 1950, George's two daughters were again presented and some memories of the day re-called. Katherine, is still very much connected with the show, through her husband, Adrian Clifton-Griffith, the current Departmental Steward of Sheep.

The show was a truly *'Royal'* event, with a massive attendance, breaking all previous records. Nearly 48,000 people paid for admission over the two days, with over 31,000 on the first day alone. The membership of the Association rose for the year to 3,000 and entries exceeded all previous totals.

The visit of Their Majesties King George VI and Queen Elizabeth, with Her Royal Highness Princess Margaret to the Royal Cornwall Show at Callington in 1950.

Seen on the right are Misses Katherine and Mary Eustice, now Katherine Clifton-Griffith, wife of Adrian Clifton-Griffith, the Shows Departmental Steward of Sheep, and Mary Tilley.

A site at Trethiggey, Newquay, was chosen for the show of 1951, with the July dates causing some concern as to any clash with the holiday season. However, no such problems materialised, and a large number of visitors attended. Red Poll cattle classes were this year to be seen for the first time, with the Blewetts of Trewhella, Goldsithney and the Trumps of Tillhouse, Exeter, taking between them, all the prizes in the two classes.

A major addition to the schedule of prizes for the year, was the inclusion of a farm competition. Such competitions had been run in earlier years on various occasions, although not for some years previously, and a new classification and points system was carefully drawn up as follows:

CATTLE	for quality and evenness of type ... 100 points		
	,, condition .... 75 ,,		
	,, pedigree status add .... 50 ,,		
	,, attested herd add .... 50 ,,		
	,, licensed T.T. or supervised herd add .... 25 ,,		300
SHEEP	for quality and evenness of flock ... 50 ,,		
	,, condition and shepherding ..... 50 ,,		
	,, registered flock add .... 25 ,,		125
PIGS	,, quality and type .... 50 ,,		
	,, condition and herdsmanship ... 50 ,,		
	,, pedigree herd add .... 25 ,,		125
BUILDINGS, YARDS, ETC.			
	for condition .... 100 ,,		
	,, cleanliness .... 50 ,,		
	,, adaptation and improvements .. 25 ,,		175
FENCES, DITCHES AND GATES			100
IMPLEMENTS — for care and maintenance ..			100
CROPS	for condition of grassland .... 125 ,,		
	,, ,, ,, arable land .... 125 ,,		
	,, management .... 50 ,,		300
GENERAL FARM MANAGEMENT			25
EXHIBITION AWARDS at 1951 Royal Cornwall Show:			
	for each Championship, add .... 6 ,,		
	,, ,, Reserve Championship, add 5 ,,		
	,, ,, 1st Prize, add .... 4 ,,		
	,, ,, 2nd Prize, add .... 3 ,,		
	,, ,, 3rd Prize, add .... 2 ,,		
	,, ,, 4th Prize, add .... 1 ,,		

The prize for this new competition was to be first, President's Challenge Cup and £20, second £15 and third £10. Entries soon came in and prior to the 1951 show the judging took place on the farms entered.

The three prize winners in that first year, are all names well-known to many today, being 1st W. C. Maddever of Looe Down, Liskeard, 2nd J. R. Dunstan of Sinns Barton, Redruth, and 3rd E. T. Hawkey, Junr., of Penpont, Amble, Wadebridge.

The competitions were not to last for a great number of years, but whilst they were in existence, great store was set by their results. This interest was no doubt enhanced by the presentation of a challenge cup for the winner of the competition, by His Majesty The King for 1952 onwards. Following the decisio to end the competitions, in 1965 this trophy was thereafter to the present day, presented to the Supreme Champion of the cattle classes.

The death of His Majesty King George VI, in early 1952, started what was to be an extremely busy year for the Association off on a sad note. Letters of condolence were sent to The Queen on behalf of the Assocation, by the President, G. P. Williams, Esq. In June, the Association were delighted to hear from Ulick Alexander, the Keeper of the Privy Purse, that Her Majesty The Queen had *'been graciously pleased to grant her Patronage to the Royal Cornwall Agricultural Association'*. The Association, were during the next year, to donate the sum of 100 guineas to the King George VI Appeal Fund, set up following the King's death.

Amongst the notable events for the Association, were during February, the staging of two evening lectures, one in Bodmin and the other in the Camborne/Redruth area. Both lectures were given by Dr. Ian Moore, Principal of Seale-Hayne College in Devon, with his subject being *'Self Sufficiency in Beef and Milk Production'*. The Association had for at least the past two years supported the college through donations of in 1951 £150 and in 1952 £300, and these lectures were a way of letting the members of the Association know how this support was being used.

The show at Redruth in June, staged at Sinns Barton, although experiencing a slight decline in entries and attendance, ran smoothly, although of course the visit of the Royal Show to Newton Abbot in July, rather overshadowed local agricultural matters.

£1,000 had been granted to the Royal Agricultural Society towards the costs of staging the show in return for free admission for members of the Royal Cornwall Agricultural Association. In addition to this, and as a joint venture with the Devon County Agricultural Association, a pavilion for the use of the members of the two Associations was erected on the showyard at Newton Abbot. The cost of this pavilion, which inclued cloakrooms, restrooms and areas for refreshments, to the R.C.A.A. amounted to just over £250.00.

1952 also saw the introduction of four new challenge cups, given by the Grampound Road Horse Show Committee and purchased with the residual funds of the show. Two cups valued at £50 and two at £25 were to be provided, with one cup to be used for the second class of the extended farm competition, to be awarded to whichever class winner of the two classes, did not win the cup given by the late King.

This year also saw the cattle classes being confined to animals from attested, supervised or licensed T.T. herds. Some moves had been made towards this end some years previously, with certain sections having already been limited by the above regulation.

Launceston saw a visit of the show in 1953, with the spectre of swine fever looming, although the pig classes were able to continue for that year.

Talk of purchasing a permanent site was again heard, although no real action was to be taken, although the increasing difficulty of providing electricity to the temporary showground each year, and the costs relating to this caused much annual debate. No doubt problems such as this added much to the arguments for basing the show on a permanent site.

Plans for 1954 saw arrangements being made for the visit of Their Royal Highnesses The Duke and Duchess of Gloucester to the show to be held at St. Austell in late May. Doubts had been expressed by the Chamber of Commerce and various other tradesmen of the town, as to the benefits of hosting the show and of having to find the necessary funds. However, such fears were overcome and a local Committee, presided over by Mrs. R. Cobbold Sawle of Penrice, St. Austell, was set up to raise the £1,000 needed for the local contribution to costs.

One of the fund raising efforts staged involved a lecture in the Public Rooms, given by *'Elephant Bill.'* Lt. Col. J. H. Williams, better known as *'Elephant Bill'* is said to have enthralled those present, with his tales of his life in the jungle covering 26 years. The event proved most successful and raised significant funds for the local committee.

Bad weather, on the day before the show, caused concern and the laying of several hundred tons of sand on the 42 acre site adjoining Cypress Avenue, near the Cornish Speedway Stadium. In addition 23 acres of car parks were prepared and detailed traffic arrangements made, including the widening of a road at St. Blazey Gate and the elimination of a dangerous corner. Other arrangements included the laying of over 2000 railway sleepers as a temporary roadway, over a site, which in parts, had been found to be quite wet.

Sir John and Lady Carew Pole acted as hosts to The Duke and Duchess of Gloucester during their stay at Antony House, Torpoint, whilst in the county, following their arrival by a Viking of the Queen's Flight at St. Eval. Their visit to the show was unfortunately, rather a muddy one, although the Duchess was described as being far more suitably dressed than many of those presented to her! A busy day of visits and presentations ensued, which included, for the Duchess a visit to the stand of the South Western Electricity Board, where such innovations as electric food mixers and cookers with electric timers were on display.

Agricultural machinery of the very latest design was said to be on show, with one newspaper commenting that farmers will *'one day manage their farms from a desk by pressing buttons'*. How right that reporter was!

A vast attendance was seen that year, although the first day, when the Royal visit was made, was not as large as that of the second day. The threat of myxomatosis had not affected the entries in the fur section, although an outbreak at Lerryn shortly before the Show had caused concern. Probably the single most important feature of the show, however, had been a great innovation made in the jumping classes. For the first time, the horses were automatically timed, by a large, four sided clock, towed into the centre of the main ring for each class. Operated by *'invisible rays'* the clock was seen as great novelty and attracted much interest, with the grandstand and the ringsiders always packed with spectators during a class. The best of the countries show jumpers were hoped to have been seen, but this spectacle had to wait for a future occasion.

Following the death of Mr. Arthur Pooley, the Director of the Showyard, after the 1954 show, the next Council meeting of the Association saw the election to this post of Mr. George Blight of Tregonning. The Blight family, have been strong supporters of the show for a great number of years, with the first member serving on the Council being elected in 1919. The family is still much involved today, although the fact that the four male members of the family to have been on Council, have been all called George, has, from time to time caused confusion!

1955 was to see a visit by the Bath & West to Launceston, and therefore no Royal Cornwall was held that year. Members of the Association received free admission to the show at Launceston, and the erection of a members pavilion for their use was no doubt much appreciated. Some of the Association's trophies were in that year lent for presentation at the show at Launceson, and the occasion of the last visit by the Bath & West to Cornwall passed off well.

Helston was chosen to play host for the show of 1956, with record numbers of trade stands (160) being displayed. An added attraction for the main ring, in addition to a display by the Band of the Scots Guards, was a P.T. display by R.N.A.S. Culdrose. Hopefully the inclement weather suffered did not create too many problems for them!

The Annual General Meeting for 1956 heard that it was the intention of the Secretary of the Association, Anthony Williams, to retire from his post. Having been in the postion for ten years the thanks of those present were minuted for all that he had done, and the necessary wheels for the findings of a replacement put into motion. The last part-time Secretary had retired and the time had come for yet another expansion of the Association's activities, with the appointment, for the first time, to a full-time Secretary.

6. A New Regime 1957–1993

Following the advertisements with regard to the position of Secretary to the Association, thirty-eight applications were received. A small sub-committee was appointed to consider the candidates and to select a few to go forward to the Council.

Only two names were in fact chosen those being J. Wesley Wilton, who was then County Organiser for the Cornwall Federation of Young Farmers' Clubs and Albert Riddle, who was then working for BOCM Silcock, after leaving the Royal Navy after some 11 years of service.

Wesley Wilton, well known to many in agricultural circles within the County, was tipped to win the post, however, each candidate had support amongst the members of the Council and no doubt, much hard campaigning was done.

Albert Riddle, had in fact been secretary of Liskeard Show for the previous five years as well as being a joint Secretary of Liskeard Fatstock Show for a couple of years. No doubt this past show experience, albeit on a much smaller scale helped tip the balance.

Mr. Jack Olliver was his proposer with a second reference being given by Mr. Jock Stanier, who was at that time Land Steward of the Duchy of Cornwall.

The final decision gave Albert Riddle the post, and Wesley Wilton is today a strong supporter of the show, acting as a Steward each year.

Along with the decision to employ a full-time Secretary, had come the need to purchase permanent office accommodation for the work of the Association. A house, which was to become well known to many over the years, was chosen, and following the gaining of planning permission for its use as offices, was subsequently purchased for the sum of £2,500.

The property in question was No. 4, Upper Lemon Villas, Truro, right at the very top of Lemon Street, one down from the forbiding figure of Landers Monument. The house, built in the early nineteenth century, had formed part of the quarters of the officers of the County Regiment, and to this day, the mounting blocks for the cavalry, can be seen along Strangways Terrace, just around the corner from the old offices of the Association. In fact, when the property was first purchased, at the end of the garden, on Barrack Lane, there still existed the stables and grooms quarters for the horses and men attached to the officers billeted in the main house. This was to be later turned into garages.

During the debate over the purchase of the house, a small committee of the Association visited the property to be shown over it, however, owing to the incapacity of an elderly lady in one of the five bedrooms, that particular room was not viewed. This typical Cornish gallantry, was however, to prove expensive with the room in question being found to need a great deal of repair and decoration. How convenient that the lady had been confined to her bed during the tour of inspection!

The new Secretary was installed in the new offices, with living accommodation being available above and behind the two main offices, formerly the formal drawing and dining rooms. A further ten rooms, spread over four upper floors comprised the rest of the rooms available plus a cellar, which was, in years to come, to need a great deal of building work. It was however, to be nearly two years before the Secretary and his family moved to Truro to live, preferring in the meantime to continue to live in their home town of Liskeard.

Albert Riddle, pictured at the time of his appointment as Secretary in 1957.

Sir John Molesworth-St. Aubyn, Bart., of Pencarrow, Bodmin, Chairman of the Royal Cornwall Agricultural Association from 1957 to 1972.

The first few years for the new Secretary were not to prove to be easy, with his first Show at Wadebridge in 1957 being hit badly by the effects of the Suez Crisis. Severe shortages of fuel not only severely cut the numbers of trade exhibitors able to travel from distant parts of the country, but also greatly hampered movement around the county to attend meetings, etc.

The year also saw a change of Chairman, with George Johnstone of Trewithen, who for so many years had led the Association, tendering his resignation. Cmdr. A. M. Williams of Werrington, was first offered the position, on a one year basis, to help see the Association through this period of re-organisation. Cmdr. Williams was however, not to take up the offer and Sir John Molesworth-St. Aubyn was duly appointed, again firstly on a one year basis only.

Sir John, was in fact to remain in office until 1972 and during those years was to help guide the Association through a period of great change. Sir John, whose family have always been so closely linked with the show, from its very earliest days, was to take a keen interest in all that went on and continued to do so until his death.

That first show under the new Secretaryship saw few changes, other than those imposed by the Suez Crisis, although Landrace pig classes were to be seen for the first time. In the Main Ring, many remember seeing Charles Fricker putting his police dogs through their paces on the showground at Wadebridge, which was to be the site eventually chosen three years later as the permanent home for the Show.

Truro, however, was to be visited for the last time in 1958, and from the moment the main showyard contractors, L. H. Woodhouse & Co. of Nottingham moved in in the early spring, the weather was against the event. The first job of the contractors when arriving at a new site was to lay trackboards, to make a temporary road surface to their yard within the showground, reversing into the site, the boards would be laid until the area of the yard was reached. In 1958 the boards as laid were disappearing into deep snow, a bad omen for the weeks to come!

Owing to the fact that the Bath & West was that year to visit Plymouth, the Council of the Association had decided to change to an earlier date than usual, with the show being held on the 21st and 22nd of May. Unfortunately during the week leading up to the show, very strong winds hit the showground at Allet on the outskirts of Truro causing severe damage to almost all the preparations made. The roof of the grandstand flapped so much that it shreaded itself into threads which covered the main ring like a heavy fall of snow. Brigadier Stephen Williams of Scorrier, sent men and machinery, and literally trailer after trailer of thread was chain harrowed to clear the ring. All tents were wrecked and the accommodation for cattle blown away.

An emergency meeting two days before the show discussed what to do and the decision to go ahead was made, after the Secretary, Albert Riddle, reported that Woodhouses and the main caterers had promised to do their best to see that the show opened on time.

Many visitors to that show can probably still remember sitting in a grandstand with no roof and ploughing through the sea of mud, and although the attendance was badly hit, still over 23,000 people paid for admission, plus the many members and ticket holders, etc.

The difficulties of that 1958 show, reinforced the argument being raised by many, that the time had come for a serious look for a permanent showground.

However, not all was gloom, for that show at Truro, brought to the county for the first time, many stars of the showjumping world, not previously seen in action in Cornwall. Although today, showjumping has lost some of its glamour during the 1950's and 60's, the sport was avidly followed and drew great interest at the show.

1958 was to see riders such as Pat Smythe on Grand Mannan and Flanagan, Ted Edgar on Mr. Pollard, Jane Summers and Scorchin, Derek Kent on Shane, Banka and Gay Romance and Fred Broome on Ballan Silver Knight. In one class Ted Edgar, following a bad fall into the sea of mud that was the main ring, had to be taken to a nearby farm, for a thorough cleaning.

That year also saw for the first time, Raymond Brooks-Ward, commentating in the main ring. That well known figure and even better known voice, was to grace the main ring of the Royal Cornwall Show, adding life to sometimes otherwise rather dull moments and filling gaps where necessary in his own style, up to and including the show of 1992. His untimely death during the summer of 1992 came as a great shock to his many friends at the Royal Cornwall, although I am sure that he will be looking over the shoulder of his great friend and colleague Tom Hudson, who will be taking over for 1993, to check on all that goes on.

Following the loss of nearly £4,000 on the running of the 1958 event, a very close look at the finances of the Association ensued. Many charges had remained static since the war, although the cost of staging the show had risen fairly dramatically. Change was needed and the Secretary together with Major Simon Bolitho of Trengwainton, were charged with the task of drawing up a clear report on the situation and suggestions for what could be done.

Various price increases for trade stands, tickets etc. were found to be necessary although the main area for debate concerned the provision of a suitable permanent site. It was decided that no delay should be made in finding the necessary land and various possibilities were examined.

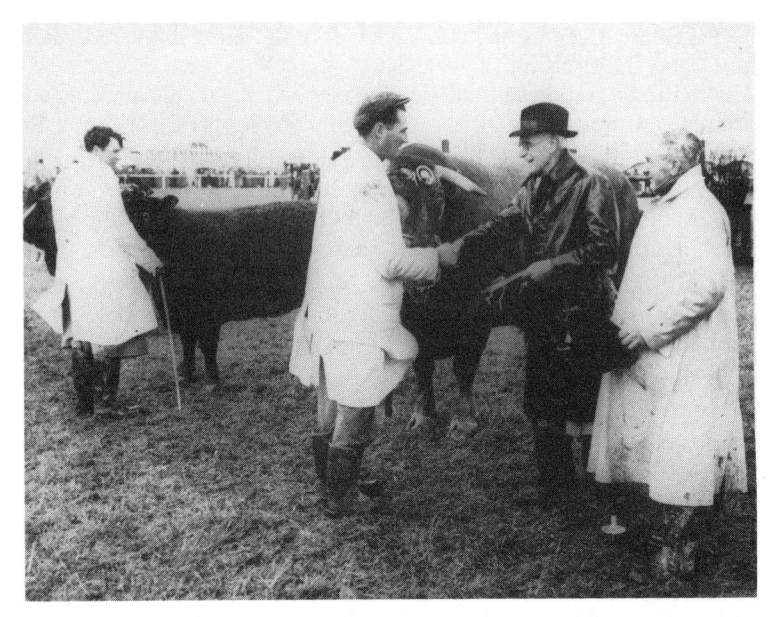

The Rt. Hon. The Viscount Falmouth, President for the 1958 Show at Truro, seen presenting Hugh Lello, owner of the Champion South Devon with his award. To the right can be seen Bill Cook, the Departmental Steward of Cattle.

Major Johnstone offered a site at Trewithen, which although suitable in many ways, was not to be available for purchase and even a long lease seemed unlikely. A site which did look like a real possibility was at Quintrell Downs near Newquay, although fears as to a proposed major road diversion scheme which would have seriously affected the site, led to its being turned down. The scheme has never actually gone ahead!

The Princes' Council of the Duchy of Cornwall, always keen supporters of the show and all that it stood for, were quick to offer their support and promised to do all that they could to help find the right site. In fact, the site used for the Wadebridge show of 1957, at Whitecross, was owned by the Duchy and in many ways it seemed to suit the purpose. Negotiations started and an agreement reached with the Duchy for the lease of the land in question for a trial period of three years, for the staging of the show in 1961, 1962 and 1963.

The support and co-operation of the tenant of the neighbouring farm, Henry Menhenitt, was of great help at this time.

Owing to the helpful nature of the Duchy throughout the period of discussion, agreement was reached earlier than anticipated and the question of whether or not the 1960 show should be held at Wadebridge was raised. That year should have seen a visit to

Falmouth, but the Mayor and Town Council kindly offered to forego any claim they may have had on the show, if moving to the semi-permanent site a year earlier would be financially advisable. This kind offer was taken up and plans laid for that show of 1960 at Wadebridge, including the purchase of three fields on the eastern side of the ground for car parking.

Her Majesty Queen Elizabeth, The Queen Mother, meeting Council Members of the Association, during a visit to the Royal Cornwall Show of 1959, held at Liskeard.
Photograph courtesy of the Western Morning News

In the meantime, that last travelling show, fondly remembered by many, saw Liskeard as its host, with the gracious presence of Her Majesty, Queen Elizabeth the Queen Mother marking the occasion. That year saw the appointment of Walter Abbiss as Advisor to the Horticultural Section of the show. Many years were to follow, with the Flower Show being staged under the guidance of a gentleman, who was to be greatly regarded both by the show and by the county who so benefitted by his enthusiasm and knowledge.

The show was once again to make a small loss in that year, emphasising the need for moving away from its itinerant existence, so welcomed just over 100 years before.

One small event during the year was a presentation of a gift to the value of £25 to Mr. Eric Westby, foreman to the Association of the contractors, L. H. Woodhouse & Co. This presentation was to mark Mr. Westby's work in erecting the show since 1925.

Taken during a Reception at her home, Trewithen, Miss Elizabeth Johnstone (*front row, seventh from right*), seen with many of the Vice Presidents and Council Members of the Association and their wives.

The show of 1960 proved a great success and although, from a financial viewpoint, the Association had no room for complacency, it could be seen that the move to what could become a permanent site, was a step in the right direction. A visit by Her Royal Highness, Princess Alexandra of Kent on the second day and the great efforts put in by all those involved in the setting up of the show, helped ensure success.

A great deal of money had been expended in preparing for the show, with for instance, £1,760 having been paid to the South Western Electricity Board for a permanent supply of electricity, with a water supply provided by Messrs. Trewartha, Gregory & Doidge of Callington, costing a further £986 12s. 3d. The laying of hardcore and the removal of hedges, etc., had cost a further £350, with the work being carried out by F. Davey & Sons of Tresillian Barton, near Newquay. It is gratifying to note, that both of these local firms have continued to work for the Association, throughout the intervening years, with Davey's only being replaced by Jack Kingdon, following their giving up of the contract side of their business prior to the 1992 show.

Albert Riddle, the Show Secretary being presented to Her Royal Highness Princess Alexandra, during a visit to the Wadebridge Show of 1960.

Trewartha, Gregory & Doidge have installed literally miles of water pipe, and are today responsible for the maintenance of dozens of toilets, various hot water and heating systems, together with the Main

Ring and Ring Two sprinkler systems. Formerly in the hands of Bill Thomson, and laterly Gervis Radford and Dick Parkes, the actual work on site is vital to the smooth running of the show, when a badly aimed tent peg can do untold damage to a water main. During the show plumbing staff sleep on site to be ready for any emergency!

Further expenditure on the site followed in preparation for the show of 1961. A pavilion, erected for the use of HRH Princess Alexandra, whilst visiting the 1960 show, was purchased for the future use of the President and any VIP visitors, with it being given the title of *'The George Johnstone Pavilion'* as a memorial to a man who had for so many years been Chairman of the Association.

The fencing in of the Showground proved to be an expensive, yet vital, exercise both to ensure the security of the trade exhibits and livestock on display, but to also ensure those visiting were paying to enter! By the end of 1961, some 750 yards of chain link fence had been erected, although the help of the Chairman, Sir John Molesworth-St. Aubyn, led to one or two delays. Always keen to be of help and realising that the Secretary was based in far-off Truro, Sir John would often visit the Showground, out of showtime, to check that all was well. During one such visit, he inspected the work of a contractor who was erecting 6 foot high fencing, but was alarmed to find that every post he tested could be easily moved from side to side.

On returning home, the Secretary, Albert Riddle, was informed of the problem and asked to investigate. The next morning, on arriving at the Showground, the Secretary was met by an irrate workman, complaining of someone who had ruined his work by moving all the posts which had recently been placed in wet concrete! Such a lack of communication was, however, rare, and as a Chairman Sir John, held the respect of all those connected with the show.

Plans progressed and the staging of a pageant as an arena attraction for the 1961 show was suggested and acted upon. The pageant, remembered by many who took part, could well have been a complete disaster if not for the dedicated work of a stalwart band of enthusiasts. With a theme of *'A Cornish Country Market and Fair and a Meet of Foxhounds'*, much scope for impromptu merryment was given, although the organisation was nothing if not professional. Directed by Francis Collingwood Selby and written by Kathleen Mackenzie, a host of characters were to take part with much help being given by schools, dramatic societies and women's institutes. Mr. G. W. Dobson, Headmaster of Wadebridge County Secondary School, chaired the organising committee, with many of the costumes being provided by Miss Joyce Hawken of Wadebridge, a well-known collector of costumes and accessories, whose name was often seen on the credits of historical television dramas, etc.

Also new was the staging of motor cycle racing in the Main Ring on the last evening of the Show. Run by the Pendennis Motor Cycle Club, this event has been an annual feature of the show ever since, only being cancelled when extreme bad weather has intervened.

Although the Schedule of classes varied little from year to year during this time, several new challenge cups were gradually presented, with the most sought after, from the point of view of horse exhibitors, making an appearance for 1961. A challenge cup presented by the Patron of the Association, Her Majesty The Queen, had been accepted by the Association for presentation to the Best Light Horse in the Show. Competition was keen, and in that first year the honours went to P. J. H. Murray Smith of Kingsbridge.

A new, more comfortable grandstand, with every seat sold, was erected that year and with such features as the *'Story of Wool'* exhibition, the show was said to be the best yet.

Following that show the decision to settle permanently at Wadebridge was made. The two *'trial'* years had proved the benefits of such a move and it was felt unnecessary to wait for the third year of the Duchy lease to pass, before the decision was made.

1962 proved a memorable year, as not only had the show a permanent home, but the decision to engage the very best in Main Ring entertainment had paid off. Few residents in Wadebridge, will forget the sight of the King's Troop, Royal Horse Artillery, trotting up Molesworth Street after arriving at Wadebridge Station by train. Although the cost, at over £2,000, was high, the visit to Cornwall of such a display created a great deal of interest and set the tone for future years.

1963 saw yet another pageant being staged, this time with the theme *'Wedding Bells'*. Linked to the legend of St. Keyne's Well, and the tradition that the first person, either bride or groom, to drink from the well after the marriage ceremony would be the dominant partner, the protrayal of a series of weddings from different ages formed the storyline.

Again written and produced by Kathleen Mackenzie, a very large cast took part with many familiar names being seen in rather unfamiliar roles. The first wedding, for instance, in the time of King Arthur, saw Melville Lawrey, Douglas and David Kellow, Henry Lobb, Henry Menhenitt, Bill Tucker, Bruce Warren, Charlie Richards and several others appearing as knights. The stage manager, Anthony Swinburne, probably deserved a medal for keeping order!

The year also saw the last classes for Red Poll cattle being staged, although with no entries being received, the classes were cancelled and the classification dropped for the following year.

Work continued on the showground, with concrete roads being laid and for the 1964 show a much enlarged, although temporary

Members' Pavilion being erected. This 'temporary' pavilion is still in use today, and although much maintenance is annually carried out, is still in very good order. Originally hired from L. H. Woodhouse & Co., the pavilion was purchased in 1972 for the sum of £6,000 and still provides a satisfactory facility today.

One item, recorded in the minutes of the Advisory Committee, on 2nd September 1963, holds some personal interest for the author of this book, in that his birth, one hour before was recorded. On a sadder note, that same meeting stood to the memory of Jack Laity, a great supporter of the show and whose family are today much involved, who had died in the preceding week.

During these first few years at Wadebridge, the various local committees that had been formed in the areas where the show had previously visited, were gradually wound up. Many of these committees had been in existence for a great number of years, in place, ready for the time when the Royal Cornwall would be invited to visit their particular district. In many cases, considerable sums of money had been accumulated by these local organisations, through their previous fund raising efforts and these sums were during the early 1960's, in the main, gradually made over to the Association. To be used for the further development of the permanent showground, these funds were to be very gratefully received and in many cases, small donations were subsequently made back to the local agricultural society of that area. For instance, in early 1960, £40 was given to the Camborne Show Society, following the loss they had sustained in 1962 owing to bad weather.

By this time a great deal of equipment such as poultry, pigeon and rabbit pens had been purchased. It was to be decided that this equipment should be loaned, free of charge, to other show societies within the county, particularly when their members were involved in the organisation of the section shows at the Royal Cornwall. This arrangement lasts to this day, with many small shows benefitting from the ongoing capital investment made and being made by the Association in such equipment.

1964 saw the Association facing the costly exercise of replacing virtually the whole of the back wall, some four storeys high, of their house at 4, Upper Lemon Villas. Early one morning, whilst sitting in the living room at the back of the house, the Secretary had heard a low rumble. Peering over the back garden wall, into that of the dentist's surgery next door, he was aghast to find nearly a trailer load of stonework had slipped from the wall down into the garden! Only the great width of the walls had saved more from collapsing and the necessary repair work took several weeks and the expenditure of a great deal of money.

Following that year's show, at which the Musical Ride of the Household Cavalry and the Band of the Royal Horse Guards were to hold centre stage in the main ring, further development of the showground was put in motion. Goats were to also see a re-introduction of classes for that year, under the guidance of the Cornwall Goatkeepers' Association.

Already some £16,000 had been spent on improvements, and the year was to see new lengths of concrete road being laid, mainly by the firm of Sidney Jewell & Sons of Wadebridge, and terraced banking built along the length of the main ring to aid viewing. The membership had now risen to almost 6,300, an increase of almost 650 on the previous year, an encouraging response to the membership drive carried out by the Secretary.

The ever increasing amount of machinery being brought to the show for display by exhibitors and indeed the larger number of firms exhibiting (241 by 1965), led to further improvements being needed. Better entrance and exit roads and the replacement of the old wooden unloading ramp, with a concrete one, were all projects undertaken prior to the 1965 event.

A new purpose built sheep shearing building, still in use, although modified, today, was also to be purchased for the YFC shearing competitions. Originally erected by the Duchy of Cornwall on the stand of the Ministry of Agriculture, it was offered for sale to the Association after the show for the magnificent sum of £150! Although having been moved and altered somewhat on various occasions since, the building still operates well and will hopefully be in use for many years to come.

The President for 1965, Sir Patrick Kingsley of the Duchy of Cornwall was to lead to the visit to the show of one of the most impressive and memorable main ring events ever staged.

During a meeting in London with Sir Patrick, the Secretary, Albert Riddle mentioned that he wondered if the Royal Mews could be persuaded to provide some sort of entertainment for that year's main ring. Sir Patrick suggested that the Secretary should visit the Mews, which was situated fairly near to his own office and find out in person. Having asked for directions of the policeman on duty at the Mews entrance and on entering the door indicated, he found he had actually entered the 'back door' of the private office of the Crown Equerry, Lt. Col. John Miller (now Sir John Miller).

Sitting behind an enormous mahogany desk, Colonel Miller, ever the perfect gentleman, welcoming the intrusion, listened patiently to the propostion. Following this meeting, it was agreed that a request would be placed before Her Majesty The Queen for some of her coaches, carriages and harness horses to appear at the 1965 show. The fact that the show was at that time still being held on Duchy property and that this therefore gave a special reason as to why the carriages should come to Cornwall, no doubt influenced the decision for permission to be granted.

The resulting display and the brilliant commentary given by Colonel Miller helped swell that year's attendance to a record beating level, with nearly 48,000 people paying for admission over the two days.

An amusing sequel to this event, which happened a few years later, came after the Royal Agricultural Society of England put in a request for a similar display to be staged at the Royal Show. Permission was subsequently granted, but when the question of what facilities should be provided on site for the coaches and horses etc. was raised, Colonel Miller told the organisers of the Royal Show to contact the Royal Cornwall. He said that the arrangements made in Cornwall had been so good, that the Royal could do no better than to ask our advice! A feather in Cornwall's cap!

The 1965 Show saw a Main Ring display featuring carriages and horses sent from the Royal Mews in London, by gracious permission of Her Majesty The Queen, the Association's Patron. Never before had such a display taken place, and the magnificent spectacle is well remembered by many.

Also seen for the first time at the show in 1965 was an overall Supreme Championship of all cattle entered in the cattle classes. Previously, no such award had existed, but the challenge cup, presented by His Majesty King George VI, which had previously been awarded to the by then defunct Farm Competitions was to be moved to this new, prestigious award. This year also saw one of the cups awarded by the Penryn Agricultural Society being offered for the *'Best Pair of Animals of a Breed, one male and one female'*. Both of these awards of course, exist to this day.

One section which was sadly to disappear from the schedule of 1966 and not to be included again until 1976 was for agricultural horses. Numbers being shown had dwindled to a level at which classes could not be justified, although today strong classes, particularly of the Shire breed are seen and will hopefully continue to grow in the future.

The Annual General Meeting for 1965, heard of the deaths of both Lady Molesworth-St. Aubyn and Captain Abbiss. Following the death of Walter Abbiss a memorial fund, operated by the *'Sou'Westers Club'*, raised money to create some lasting mark of his work in the county as Horticultural Advisor. It was decided to create a garden at the Royal Cornwall Showground, and the area surrounding the Members' and President's Pavilions was chosen and to this day reminds visitors to the show of the work of this man.

1966 again saw the membership of the Association rise to a staggering 7,500, with one of the main improvements being the installation of flush toilets in two of the main blocks! Today, of course, all the shows toilets flush, but this move must have been welcomed by many. The days of the old 'thunderboxes' were numbered!

'Burncoose No Use 5th', a Devon bull owned by S. J. Skinner of Winkleigh, Devon, pictured at the 1966 Show after winning, amongst other prizes, 'The King George VI Perpetual Challenge Cup' for the Supreme Champion Animal in the Cattle Classes.

F. Davey & Sons exhibit of 1966, with a dramatic message to visitors to the Show.

At around this time, the showground at Wadebridge began to see the staging of a few other events, although the full programme of activities seen today, was still some way off. The Four Burrow Pony Club for several years held their summer camps on the showground, with the Members' Pavilion being used as a large dormitory and the President's Pavilion or 'Buck House' as it was known for this particular event, being headquarters to the organising team.

Such well-known figures as Winifred Helm and Captain Dick Micklem held sway and an enjoyable and educational holiday was had by many.

1967 again saw the membership of the Association rise, this time to nearly 8,000. An *'Equestrian Cavalcade'*, featuring a host of famous names, both two and four legged, from the equine world, held centre stage in the main ring. Plans were already in hand for filling an area of land at the Eastern end of the showground to create a level site for a much enlarged flower show to be in place for 1969.

Trotting races were to make an appearance for 1968, following a visit by the Secretary and Mr. Nelson Hosking, the Vice-Chairman of the Association to the Royal Welsh Show in the previous year. Popular in Wales and the Border Counties, the sport had previously not been seen in Cornwall. However, what should have proved to have been an exciting and breathtaking show, had lost some of its sparkle by the time it had crossed the Tamar.

Those competing, whilst keen to win in their own part of the country, were not quite so prepared to risk life and limb for the entertainment of the visitors to that year's show, and the end result was somewhat tame. However, a change is as good as a rest, and most of those watching enjoyed the spectacle.

The year also saw the purchase, by the Association of a further 7½ acres of land for car parking at a cost of £1,500. It should be remembered that at this time the actual showground and some of the car parks were still in the possession of the Duchy of Cornwall.

Throughout these years, much help was given by the Association to the other, smaller agricultural and horse shows from within the county. 1968 had, for instance, seen the sum of £200 donated to Camborne Show to help them over a difficult period brought about by bad weather at their shows. 1969 was to see £140 set aside to repair two trailers, used in the county to transport show jumps from one show to another. Such help, always well received, did much to keep some of the smaller shows going through difficult times and today most of them flourish and are financially secure. To show how generous the Association could be, a donation was even made to a show in Devon! To assist with the building of a hardcore road, £100 was donated in 1969 to Holsworthy Show.

1969 proved to be a somewhat disappointing show from a financial aspect, although the main ring theme of *'This World of Sport'*, provided a new type of entertainment. Stars of many sporting fields paraded and events such as caber tossing and volley-ball (a novelty at the time) drew much interest. A new sponsorship from Whitbread, enabled the sum of £500 to be offered as prizes in a showjumping class, and the team of Whitbread Shires were a popular extra attraction.

The attendance that year owing to very bad weather, dropped by over 5,000, and this coupled with the heavy on-going capital expenditure on the site, resulted in a deficit on the year of just over £1,000, after writing off for depreciation of some £4,000 for the year.

On a brighter note, the Musical Ride of the Household Cavalry were to be looked forward to in 1970 as was a visit to the show by His Royal Highness The Prince of Wales. This was to be the first of three visits so far made by the Prince, when great interest has been shown by him in all that has been going on. Even further ahead, a 14 day cruise to the Mediterranean was planned and organised by the secretary for members in late 1971.

The cruise to the Greek Islands and the Black Sea, for over 300 members, was a great success and although a great deal of organisation was entailed, the event went well. Sailing from Venice on the 12,000 ton Greek owned, Regina, the cruise included visits to places such as Istanbul, Constanta in Rumania, Athens and Dubrovnik in Yugoslavia. A slight hiccup in the plans was to occur when a small

Accepting a Show Catalogue from the Show Secretary, His Royal Highness, The Prince of Wales, Duke of Cornwall, on a visit to the Show in 1970. To the left is Sir John Molesworth-St. Aubyn, Bart., President for that year as well as being the Chairman of the Association. Behind Sir John can be seen Major Simon Bolitho of Trengwainton, Penzance.

number of the group, on an excursion to Moscow, following a call by the ship at Odessa, were unable to rejoin the liner owing to bad weather closing Moscow airport. It was not until the ship docked at Corfu, some days later that the group were to be all back together again. A second cruise, this time along the coast to Yugoslavia, was also to be organised in later years, again proving to be a great hit with those taking part.

Once again, in the Annual Report for 1969, reference was made to the problems experienced in connection with the traffic either coming to or from the show. Annual changes to systems and routes were made to help ease the problem, and today, hopefully most of the difficulties have been overcome. The one single greatest improvement, which has been awaited since the show first moved to Wadebridge is the construction of the Wadebridge By Pass. Now nearing completion, the new road bridge will enable traffic to miss travelling through Wadebridge and hopefully far fewer problems should be experienced.

As far back as 1960, the Assocation was advised to build its main entrance building set back from the A39, owing to the impending road improvements. Thirty years later, those building works are nearly finished!

March of 1972 saw two of the Association's longest serving supporters retiring from office. Sir John Molesworth-St. Aubyn, was to retire as Chairman of the Council after being in office since 1957 and George Blight was to retire from his post as Director of the Showyard.

The new Chairman of the Association, who has so successfully led the show for over twenty years, was to be The Rt. Hon. the Viscount Falmouth. Still in office today, it is particularly appropriate that the Boscawen family should be still represented in this way, as way back in 1793, it was the 3rd Viscount Falmouth who was to be elected the first President of the then Cornwall Agricultural Society. Having a unique interest in not only one, but two county shows, as for many years Lord Falmouth was also Chairman of the Kent County Agricultural Society, the families care and support for all things Cornish is today greatly appreciated.

It was to be Nelson Hosking of Penzance who was to take over as Director of the Showyard. A well known figure in the show world, Nelson was admirably suited to the role.

By this time, the show had grown to include much more than that seen in its early years. Although still very much an agricultural show, various other aspects, such as a forestry section, had been added, with a far wider range of trade exhibits by this time being seen. Virtually any product could be seen on the site, although the total number of trade exhibits, at just under 300, was still less than half of what is present today.

1971 was to see the first of two holidays organised by the Secretary, Albert Riddle for members of the Association. Seen here are some of the 350 who took part in that first cruise around the Mediterranean, prior to take off from RAF St. Mawgan.

Ever keen to take up a challenge, 1971 saw the Cornwall Federation of Women's Institutes Keeping Britain Tidy.

For several years, the question of whether or not to stage a three day as opposed to a two day show had been debated. Many felt that the exhibitors of livestock would not be able to attend the show for a third day, whilst those taking part in other sections would find it extremely difficult to find the extra time. However, the decision to try the venture was to be taken for the show of 1974, with all sections, including the livestock to be expected to be in attendance throughout the show period.

Pig classes, cancelled for the show of 1973, owing to outbreaks of disease were, unfortunately, not to see competitive classes at the show again. The support for the section had significantly decreased over the years and although discussions for their re-instatement for 1974 were had, no further action was taken. Over the years, pigs have been represented in various other ways, with a *'Pig Fair'*, displaying the many aspects of pig production being a feature for quite some time.

1973 will also be remembered by many as the year when several cases of severe food poisoning hit the show. Stemming from food served in the members catering area, the incident highlighted the need for strict measures at events such as a show and thankfully no such problem has been experienced since.

Throughout this period, great thought was being given to the further development of the site as an ever more popular show, meant an ever increasing range of facilities being provided. An appeal fund

The horses of the Kings Troop, taking time off from their displays in the Main Ring, of the Show. Ever popular, this spectacle has been a major feature twice at the Show.

to help cover the costs of the development of the showground was launched and substantial support received both from individuals and organisations who wished to help the show.

The one gift to the Association, however, that was to really be of help at this time, was the donation of a house by Dean Developments. For some time, the question of moving the offices of the Association from Truro to the showground had been discussed, although the cost of building a suitable house and offices had been a great stumbling block. The offer by Dean Developments to provide this accommodation, with the offices being built at cost and the house, free of charge, was to be an offer too good to miss. Planning consents were obtained, and work commenced early in 1974, with the offices being occupied by the show and the house later in the summer.

The additional offer by West Country Oils Ltd. to provide a central heating system for the house and offices, again totally free of charge, was also to be gratefully received. West Country Oils were in fact to later provide a similar offer for a heating system for the members' pavilion and to this day continue as great supporters of the show.

The experiment of holding a three day show proved to be success, with traffic problems being eased. The year was to see the staging of the Women's Weekly exhibition for the first time, with fashion parades and cookery demonstrations proving a popular attraction. Although fairly costly to stage, the exhibit was to prove successful for a number of years, and only ceased to exist following a change of policy by IPC magazines, some years later.

Molesworth House, named after the family who have been so involved with the show since its founding in 1793. Seen being built in 1974, the house was generously presented to the Association by Dean Developments.

1974 also saw donkey classes being staged and the provision of a caravan park for exhibitors wishing to bring caravans to the show.

It was on the last evening of this show that Albert Riddle broke his right leg and along with the other various show reports, local newspapers were to include photographs of the Secretary, with his leg in plaster, propped up in a hospital bed!

The rest of 1974 and 1975 were to prove to be exciting times for the Association. Not only did the Duchy of Cornwall make the purchase of the site by the Association possible at the very reasonable figure of £15,000 for 76 acres, but plans were also put in motion for the erection of a large new cattle hall. A steel-framed exhibition hall, measuring 240' × 100' was, after much discussion to be built, ready to house the cattle for the Royal Cornwall Show of 1975. This building has proved to be a great asset to the Association and has been used for a host of events, other than the show, throughout its life.

Dog and poultry shows, dances, exhibitions and religious gatherings have all taken place, with such highlights as a weeks visit by the Royal Shakespeare Company being particularly memorable. The Caravan Club, during their annual summer rally, have used the building as an evening entertainment centre, with several thousand people dancing on a specially constructed dance floor.

However, without the generosity of the Duchy of Cornwall, none of this would have been possible. The purchase of the site in 1975, which included the site and some of the car parks, has since enabled the Association to develop the area as funds have allowed to a showground the envy of most other shows.

Adding that Continental 'note', Les Tromps De Chasse, representing several French hunts, seen in the Main Ring in 1974.

Michael Rosenberg (*right*), a major instigator of the Rare Breeds Survival Trust, seen with one of the Manz Loghtan sheep, in the 'Animal Land' exhibit at the 1975 Show. The Royal Cornwall was the first Show that the Trust ever exhibited at, and the Association still remains a strong supporter today.

His Royal Highness The Prince of Wales, Duke of Cornwall, visiting the Flower Show during the 1977 Show. Seen to the Prince's right is Neil Treseder, then Chairman of the Flower Show.

Photograph by James Rusbridger

His Royal Highness The Prince of Wales, Duke of Cornwall, whilst visiting the show in 1977.

The use of the showground out-of-season, continued to grow, although 1977 was to see one of the largest single events yet staged. The year, having a truly Royal flavour, included a visit by His Royal Highness The Prince of Wales on the second day of the show, on an occasion when the 150th year since the formation of the Association in 1827 was celebrated.

However, as 1977 was the Silver Jubilee Year of Her Majesty The Queen's reign, a rather special event was to take place on the showground during August. As part of a tour of the Westcountry, using the Royal Yacht Britannia, as a base, a visit to the ground to inspect a massed rally of the youth of the county was planned. Several thousand youngsters were gathered, representing a host of youth organisations and a memorable day was had. For Albert Riddle and his wife Betty, who has been so involved with the show for a great many years, the day was to end particularly well, when they attended a reception given by Her Majesty The Queen on board HMY Britannia. To commemorate the visit by Her Majesty to the showground, a garden seat with a suitably inscribed plaque, was presented, to be placed in the grounds at Sandringham.

A major improvement for the show of 1978 was the erection of the herdsmen's cubicles, where those attending cattle and sheep could sleep during the show. Purchased from the Royal Showground at Stoneleigh, the cubicles, although secondhand, have proved a worthwhile investment and provide weatherproof, albeit, basic accommodation.

At this time, the Association were to purchase a further small piece of land from the Duchy of Cornwall. Amounting to approximately 3 acres, the old Tredinnick farmyard, situated on the south eastern side of the ground, was all that remained of the original farm serving the land already purchased. As many will know, the one up, one down cottage, with adjoining cowshed and hay barn, has now been converted into a house for the past Secretary, Mr. Albert Riddle, who now acts in a capacity of consultant. The other buildings have been converted into toilet blocks and a store, with the land being now part of the main showground.

Further building work was to be proposed at this time, with the National Farmers' Union/NFUM, looking at the possibility of building a permanent stand on the showground. This building was of course to eventually be erected and now stands opposite to the Members' Pavilion on the central concourse.

Gradual changes to the layout and management of the show, were to be seen at each successive show, with more space being taken in at the Western end to allow room for horse box parking, etc. Classifications gradually changed with Shorthorn classes disappearing and in 1979 the first classes for Charolais cattle being staged. Private Driving had by this time been introduced and was to become extremely popular both with exhibitors and those visiting as was the introduction of double harness scurry pony classes in 1980.

During 1977, the Silver Jubilee Year, Her Majesty The Queen and His Royal Highness Prince Philip, Duke of Edinburgh, visited the Showground during August to review a gathering of the youth of the County.

Photograph by James Rusbridger

May of 1980 saw the Wadebridge Agricultural Education Unit being declared open by HRH The Prince of Wales, Duke of Cornwall. Constructed on a site leased to the Cornwall County Council by the Association, the Centre has provided part-time and evening classes in many subjects linked to agriculture and is adjacent to the permanent exhibition building built by the County Council for their use during the show.

May of 1980 saw the Wadebridge Agricultural Education Unit being declared open by HRH The Prince of Wales, Duke of Cornwall. Constructed on a site leased to the Cornwall County Council by the Association, the Centre has provided part-time and evening classes in many subjects linked to agriculture and is adjacent to the permanent exhibition building built by the County Council for their use during the show.

1981 was to be Albert Riddle's twenty-fifth show and with a visit by Her Royal Highness, Princess Alexandra and record entries in the cattle, sheep and horse sections, a memorable show was to be staged. The main ring was graced by a parade of champions, including 'Aldiniti' winner of the 1981 Grand National and a massed parade of hounds. For the first time the show's attendance, including the paid gate, members and ticket holders, was to pass the 100,000 mark, with a total of 100,255 being recorded.

The year had seen the Association venture for the first time into farming, with the fattening of over 1,000 store lambs, bought in from Kent. Although previously only responsible for the taking off of grass for sale for silage or the letting of grass keep for sheep, this venture marked a move into the actual farming of livestock. The buying in of store lambs, was to continue for a few years, with for a while an experimental project of indoor fattening undertaken.

A gradual decrease in the finished price for lambs was to put an end to this farming project however, with the letting of grass keep only, becoming a more financially viable option.

During 1981, following the retirement of Henry Menhenitt, the tenant, a further portion of Tredinnick farm became available, and agreement was reached between the Association and the Duchy of Cornwall for its purchase. The land, amounting to just over 51 acres, comprised what is now the large Southern car park, the livestock exhibitors caravan park and the various areas in between, such as the empty lorry park etc. At a cost of £65,000, this purchase represented a major capital investment for the show, but has proved well worthwhile.

In recognition of his valuable services to the show, Henry Menhenitt and his wife Mary, who is a daughter to that show stalwart George Blight, were both made Honorary Life Members of the Association. A lasting memorial to the work of George Blight can of course be seen over the entrance of the Members' Pavilion where the old station clock from Wadebridge Station, presented to the Association by the Blight family stands in his memory.

Bill Cook, who had for very many years served as the Departmental Steward of Cattle, was to be also so honoured with an Honorary Life Membership, following his retirement from office that year.

At this time, I was to start work on a full time basis with the Association, whilst attending two day release courses, one in business management and the other in agriculture at the day release centre on the Showground itself. Having worked in a part time capacity for some years, both in preparation for the show and on the caravan site that was then being operated during the summer months, the natural step involved being employed on a more regular basis.

1983 saw a major project placed before the Council of the Association for the development of a permanent grandstand, incorporating a members' pavilion and several permanent trade stands. The total estimated cost of the project was to be £1.3 million, which it was hoped could be funded from the leasing of the permanent stands to exhibitors. To be built into the side of the hill, near the cattle judging rings, the new stand would have provided a wonderful facility both for the show and the county, but unfortunately the financial arrangements were not to come about. A possibility for the future perhaps?

Mary Chipperfield's Racing Camels, appearing in the Main Ring at the 1982 show.

The Royal Navy Display Team 'Mast Manning' at the Royal Cornwall Show of 1983.

That year was to also see the show opening on the day of the General Election, although thankfully, attendances were not affected. Radio Cornwall were to be seen that year, with live radio broadcasts coming from the show for the first time. A packed programme of entertainment was to fill the Radio Cornwall Show Theatre over the three days, with live traffic reports and a host of show information being transmitted. The happy relationship which has grown up with Cornwall's first radio station has been mutually beneficial and will hopefully continue over many years to come.

A popular feature in the main ring was the Royal Navy Display Team with their mast manning and window ladder displays. The performances thrilled those watching and together with the Royal

Military Police Display Team, the 'Redcaps', a full main ring programme was seen. Today military cut-backs restrict the type of displays available, making the annual task of filling the programme ever more difficult.

A new feature in the cattle section had been the employment of a cattle commentator, Mike Tucker, to cover the judging of the classes and the presentation of trophies. In 1991 the idea was extended to the sheep section with Charles Emery taking on the task, who with Ashton White, now covering the cattle, make a very talented team.

Once again a Royal president was to officiate, with for 1984, His Royal Highness, The Prince of Wales, Duke of Cornwall taking office. The show that year was to be the twenty-fifth held since moving to the Wadebridge site in 1960. The year was also to be the first to see the office using a computer to complete many of the tasks necessary to the running of the show. All members records, trade stand bookings and livestock entry details were to be dealt with in this way, with the writing of cheques being added the following year. At a cost of nearly £8,000, the installation of such equipment was a great step forward and today, although the original equipment has now been replaced, running the show without it, could not be imagined.

The Secretary appearing to have a little difficulty reading the label, during the presentation of Long Service Awards, by His Royal Highness The Prince of Wales, Duke of Cornwall, President of the Royal Cornwall Show of 1984.

By this time some £240,000 had been spent on the development of the showground with an on going programme of improvements and changes still to come. The year was to see the introduction of the coaching marathon, an event, which has now become a great part of the show and one which attracts entries from throughout the country. Only 3 entries appeared for that first year, with of course, Bill Tucker being one of them. Bill, of Trewornan, near Wadebridge, was also able to transport the president, HRH The Duke of Cornwall around the site in his landau and pair.

The year also saw texel sheep classes for the first time, with an entry of 12. Today much larger numbers are seen and the sheep section has grown dramatically from that seen even less than ten years ago.

The following year, 1985, saw the election of a president, who was to prove to be one of the most popular ever. Never before had a Bishop held the office, but Peter Mumford, The Rt. Revd. The Lord Bishop of Truro, was soon to be a well loved addition to the show team. The show was to be graced by a visit by Her Majesty, Queen Elizabeth the Queen Mother, who was to also exhibit her Buff Orpington poultry in the poultry section that year.

Retiring as Honorary Show Veterinary Surgeon, after 25 years of service to the Association, was Mr. Alan Abbott, who together with his wife were that year to be made Honorary Life Members. Having been on call, in his caravan near the horse lines, for so many years, Mr. Abbott had become a familiar part of the show scene and was to be missed by many of those attending.

Today the Royal Cornwall Show relies on members of the Cornwall Veterinary Association for the supply of vets to cover the show period, with at most times four people needing to be on duty. This help is vital to the livestock sections of the show and the Association are extremely grateful for all that is done. The current show team is headed by Owen Pearce, who practices in St. Columb and Alan Abbott's son Peter, as one of those involved, is helping to carry on the family tradition.

Often working side by side with the vets is the show's farrier. Harold Rich of Blisland, has been working in this capacity since attending a call at the first show at Wadebridge in 1960. Ever since he has been available on site to deal with any problems that may arise with horses shoes whilst at the ground. In the last couple of years Harold has passed his business to his assistant, Anthony Ball but still remains available for advice and continues to act as Stable Manager to the Association.

1985 was also to see Christopher Riddle, appointed as Assistant Secretary to his father. Unique in the world of the county shows, this arrangement was to continue until late 1989, when on the retirement of his father, Christopher was to be appointed to the Secretaryship.

Her Majesty Queen Elizabeth, The Queen Mother, during a visit to the 1985 Royal Cornwall Show. During her visit, The Queen Mother took great interest in an album, compiled by the Association to commemorate her visit together with the King and Princess Margaret in 1950. Her Majesty also inspected the Buff Orpington poultry, she was exhibiting at the Show.

Her Majesty Queen Elizabeth, The Queen Mother visiting the Show of 1985. Seen with Her Majesty is Arthur Daniel, Director of the Showyard, inspecting H. V. Benney's Champion Jersey 'Trevorlis Harvester's Janet'. In the background, to the right of Lord Falmouth, is The Right Revd. The Lord Bishop of Truro, Peter Mumford, Show President for that year.

Plans for 1986 progressed with the news that the *'JCB Dancing Diggers'* would be appearing causing much interest. With the year having been launched by HRH Prince Philip as *'Industry year'*, it was felt appropriate to recognise this fact at the Show. A large indoor exhibition was staged, covering a wide range of areas of industry, and although support for the project was not easy to find, it was felt that the end result was worthwhile.

The year also saw a presentation to Mrs. Peggy Colenso, who had for a very great number of years worked as the Secretary's secretary! Peggy Colenso and her husband Norman, have comprised a vital part of the team behind the scenes for a long period. The presentation took the form of a silver coffee pot which was presented as a small memento of all her years' work, at a reception given by that year's President, Mr. Michael Galsworthy at his home Trewithen.

First started in that year and continuing today, was a scheme whereby with the assistance of Age Concern Cornwall, 500 elderly people, from the various districts of the county, could visit the show each year, on a specially run coach trip, at a nominal price of £1 each, including transport, admission, lunch and tea. The trips still run today and have enabled several thousand people, to visit the show, who perhaps would otherwise have been unable to do so.

A further addition of land to that owned by the Association was made during 1986 with 14 acres being purchased, at auction, from the Duchy of Cornwall. The field, situated at Whitecross Corner, had in most years been used for car parking and its purchase secured the necessary area for this end of the ground for future years.

A further extension to the showground was also to be made during the year, with the Western boundary of the showyard, being moved to meet Tredustan Lane. A brand new horse section, complete with lorry parking, a separate entrance and a ring equivalent in size to the main ring was constructed. The scheme also included the construction of a lake and *'Countryside Conservation Area'* which today forms a tranquil area where conservation matters are depicted and flycasting, falconry and other country pursuits are demonstrated.

A beautiful pair of gates, were presented during the 1987 show to the Association by the Eustice family, in memory of George Eustice, former Director of the Showyard. The gates, which were accepted by the President, Lt. Col. G. T. G. Williams, stand at the entrance to the Conservation area.

During the later part of 1986, a year which marked Albert Riddle's 30th as Secretary, a dinner was held at the Hotel Bristol at Newquay to commemorate this achievement. During the evening, Mr. and Mrs. Riddle were presented with a pair of silver candlesticks, and those present were entertained by the late, Raymond Brooks-Ward, with a speech given in his own, special way.

1987 saw the sheep section, receiving a great boost, with a completely new layout and for the first time, classes for commercial sheep included. These classes were for ewes with cross-bred lambs, bred for the butcher rather than for the breeder. Judged in the pen, the classes enabled those without pedigree flocks to exhibit and over the years, have proved a popular addition. The change from penning the sheep under individual runs of canvas, to under one roof in a large marquee has also pleased exhibitors, with the sheep section having grown from 208 entries in 1987 to 634 in 1992.

During the year, a further small purchase of land was made, with 5½ acres being purchased to add to an existing car park. Work also started on the conversion of the barn/cottage in the old Tredinnick farmyard. Now the home of the past Secretary, Albert Riddle, consultant to the Association, the house is known as Boscawen House, after the family name of Lord Falmouth.

The Flower Show, for so many years such a feature of the Show and the envy of many other shows, was to move to a new home for 1988. Having outgrown its original position at the Eastern end of the showground, a new site, near the Countryside Conservation Area was to be chosen from 1988. An enlarged marquee, in an L-shape, was erected, and the section proved to be a great draw to that area of the ground. The move enabled those exhibiting to have their own car and caravan park, near the marquee.

The Countryside Conservation Area was also to be officially opened in that year, by Her Royal Highness Princess Margaret, Countess of Snowdon, who was to plant a tree to commemorate the event, whilst on a visit to the show.

Pictured at a Dinner held at the Hotel Bristol, Newquay, in 1986, to celebrate Albert Riddle's 30th Show as Secretary are Albert and Betty Riddle, together with some of the team, so vital in running the Show. Bob Gregson, front row—left, back row from left, Richard Blake, Barbara Hewitt and her husband Ivan, Jack Gard and Geoff Eddy.
To mark the occasion, the Secretary was presented with a pair of silver candlesticks.

In the cattle section, classes for Limousin cattle were introduced for that year and have since received excellent entries. As far as horses were concerned, the year was to see the first of two major expansions in the classes offered for Welsh Ponies and Cobs. Originally only having five classes, a total of twenty classes are now staged, five for each of the four sections of the breed.

During the year, a special committee was formed to look into the question of the retirement of Albert Riddle as Secretary, in August 1989. I am glad to say that following various interviews, and the Council's acceptance of the recommendation made by the Committee, I was offered the position. It was to be particularly pleasing that the gentleman to recommend my promotion, Jack Olliver, had been the proposer of my Father to the position way back in 1957!

The President for 1987, Lt. Col. G. T. G. Williams, DL., seen accepting on behalf of the Association, a magnificent pair of gates, erected at the entrance to the Countryside Conservation Area. The gates were presented by the family of the Late George Eustice, who had served on Council for 45 years and had been Director of the Showyard for 10 years prior to his death. Seen presenting the gates is Mrs. George Eustice and George's son Paul.

The following year, leading up to the change in Secretary, was to be a particularly busy one, with not only Her Majesty The Queen and His Royal Highness Prince Philip visiting the show, but the year was also chosen to celebrate British Food and Farming Year.

The Association were to early on agree to support the year as far as was possible, and I was to represent the Association on the county committee. A large indoor exhibition was staged at the Royal Cornwall Show, covering a host of types of agriculture, horticulture and food production and processing. Visited by The Queen and Prince Philip, as part of their showground tour, I was honoured to present a very large hamper of Cornish foods to Her Majesty, on behalf of the food producers of the county. The gift was much appreciated and it was later heard that on the Sunday after the show, The Queen and Prince Philip personally sorted through the hamper. Both to commemorate the visit and to mark Prince Philip's birthday on the following day, a garden seat was presented to The Queen and a set of five horse rugs to Prince Philip.

The day holds very happy memories for many of those present, and I particularly remember the great interest shown by Her Majesty in an egg grading machine, being demonstrated in the Food and Farming Exhibit. Simmental cattle and Jacob sheep were both to make an appearance in 1989, with specialist classes, attracting good entries, being staged.

An aerial view of the Royal Cornwall Show, taken in 1988.

Her Royal Highness Princess Margaret, The Countess of Snowdon on a visit to the Show in 1988. Accompanying her are Lady St. Levan, the Show President and The Rt. Hon. The Viscount Falmouth, Chairman of the Association.

A right royal curtsy! Grove Marlborough, Derrick Eustice's Supreme Champion greeting Her Royal Highness Princess Margaret, The Countess of Snowdon at the 1988 Show.

During the visit of Her Majesty The Queen and His Royal Highness Prince Philip, The Duke of Edinburgh, the then Assistant Secretary, Christopher Riddle, presenting a hamper of Cornish foods on behalf of the food producers of the County in recognition of British Food and Farming Year 1989.

Show Secretary, Albert Riddle (*left*), Lady Molesworth-St. Aubyn, Show President, and Arthur Daniel, (*right*), Director of the Showyard, with the Show's Royal guests in 1989. This was to be Albert Riddle's last year as Secretary, after organising 33 Shows.

The Late Bruce Warren (*left*) and Paul Hancock, parading the Four Burrow Hounds at the 1989 Show.

Albert and Betty Riddle, pictured in the Main Ring at the 1989 Royal Cornwall Show, surrounded by Show Stewards at the end of Albert Riddle's last show as Secretary.

A further area of land, of some 14 acres was to be purchased by the Association from the Church Commissioners during the year, with in addition 18 acres being leased. This land, representing part of the glebe land, attached to St. Breock Church, had in past years been used for car parking, and the purchase helped secure an area for the Association, which was needed as parking once again.

Part of this land together with an area of the showground itself, was to be lost when the long awaited construction of the Wadebridge By-Pass began in late 1991. Excess fill from the construction work has enabled the Association to fill a gulley on the Eastern boundary of the showground, bringing a virtually unusable area into use as showground, and replacing the land lost to the by-pass work.

For the new Secretary, a massive show of livestock was to mark his first year. Records in the horse, cattle and sheep sections were all broken, with in the Jersey classes a judge officiating who was to take everyone by storm. Visiting the show from her home in Montevideo in Uruguay, Senora Cecelia Gallinal de De Haedo, did much to brighten up a show, dogged by rather wet weather. Cecelia still keeps in contact with the Association and many of her Cornish friends, and no doubt will be seen in the county again in years to come.

Making a first visit to the show in 1991 was Her Royal Highness, The Duchess of Kent. During a long, but informal day at the show, Her Royal Highness endeared herself to those attending during visits to most areas of the showground. President for the year, and therefore

Visiting from Montevideo in Uruguay in 1990, Senora Cecelia Gallinal de De Haedo, added her own style to the Jersey cattle judging of that year.

The victorious Cornish rugby team, seen at the Show of 1991, with Show President, Peter Prideaux-Brune.

Two of the Show's most stalwart supporters, Mr. F. Julian Williams (*left*), the Association's Vice Chairman and The Rt. Hon. The Viscount Falmouth its Chairman.

Christopher Riddle, Secretary of the Royal Cornwall Agricultural Association since September 1989.

host to the Royal visitor, Peter Prideaux-Brune of Padstow made a popular figure and one who could trace his families involvement with the show back to its earliest days. The Revd. C. P. Brune, who as Charles Prideaux was to marry into the wealthy Brune family and add the name to inherit estates in Hampshire and Dorset, was great, great, great, great, great Grandfather to Peter Prideaux-Brune and a strong supporter of the original Cornwall Agricultural Society.

Charolais sheep were to be awarded their own classes at that show in 1991, after their debut in the *'Any Other Pure Breed'* section, and with the British Bleu Du Maine breed being added for 1992, the continental influence is ever more apparent.

1991 also saw the Association making a substantial donation towards the cost of a new R.S.P.C.A. Horse Ambulance. Needed to replace an existing, worn-out trailer, the new ambulance, which can be seen at a great number of horse shows etc. throughout the South West, has been much praised by those showing horses and hopefully provides the R.S.P.C.A. with a much needed piece of equipment. This type of donation has been seen throughout the history of the show, and when the funds have been available, suitable worthy causes have been assisted.

The end of 1991, saw both Albert and Betty Riddle, created Honorary Life Members of the Association for their work over so many years. At the same time, Neil Treseder, who had recently retired as Chairman of the Horticultural Section of the Show, was so honoured

Her Royal Highness The Duchess of Kent at the Show of 1991, presenting The Royal Cornwall Cup for the Best Exhibit in the Women's Institute Club Competition to Lostwithiel W.I.

in recognition of 26 years in office. During that year's show, a framed certificate recording this achievement had been presented on behalf of the Association, to Neil Treseder, by HRH The Duchess of Kent. A similar presentation, again by the Duchess, was also made to Mrs. Beck Chapman, whose husband Ron, who had sadly died only shortly before the show, had for over 30 years acted as Honorary Jeweller and Cup Steward to the Association. A very familiar figure to visitors to the show, Ron was to be succeeded by my wife, Thelma Riddle, whose family took over the Chapman family's jewellers business in Wadebridge.

1992 saw the President, Michael Latham of Trebartha, near Launceston tackling a particularly busy twelve months. Not only President of the Royal Cornwall Agricultural Association, but also High Sheriff, a hectic year was to be had. The year saw the dog show and the Craftsmen of Cornwall Exhibit both staged on a temporary hard-surface, at the Eastern end of the showground, in part of the area under development in connection with the construction of the Wadebridge by-pass. The area, now completed will make a useful and attractive addition to the showyard and for 1993 will house, amongst other displays a yet-again enlarged Flower Show, moving from the Western area.

The man with the white gloves, Ron Chapman who for so many years so carefully tended to the Association's vast collection of trophies. See here with the Secretary's wife, Thelma Riddle, who following Ron's untimely death, took over as the Association's Jeweller.

As the Association moves into 1993, plans for the Bi-Centenary Show take shape under the Presidency of Her Royal Highness The Princess Royal, who will no doubt take an active interest in all that goes on.

However, the plans made by the Secretary and the Officers of the Association can only come to fruition with the help of the host of volunteers and organisations who annually give so much time and effort to make the show the success it is. This support over the years has ensured the future of the *'Royal Cornwall'* and as it moves towards the start of its third century, those involved today, whether exhibiting, stewarding or helping in any one of a thousand ways should be proud of playing their part, one and all.

Cornwall forever! (Royal Cornwall, of course).

'AROUND THE SHOW IN 1992'

Arthur Daniel, Director of the Showyard and Hugh Lello, Departmental Steward of Cattle, sharing a joke during the 1992 Show.

Charlie Richards, left, Departmental Steward of Horses and Ernie Bowden, conferring during one of the Horse Parades at the 1992 Show.

Adrian Clifton-Griffith, Departmental Steward of Sheep, during the Grand Parade at the 1992 Show.

Playing a vital role, members of Cornwall Girl Guides who operate the Show Messenger Service, seen with Donald Dunstan, the Departmental Steward of the Grandstand, during the 1992 event.

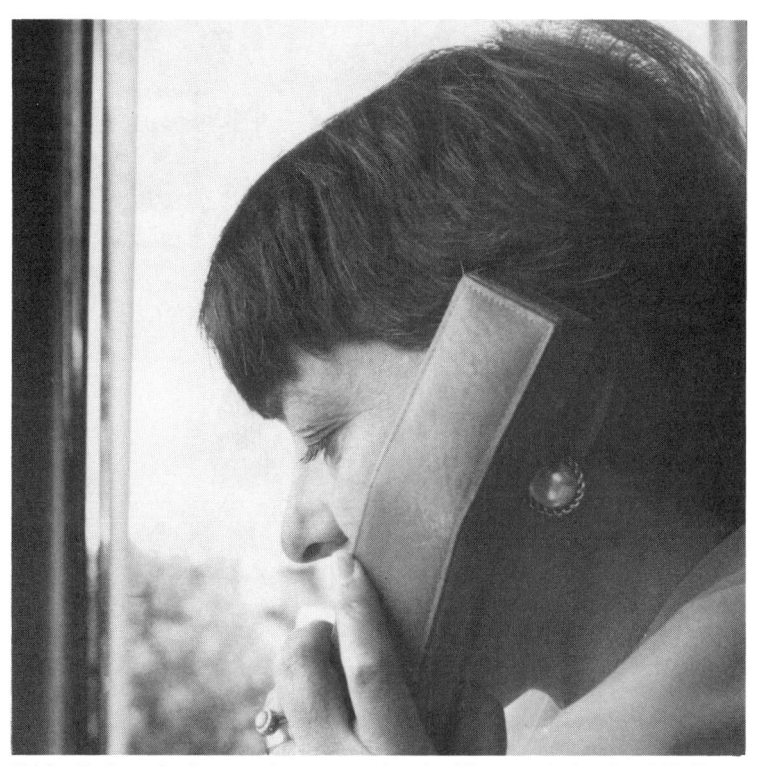

Shirley Parkyn, the Secretary's secretary, in a familiar pose during the 1992 Show.

Roy Meagor and Mike Wilson, two of the Association's gallant team of helpers, manning the Reception of the Secretary's Office.

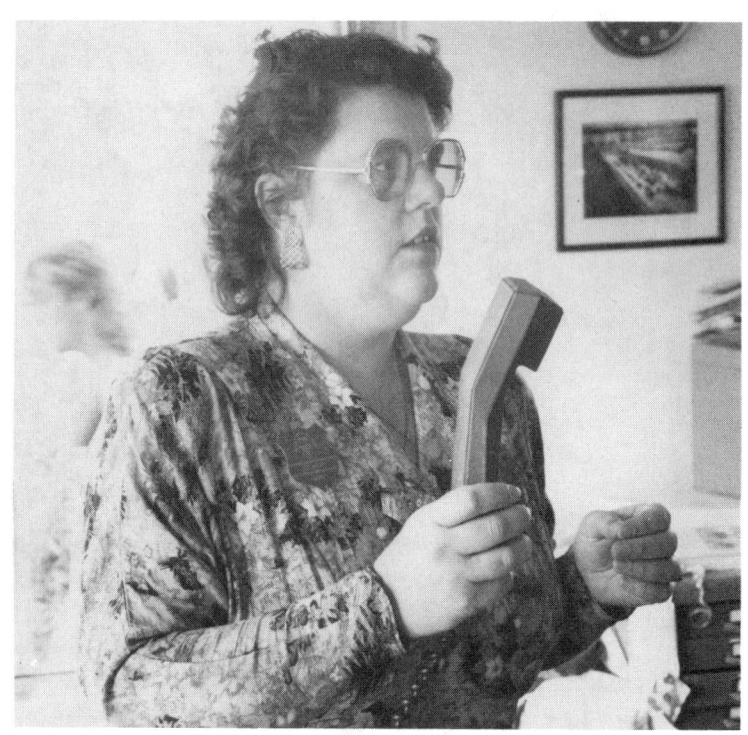

Myra Langdon, one of the Association's small team of office staff.

Cattle commentator, Ashton White, in action during the judging of cattle in 1992.

For so long such a familiar part of the Royal Cornwall's Main Ring, Raymond Brooks-Ward will be sadly missed, following his untimely death during the summer of 1992.

Behind bars at last! Reg Roberts (*left*), Chief Show Jumping Judge and Alan Oliver, Course Builder, pictured at the 1992 Show.

Antony Ball, who has now followed Harold Rich as the Association's Farrier, seen in action with one of Rothmans of Pall Mall's pair of greys, at the 1992 Show.

Owen Pearce of St. Columb, one of the Association's Veterinary Surgeons, who on behalf of the Cornwall Veterinary Association, tend to the needs of the livestock whilst at the Show.

Don Waterhouse (*right*), Chairman of the Flower Committee, seen with TSW's gardening expert, Terry Underhill in 1992.

True perfectionists, part of the hard working Flower Show Committee, pictured at the Royal Cornwall Show of 1992.

Maurice Lobb, Secretary of the Cage Birds Section of the Show.

The Rt. Hon. The Viscount Falmouth (*left*), and Mr. Michael Latham, Show President, (*centre*), during a visit to the Poultry Section of the 1992 Show. Seen on the right is Kevin Dowrick, Secretary of the Poultry Section.

The Secretary of the Show's Fur Section, Marion Dyer, seen with one of the Section's larger exhibits!

John Gregory (*left*), Sheila Sleeman, Section Secretary and Morley Kingdon (*right*), of the Bees and Honey Section, pictured outside their marquee in 1992.

John Robilliard, Chairman of the Pigeon Section, with a prize exhibit.

Sue Emrys-Jones, Chairman of the Association's Canine Section, seen with some of her Committee at the 1992 Show.

Unseen by many, Bill Vernon, the Association's Milk Steward, for a great number of years has directed operations in the Showground Milking Parlour, along with his hard working team of assistants.

Gerald Radford and Dick Parkes, who on behalf of Trewartha, Gregory and Doidge of Callington, keep the Shows plumbing and heating systems in order.

Mrs. Joan Latham and the President, Mr. Michael Latham, seen visiting the Children's Creche during the 1992 Show. The Creche, which can cope with dozens of children daily is run by the Women's Royal Voluntary Service and the Playgroups Association (Cornwall County) for the Pre-School Child.

SHOW VENUES, ATTENDANCES AND SHOW PRESIDENTS 1793 TO 1993

Please Note: The venues listed do not include the venues of ploughing matches.

Year	Show Venue(s)	Attendance	President
1793			Lord Viscount Falmouth, Tregothnan, Truro
1794	Bodmin		Lord Viscount Falmouth, Tregothnan, Truro
1795	Bodmin		Lord Viscount Falmouth, Tregothnan, Truro
1796	Bodmin		
1797	Bodmin		Philip Rashleigh Esq
1798	Bodmin		John Coryton Esq, Pentillie Castle
1799	Bodmin		
1800	Bodmin		Lord De Dunstanville, Tehidy
1801	Bodmin		
1802	Bodmin		
1803	Bodmin		Edmund John Glynn Esq, Glynn
1804	Bodmin		Francis Glanville Esq, Catchfrench
1805	Bodmin and Helston		Lt General Morshead
1806	Bodmin and Helston		Francis Hearle Rodd Esq, Trebartha Hall, Launceston
1807	Bodmin and Stratton		John Hearle Tremayne Esq, Heligan, St Austell
1808	Bodmin and Helston		Vyel Vyvyan Esq, Trelowarren
1809	Bodmin and Stratton		
1810	Bodmin and Helston		
1811	Bodmin and Stratton		
1812	Bodmin and Helston		
1813	Bodmin and Callington		J Vivian Esq, Pencalenick
1814	Bodmin and Helston		Sir Arscot Ourry Molesworth, Bt
1815	Bodmin and Stratton		Henry Peter Esq, Harlyn
1816	Bodmin and Helston		Joseph Thomas Austen Esq, Place House, Fowey
1817	Bodmin and Liskeard		Sir Arscott Ourry Molesworth, Bt
1818	Bodmin and Stratton		John Penhallow Peters Esq, Crigmurrion
1819	Bodmin		W L S Trelawney Esq
1820	Bodmin		John Buller Esq, Morval
1821	Bodmin		
1822	Bodmin		Richard Gully Bennet Esq

Year	Show Venue(s)	Attendance	President
1823	Bodmin		Humphry Grylls Esq
1824	Bodmin		
1825	Bodmin		
1828	Truro		
1829	Truro		
1830	Truro		
1831	Truro		
1832	Truro		
1833	Truro		
1834	Truro		Richard Johns Trewince
1835	Truro		J H Tremayne Esq Heligan, St Austell
1836	Truro		
1837	Truro		G W F Gregor Esq Trewarthenick
1838	Truro		The Rt Hon The Earl of Falmouth Tregothnan, Truro
1839	Truro		
1840	Truro		The Rt Hon The Earl of Falmouth Tregothnan, Truro
1841	Truro		E W Wynn Pendarves Pendarves
1842	Truro		J Vivian Esq Pencalenick, Truro
1843	Truro		The Rt Hon The Earl of Falmouth Tregothnan, Truro
1844	Truro		P Andrews Esq Bodmin
1845	Truro		Capt B Reynolds Penair
1846	Truro		Humphry Willyams Esq Truro
1847	Truro		J D Gilbert Esq Trelissick, Truro
1848	Truro		Glanville Gregor Esq Trewarthenick, Tregony
1849	Truro		John Ennis Vivian Esq Truro
1850	Truro		The Rt Hon The Earl of Falmouth Tregothnan, Truro
1851	Truro		William Williams Esq Tregullow
1852	Truro		Richard Davey Esq Redruth
1853	Truro		John Tremayne Esq Heligan

Year	Show Venue(s)	Attendance	President
1854	Truro		Michael Williams Esq MP
Trevince
1855	Truro		The Rt Hon The Earl of Falmouth
Tregothnan, Truro
1856	Truro		William Rashleigh Esq
Menabilly
1857	Truro		R Gully Bennet Esq
Tresillian House
1858	Camborne		Richard Davey Esq MP
Polsue, Philleigh
1859	St Austell		T J Agar Robartes Esq MP
Lanhydrock, Bodmin
1860	Penzance		John St Aubyn Esq MP
1861	Merged Show
— Bath & West at Truro		N Kendall Esq MP
1862	Liskeard		W H Pole Carew Esq
1863	Truro		J F Basset Esq
Tehidy
1864	Saltash		Earl of Mount Edgcumbe
Mount Edgcumbe, Devonport
1865	Falmouth		John Jope Rogers Esq MP
1866	No Show — Cattle Plague		E Brydges Willyams Esq MP
Carnanton, St Columb
1867	Launceston		Sir Colman Rashleigh Bart.
Prideaux, Par Station
1868	Merged Show
— Bath & West at Falmouth		Sir William Williams Bart.
1869	Penzance		Colonel Tremayne
Carclew, Perranarworthal
1870	Launceston		Captain Fortescue
1871	Truro	17,231	Colonel Archer
Trelaske, Launceston
1872	Bodmin	17,732	G L Basset Esq
1873	Penryn	11,812	Lord Eliot
1874	St Austell	17,022	A Pendarves Vivian Esq
Bosahan, Helston
1875	Truro	24,212	HRH The Prince of Wales
1876	Liskeard	14,228	Colonel Hon E Boscawen
1877	Camborne	17,186	Johathan Rashleigh Esq
Menabilly, Par Station
1878	Saltash	13,574	Sir F M Williams Bart, MP
1879	Falmouth	17,027	Hon T C Agar Robartes
Lanhydrock, Bodmin
1880	Lostwithiel	11,408	Digby Collins Esq
Newton Ferrers, Callington
1881	Redruth	20,105	Richard Foster Esq
Lanwithan, Lostwithiel
1882	Launceston	16,409	T Robins Bolitho Esq
Pendrea, Penzance

Year	Show Venue(s)	Attendance	President
1883	Truro	21,871	John Tremayne Esq Heligan, St Austell
1884	Bodmin	16,057	Rev Sir Vyell Vyvyan Bt Trelowarren, Helston
1885	Penzance	15,732	Sir Charles Sawle Bart Penrice, St Austell
1886	St Austell	16,551	W Cole Pendarves Esq Pendarves, Camborne
1887	Camborne	17,669	The Earl of St Germans Port Eliot
1888	Newquay	11,711	John C Williams Esq Caerhayes Castle, St Austell
1889	Helston	12,974	Sir W L Salusbury Trelawny Bt Trelawne, Duloe
1890	Truro	12,118	T Bedford Bolitho Esq Trewidden, Penzance
1891	Par	13,484	Viscount Valletort, Mount Edgcumbe
1892	Redruth	18,793	C Davies Gilbert Esq Trelissick, Truro
1893	Liskeard	13,047	J B Fortescue Esq Boconnoc, Lostwithiel
1894	Falmouth	18,316	J Claude Tremayne Esq Heligan, St Austell
1895	Wadebridge	16,342	John Claude Daubuz Killiow, Truro
1896	St Ives	13,040	Hon John Boscawen Tregye, Perranwell
1897	Lostwithiel	9,498	F Buller Howell Esq Ethy, Lostwithiel
1898	Penzance	17,688	Richard Carlyon Coode Esq Polapit Tamar, Launceston
1899	Launceston	12,822	A F Basset Esq Tehidy
1900	Truro	14,550	John Tremayne Esq Heligan, St Austell
1901	Bodmin	14,888	Sir Lewis Molesworth MP Trewarthenick
1902	Camborne	9,506	Maj The Hon John St Aubyn St Michael's Mount
1903	St Austell	19,270	Col The Hon Charles Byng Edymead House, Launceston
1904	Falmouth	16,287	John Williams Esq Scorrier House
1905	Newquay	15,065	Sir Colman Rashleigh, Bart Prideaux, Par Station
1906	Redruth	20,318	Roger W G Tyringham Esq Trevethow, Lelant
1907	Liskeard	12,648	Horace B Grylls Esq Lewarne, St Neot

Year	Show Venue(s)	Attendance	President
1908	Helston	17,370	His Grace The Duke of Leeds Hornby Castle, Yorkshire
1909	St Columb	15,757	Capt W F Tremayne Carclew, Perranarworthal
1910	St Ives	14,250	Col Courtenay Vyvyan CB Tremayne, St Martin
1911	St Austell	18,157	The Hon T C Agar-Robartes Lanhydrock
1912	Penzance	21,454	H Harcourt Williams Esq Pencalenick
1913	No Show — Bath & West at Truro		
1914	Fowey	14,094	J De Cressy Treffry Esq Penarwyn, Par Station
1915	Camborne	18,151	Sir Hugh Molesworth-St Aubyn, Bart Clowance, Praze
1916 to 1918	No Shows — First World War		
1919	Truro	29,768	John Charles Williams Esq Caerhays Castle, Gorran
1920	Callington	17,682	HRH The Prince of Wales KG Duke of Cornwall
1921	Falmouth	16,372	Col The Hon H W F Trefusis Trefusis, Falmouth
1922	Newquay	14,128	The Rt Hon Viscount Falmouth Tregothnan, Truro
1923	Camborne	17,231	P D Williams Esq Lanarth, St Keverne
1924	Wadebridge	13,373	Col C R Prideaux-Brune DL JP Prideaux Place, Padstow
1925	Helston	16,015	H Montague Rogers Esq Nansloe, Helston
1926	Launceston	17,132	E G Baron Lethbridge Esq Tregeare, Egloskerry
1927	Truro	26,277	HRH The Prince of Wales, KG Duke of Cornwall
1928	Bodmin	19,937	George H Johnstone Esq Trewithen, Grampound Road
1929	Penzance	18,346	Lt Col E H W Bolitho, DSO Trengwainton, Penzance
1930	Liskeard	16,040	The Hon Montague Eliot Port Eliot, St Germans
1931	St Columb	12,402	Lt Col E N Willyams, DSO Carnanton, Newquay
1932	Penryn	17,363	Capt C H Tremayne Carclew, Perran-ar-Worthal
1933	St Austell	24,187	Col Lord Vivian, DSO, TD, ADC Glynn, Bodmin
1934	Camborne	19,910	J C Bickford-Smith Esq Trevarno, Helston

Year	Show Venue(s)	Attendance	President
1935	Newquay	19,431	Col Treffry, CMG, OBE, TD, DL, ADC, Place, Fowey
1936	St Ives	17,671	Mrs Charles Williams Trewidden, Penzance
1937	Wadebridge	19,469	The Rt Hon Viscount Clifden Lanhydrock, Bodmin
1938	Helston	21,181	Capt J Lionel Rogers Penrose, Helston
1939	Bude	17,825	George A C Thynne Esq Trelana, Bude
1940 to 1946	No Shows — Second World War		
1947	Truro	44,612	Maj The Hon The Lord Seaton JP Beechwood, Plympton
1948	Bodmin	38,659	Sir John Molesworth-St Aubyn Bt Pencarrow, Bodmin
1949	Falmouth	32,960	Brig Stephen Williams, DSO, MC Tregullow House, Scorrier
1950	Callington	47,732	Col Sir John Carew-Pole Antony House, Torpoint
1951	Newquay	34,551	George H Johnstone Esq OBE Trewithen, Grampound Road
1953	Launceston	29,582	Cmdr A M Williams DSC Werrington, Launceston
1952	Redruth/Camborne	25,628	G P Williams Esq MFH Tregullow House, Scorrier
1954	St Austell	34,395	The Rt Hon Charles Williams MP
1955	No Show — Bath & West at Launceston		
1956	Helston	26,786	J Lionel Rogers Esq Penrose, Helston
1957	Wadebridge	25,397	Lady Molesworth-St Aubyn JP Pencarrow, Bodmin
1958	Truro	23,897	The Rt Hon Viscount Falmouth Tregothnan, Truro
1959	Liskeard	29,985	Lady Cynthia Carew-Pole Antony House, Torpoint
1960	Wadebridge	29,848	Miss E Johnstone Trewithen, Grampound Road
1961	Wadebridge	32,141	M Bickford-Smith Esq Trevarno, Helston
1962	Wadebridge	34,439	Lord St Levan St Michael's Mount, Marazion
1963	Wadebridge	38,762	P M Williams Esq OBE Burncoose, Redruth
1964	Wadebridge	36,522	The Rt Hon Viscount Falmouth Tregothnan, Truro
1965	Wadebridge	47,778	Sir Patrick Kingsley KCVO Duchy of Cornwall, London
1966	Wadebridge	38,439	Major J C B Lethbridge MFH Tregeare, Launceston
1967	Wadebridge	40,464	H R Graham-Vivian Esq TD Bosahan, Manaccan, Helston

Year	Show Venue(s)	Attendance	President
1968	Wadebridge	46,137	J D G Fortescue Esq Boconnoc, Lostwithiel
1969	Wadebridge	41,843	F Julian Williams Esq Caerhayes Castle, Gorran
1970	Wadebridge	39,212	Sir John Molesworth-St Aubyn Bt CBE Pencarrow, Bodmin
1971	Wadebridge	36,624	Mrs Stephen Williams Scorrier, Redruth
1972	Wadebridge	35,605	Major S E Bolitho MC Trengwainton, Penzance
1973	Wadebridge	42,409	Viscountess Falmouth Tregothnan, Truro
1974	Wadebridge	76,147	Alderman K G Foster, CBE, JP Trewin, Sheviock, Torpoint
1975	Wadebridge	72,249	Major J Coryton, MC Pentillie Castle, St Mellion
1976	Wadebridge	85,593	Lt Col J A Molesworth-St Aubyn MBE The Manor, Tetcott, Holsworthy
1977	Wadebridge	86,200	The Rt Hon Viscount Falmouth Tregothnan, Truro
1978	Wadebridge	80,600	The Hon John St Aubyn St Michael's Mount, Marazion
1979	Wadebridge	85,700	The Rt Hon The Earl of Morley Pound House, Yelverton
1980	Wadebridge	95,000	Mrs Simon Bolitho Trengwainton, Penzance
1981	Wadebridge	100,255	Richard Carew-Pole Esq Erth Barton, Elm Gate, Saltash
1982	Wadebridge	99,076	The Most Honourable The Marquess of Lothian, 54 Upper Cheyne Row, London
1983	Wadebridge	99,655	Major E W M Magor, CMG, OBE, DL Lamellen, St Tudy, Bodmin
1984	Wadebridge	103,950	HRH The Prince of Wales, Duke of Cornwall
1985	Wadebridge	104,137	The Rt Revd The Lord Bishop of Truro Lis Escop, Feock
1986	Wadebridge	105,348	A M J Galsworthy Esq Trewithen, Grampound Road
1987	Wadebridge	98,452	Lt Col G T G Williams, DL Menkee, St Mabyn, Bodmin
1988	Wadebridge	106,886	Lady St Levan St Michael's Mount, Marazion
1989	Wadebridge	111,789	Lady Molesworth-St Aubyn Pencarrow, Bodmin
1990	Wadebridge	107,890	P M Bickford-Smith Esq Trevarno, Helston
1991	Wadebridge	104,228	P J N Prideaux-Brune Esq Prideaux Place, Padstow
1992	Wadebridge	105,778	E M L Latham Esq Trebartha, Launceston
1993	Wadebridge		HRH The Princess Royal GCVO

CHAIRMEN OF COUNCIL OF THE ASSOCIATION

Prior to 1950 no official post of Chairman existed, with the chair at meetings of the Council being taken either by the President for that year, the Chairman of one of the main Committees or a senior Member of Council or Vice President.

1950 to 1957
George H Johnstone
Trewithen, Grampound Road

1957 to 1972
Sir John Molesworth-St Aubyn, Bart.,
Pencarrow, Bodmin

1972 to date
The Right Hon The Viscount Falmouth
Tregothnan, Truro

DIRECTORS OF THE SHOWYARD

1879 to 1891
W Trethewy
Tregoose, Probus

1892 to 1898
Richard Olver
Trescowe, St Mabyn

1899 and 1901
John Tucker
Menheniot, Liskeard

1900 and 1902 to 1907
J De Cressy Treffry
Penarwyn, Par Station

1908 to 1933
Brooking Trant
Liskeard

1934 to 1936
Capt C H Tremayne
Carclew, Perranarworthal

1937 to 1939
J Curnow
Rejarden, Marazion

1947 to 1950
Edwin Baker
Treniffle, Launceston

1951 to 1954
J A Pooley
Chacewater, Truro

1956 to 1971
George Blight
Tregonning, Breage

1972 to 1976
A Nelson Hosking
Penzance

1977 to 1985
George Eustice
Gwinear, Hayle

1986 to date
Arthur Daniel
Trescowe, St Mabyn and Wadebridge

SECRETARY'S OF THE SOCIETY/ASSOCIATION

1793 to 1827
John Wallis, Attorney, Bodmin

1828 to 1838
John C Downe

1839 to 1858
William Floyd Karkeek
Veterinary Surgeon, Truro

1859 to 1909
Henry Tresawna
Probus, Truro

1910 to 1922
J F Crewes

1923 to 1946
W Courtney Hocking
Truro

1947 to 1956
Anthony Williams
Helston

1957 to 1989
Albert H Riddle

1989 to date
Christopher P Riddle

AN INDEX OF COUNCIL MEMBERS 1881 TO DATE

ALFORD W, Tregarrick, St Tudy 1957-1960, 1961-1964, 1965-1968, 1969-1972
ALSEY J, County Organiser, YFC, Truro 1972-1975, 1976-1979, 1980-1983, 1984-1987
ALLCHIN F, Lanteglos-by-Fowey 1892-1895
ANDERSON W H, Boconnoc, Lostwithiel 1934-1937
ANDERTON A C, Falmouth 1901-1904, 1907-1910, 1913-1919
ANDERTON E D, Falmouth 1887-1890, 1893-1896
ANDREW W J, Redruth 1905-1908
ANNAND D, Whiteford Hm Fm, Stokeclimsland 1920-1923
ARCHER C, Trelaske, Launceston 1886-1889
ARTHUR B, Trengrove, Merrymeet, Liskeard 1990-1993
AYRE E E, Trevanion, Newquay 1928-1931

BADCOCK W C, St Hilary 1949-1952
BAIN D W, Portreath 1884-1887, 1892-1895
BAIN F D, Portreath 1905-1908
BAKER E, Treriffle, Launceston 1933-1936, 1939-
BAKER RUSSELL, Worthyvale, Camelford 1924-1927, 1930-1933, 1936-1939, 1947-1950
BARON THOMAS, Lidcutt, Bodmin 1883-1886, 1893-1896, 1900-1903
BARON LETHBRIDGE E C, Tregeare, Launceston 1913-1919 Deceased
BASSETT F W, Mitchell 1887-1890, 1895-1898
BASTARD JOHN, Tinten, St Tudy 1899-1902, 1905-1908
BASTARD H E, Lemail, Egloshayle 1923-1926, 1929-1932, 1935-1938, 1947-1950, 1954-1957, 1958-1961
BATE F, Launceston 1930-1933
BAX C, Tregorden, Wadebridge 1988-1991
BAX T, Tregorden, Wadebridge 1959-1962, 1963-1966, 1967-1970, 1971-1974, 1975-1978, 1981-1984, 1985-1988
BEAK H, Praze Farm, Sithney, Helston 1959-1962
BEAUCHAMP C, Trevince, Redruth 1907-1910, 1914-1920
BENNETT A J, Carne Beach, Veryan, Truro 1979-1982, 1983-1986, 1987-1990, 1991-1994
BENNETT S E, Calendra, Ruanhighlanes 1930-1933, 1937-
BENNETTS F G, Sticker, St Austell 1952-1955, 1956-1959, 1960-1963, 1964-1967, 1968-1971, 1972-1975, 1976-1979, 1980-1983, 1984-1987, 1988-1991
BENNEY G, Namplough, Cury, Helston 1970-1973
BENNEY T J P, Falmouth 1949-1952
BENNEY W, Alma House, Mullion, Helston 1962-1965, 1966-1969
BENNEY W H, Mullion 1949-1952
BENNEY W T G, Poldown Farm, Nancegollan 1978-1981, 1982-1985, 1986-1989, 1990-1993
BERRYMAN J, Tremedda Farm, St Ives 1970-1973, 1974-1977
BERSEY N, Minard Farm, Polbathic, Torpoint 1985-1988, 1989-1992
BETTENS W J, Perranuthnoe, Penzance 1939-1949, 1952-1955, 1956-1959, 1960-1963
BETTENS W J, Rosper, Perranuthnoe, Penzance 1976-1979, 1980-1983, 1984-1987, 1988-1991
BICE CHARLES, Tregassow, St Erme 1898-1901, 1909-1912
BICE FRANK, Newlyn 1897-1900
BICE JAMES, Burthy, St Enoder 1902-1907
BICE LUKE, Nanswhyden, St Columb 1883-1886, 1896-1899

BICE L V, Newquay	1911-1914
BICE W C, Nanswhyden, St Columb	1906-1909, 1920-1923, 1927-1930, 1934-1937
BICE W F R, Nanswhyden, St Columb	1947-1950, 1953-1956
BICE ZACCHEUS, St Columb Minor	1887-1890, 1904-1907
BIDDICK JAMES, Tregolds, St Wenn	1898-1901
BISHOP H H, Tregellas, Grampound Road	1906-1909
BLACKBURN R, Highfield, Marazion	1926-1929 Deceased
BLAMEY AUG., Veryan	1884-1887
BLAMEY C, Veryan	1881-1884, 1889-1892, 1896-1899
BLAMEY C, Trewithen, Liskeard	1927-
BLAMEY M L, Camels, Veryan	1903-1906
BLAMEY THOMAS, Trethennal, Veryan	1887-1890, 1893-1896
BLAMEY TOM, Camels, Veryan	1900-1903
BLAMEY T WALLIS, Pennare, Veryan	1902-1905, 1908-1911, 1914-1920
BLAMEY W C, Pennare, Veryan	1905-1908
BLENKINSOP A, Truro	1911-1914
BLIGHT F, Trolvis, Stithians	1948-1951
BLIGHT GEORGE, Tregonning, Breage, Helston	1919-1922 Deceased
BLIGHT GEORGE, Tregonning, Breage, Helston	1923-1926, 1929-1932, 1935-1938, 1947-1950, 1950-1953
BLIGHT GEORGE, Chytodden Fm, Breage, Helston	1960-1963, 1964-1967, 1968-1971, 1972-1975, 1976-1979, 1980-1983, 1984-1987, 1988-1991
BLIGHT GEORGE, Tregonning, Breage, Helston	1983-1986, 1987-1990
BODY E, Twelvewood, Liskeard	1885-1888
BODY E H, Twelvewood, Liskeard	1912-1915, 1921-1924
BODY E W, Twelvewood, Liskeard	1900-1903, 1907-1910, 1913—Deceased
BODY J, Twelvewood, Liskeard	1894-1897
BOGER WALTER, Wolsden, Antony	1896-1899
BOLITHO, Col E H W	1925-1928
BOSCAWEN Revd A, Gulval	1905-1908
BOSCAWEN The Hon EVELYN, St Michael Penkivel	1981-1984, 1985-1988, 1989-1992
BOWDEN E, Gwealavellan Fm, Gwithian, Hayle	1974-1977, 1978-1981, 1982-1985, 1986-1989, 1990-1993
BOWDEN E W, Tregenna Lane, Camborne	1953-1956, 1958-1961, 1963-1966, 1967-1970, 1971-1974, 1975-1978, 1979-1982
BOWDEN M, South Tehidy, Camborne	1989-1992
BOWDEN S, Tresawle, St Columb	1965-1968
BRAY FRANK, Padstow	1897-1898 (Resigned)
BRENDON GEORGE, Bude	1890-1893
BRENT C, Clampit, Callington	1947-1950, 1953-1956
BRENT D, Clampit, Callington	1884-1887
BRENT, W, Brentholme, Linkinhorne	1980-1983, 1984-1987, 1988-1991
BRENT W G, Warrens Park, Coads Green	1926-1929, 1935-1938
BRENTON G H, St Germans	1890-1893
BRENTON WILLIAM, St Germans	1887-1890, 1893-1896, 1899-1902, 1907-1910
BRENTON WILLIAM HAROLD, St Germans	1923-1926, 1931-1934, 1938-
BREWER A D, Clifton Gardens, Truro	1921-1924
BREWER D, Grampound Road	1888-1891
BREWER J V, Meadowside, Grampound Road	1936-1939

BROCK J, Chy-an-Mor, Coverack, Helston   1975-1978, 1979-1982
BROOKHAM D, West Druxton,
Launceston   1975-1978, 1979-1982, 1983-1986, 1987-1990,
                                         1991-1994
BROWN G, Tresmarrow, Launceston   1957-1960
BROWN P G, Tremadart, Duloe   1901-1904, 1907-1910, 1913-1919, 1922-1925,
                                         1929—Deceased
BROWN T R, Tremadart, Duloe   1931-1934
BRYANT N, Truro   1911-1914
BUCHER N H, Boscawen St, Truro   1911-1913
BULLMORE GEORGE, Newquay   1888-1891, 1900-1903, 1904-1907, 1910-1913
BULLMORE J C, Tregair, Newlyn East   1905-1908, 1915-1921, 1924-1927 Deceased
BUNT A, Lampen Farm, St Neot, Liskeard   1988-1991
BUNT C B, MFH, Woodlands, St Neot   1983-1986, 1987-1990, 1991-1994
BUNT W P, Lanteway, St Neot, Liskeard   1977-1980, 1981-1984
BURLEY S J, Higher Newham Farm,
Truro   1981-1984, 1985-1988, 1989-1992
BURLEY W H, Higher Newham Farm,
Truro   1966-1969, 1970-1973, 1974-1977, 1978-1981,
                                         1982-1985, 1986-1989
BURLEY W T, Gwarnick Mill Farm,
St Allen   1988-1991
BUTTON JOHN, Michaelstow   1899-1902
BUTTON S J, Town Farm, St Tudy   1924-1927, 1933-1936, 1939-
BYNG The Hon Col C, Launceston   1898-1901
CARDELL JOHN, Trebelsue, St Columb
Minor   1900-1903
CARDELL RICHARD, St Columb Minor   1891-1894
CARVETH A, Boswellick, St Allen   1913-1919
CARVETH W J, St Allen   1897-1900, 1905-1908
CAWRSE R, Penwarne, Mevagissey   1947-1950
CHAPMAN G, Gerrans   1885-1888, 1891-1894
CHAPMAN WILLIAM, Gerrans   1894-1897
CHAPPELL W R, Park Farm, Carn Brea   1938-
CHELLEW P J, St Ives   1936-1939
CHELLEW S H, Polkinghorn, Gulval   1954-1957, 1959-1962, 1963-1966
CHINN W S I, Treliver, Penryn   1948-1951, 1954-1957
CHIRGWIN Mr, Truro   1885-1888
CHOAK H L, Menadew, St Clement   1920-1923, 1926-1929
CHRISTOPHER S, Eugew, Gwithian, Hayle   1926-1929
CHRISTOPHERS T W R, Truro   1967-1970, 1971-1974
CHUDLEIGH C, The Old Rectory,
Lawhitton   1949-1952, 1955-1958, 1960-1963, 1968-1971,
                                         1972-1975, 1976-1979, 1980-1983, 1984-1987,
                                         1988-1991
CLARK W, Tregavethan, Truro   1931-1934
CLARKE RICHARD, Luney, St Ewe   1890-1893, 1898-1901, 1906-1909, 1912-1915
CLIFTON GRIFFITH A, Porthtowan,
Truro   1982-1985, 1986-1989, 1990-1993
CLUNES FREDERICK, St Austell   1902-1905
CLYMO J, Penpont, Altarnun   1930-1933
COAD E J, Duchy Home Fm,
Stokeclimsland   1955-1958, 1959-1962
COAD JOHN, St Keverne   1891-1894, 1900-1903
COAD J, Mawgan   1932-1935
COAD R, St Keverne   1907-1910, 1913-1919 Deceased
COAD R, Treleague, St Keverne   1939-1949
COAD R, Little Tregear, Mawgan, Helston 1957-1960
COATH GEORGE, Raphael, Polperro   1898-1900

COLLINS J, Gweek, Helston	1913-1919, 1924-1927, 1934-1937
COLLINS JOSE, Gweek	1930-1933
COLLINS T, Trewothack, Manaccan	1936-1939
COLLINGS F L, Trewolland, Liskeard	1948-1951, 1956-1959, 1963-1966
COLWILL W, Delabole Barton, Delabole	1951-1954
CONGDON T E, Pelyn Fm, Lostwithiel	1889-1892
COOK W R, Creathorne, Bude	1948-1951, 1954-1957, 1958-1961, 1962-1965, 1966-1969, 1970-1973, 1974-1977, 1978-1981
COON F A, St Austell	1908-1911
COPPLESTONE H, Shenstone, Lostwithiel	1933-1936, 1939-
COPPLESTONE W, Lostwithiel	1908-1911
CORYTON Capt J L, Pentillie Castle, Saltash	1922-1923, 1928-1931
CORYTON W, Pentillie Castle, Saltash	1884-1887
CRAPP M, Treworrick, St Cleer, Liskeard	1921-1924
CRAVEN W G, Trewen, Camelford	1889-1892
CRAZE A, Rosewarne, Camborne	1908-1911
CRAZE F, Lelant	1894-1897
CROGGON JOHN, Creed	1885-1888
CURGENVEN E, Pulla, Gwennap, Redruth	1906-1909, 1914-1920
CURNOW J, Rejarden, Marazion	1925-1928, 1931-1934
CURNOW S W, Pollard Farm, Helston	1952-1955
DANIEL A G, Tredannen, Wadebridge	1957-1960, 1961-1964, 1965-1968, 1969-1972, 1973-1976, 1977-1980, 1981-1984, 1985-1988, 1989-1992
DANIEL J H, Wellington Road, Camborne	1939-
DANIEL W L, Ley, Upton Cross, Liskeard	1961-1964, 1965-1968
DAUBUZ J CLAUDE, Killiow, Truro	1882-1885
DAVEY F B, Trevarthian Fm, St Newlyn East	1979-1982, 1983-1986, 1987-1990, 1991-1994
DAVEY J V, Kellywell, St Mabyn, Bodmin	1979-1982, 1983-1986, 1987-1990, 1991-1994
DAVIES J N, Gweleath, Cury	1890-1893, 1898-1901, 1910-1913
DAWE J H, Parc Mean, Helston	1959-1962
DINGLE J, Linkinhorne	1881-1884
DINGLE JOHN, Darley, Callington	1891-1894
DINGLE J K, Trezare, Fowey	1936-1939
DONEY Jos, Lerryn	1901-1904, 1908-1911
DOWMAN Capt W A H, Trevissome, Flushing	1925-1928
DREW ROBERT, Mawgan	1898-1901, 1909-1912
DUNSTAN D, Blankednick, Perranarworthal	1950-1953
DUNSTAN D H, Crosswinds, Mullion, Helston	1963-1966, 1967-1970, 1971-1974, 1975-1978, 1979-1982, 1983-1986, 1987-1990, 1991-1994
DUNSTAN H, Towednack, St Ives	1925-1928
DUNSTAN J R, Sinns Barton, Redruth	1947-1950, 1953-1956, 1960-1963
DUNSTAN R, Sinns Barton, Redruth	1973-1976, 1977-1980, 1981-1984
DUNSTAN R S, Trannack Fm, Sithney, Helston	1959-1962
DUNSTAN W T, Sinns Barton, Redruth	1991-1994
DYER, F J, Penventinnie Hse, Kenwyn, Truro	1974-1977, 1978-1981, 1982-1985, 1986-1989, 1990-1993
DYER J, Treworder, Kenwyn, Truro	1969-1972
DYER T, Tencreek, Liskeard	1888-1891, 1901-1904, 1912-1915

EDDY A C, Trendrine, St Neot, Liskeard 1939-1949, 1952-1955, 1956-1959, 1960-1963, 1964-1967, 1968-1971, 1972-1975, 1976-1979, 1980-1983, 1984-1987
EDDY J W, Trevalgan, St Ives 1930-1933, 1936-1939
EDDY OAKLEY, Treloweth, St Erth 1922-1925
EDDY P O, Charmon Hurst, Probus, Truro 1948-1951, 1954-1957, 1958-1961, 1962-1965, 1966-1969, 1970-1973
EDDY P Q, Bosigran, Dobwalls, Liskeard 1975-1978, 1979-1982, 1983-1986, 1987-1990, 1991-1994
EDDY W J, Trewidden, Penzance 1947-1950, 1953-1956, 1957-1960
EDE F, Carludden, St Austell 1932-1935, 1938-
EDWARD-COLLINS E C, Trewardale, Bodmin 1883-1886
EDWARDS B C, Marazion 1923-1926, 1931-1934, 1937-
ELERS Maj C G E, Mayfield Cott, Torpoint 1926 — Deceased
ELFORD R K, St Tudy 1896-1899
ELLIOTT J R, Boswague, Tregony 1936-1939
EUSTICE D, Bezurrell, Gwinear, Hayle 1964-1967, 1968-1971, 1972-1975, 1976-1979, 1980-1983, 1984-1987, 1988-1991
EUSTICE G, Bezurrell, Gwinear, Hayle 1924-1927, 1931-1934, 1937-
EUSTICE G, Tregotha, Gwinear, Hayle 1948-1951, 1955-1958, 1959-1962, 1978-1981, 1982-1985
EUSTICE G H, Bezurrell, Gwinear, Hayle 1950-1953
EUSTICE J, Roslyn, Wadebridge 1953-1956, 1958-1961, 1967-1970, 1971-1974, 1975-1978, 1980-1983, 1984-1987
EUSTICE J M, Hay Farm, Wadebridge 1987-1990, 1991-1994
EUSTICE J W, Treglinnick, St Ervan 1979-1982, 1983-1986, 1987-1990, 1991-1994
EUSTICE P, Trevaskis Fm, Gwinear, Hayle 1979-1982, 1983-1986, 1987-1990, 1991-1994
EVA HENRY, Croft West, Truro 1922-1925, 1932-1935
EVA S H, Roseworthy, Camborne 1924-1927, 1934-1937

FALMOUTH Rt Hon Viscount, Tregothnan, Truro 1920-1923
FAULL A, Tresawle, Probus 1889-1892, 1897-1900, 1906-1909
FFRENCH-BLAKE Lt-Col R L V, Pensilva 1951-1954
FILKINS J, Creed 1894-1897
FORTESCUE G, Boconnoc, Lostwithiel 1925-1928, 1935-1938
FRAYNE JAMES, Pipers Pool, Launceston 1902-1905
FRAYNE JOHN, St Stephens, Launceston 1906-1909
FREETHY J F, Sowenna, Cury Cross Lanes 1925-1928
FREETHY W G, Burnow, Helston 1922-1925, 1928-1931

GARLAND L V, Greenbank, Hayle 1932 — Deceased
GILES H, Polsue, Philleigh 1902-1905
GILL JOHN, Ladock 1886-1889
GILL RICHARD, Clifton Gdns, Truro 1893-1896, 1902-1905, 1909-1912, 1915-1921
GILL R E, 14 Market St, Falmouth 1920-1923
GILLARD Mr, St Ewe 1885-1888
GOLDSWORTHY THOMAS, Trellesick, Hayle 1892 — Deceased
GRAHAM-VIVIAN Capt H, Bosahan, Manaccan 1951-1954
GRIGG JOSEPH, Creed 1887-1890, 1895-1898
GRIGG R, Porthmissen, Padstow 1958-1961
GRIGGS M, Tremedda, Zennor, St Ives 1932-1935, 1938-1948, 1951-1954, 1957-1960, 1961-1964, 1965-1968, 1969-1972, 1973-1976
GROSE WESLEY, Penpont, St Kew 1881-1884, 1886-1889, 1893-1896
GRYLLS W M, Falmouth 1888-1891, 1901-1903, 1906-1909

GUNDRY ISAAC, Bosveal, Mawnan 1929-1932, 1935-1938, 1948-1951
GUNDRY Jos, Trebah, Constantine 1907-1910, 1913-1919, 1923-1926
GYNN J C, Boscastle 1949-1952, 1954-1957
GYNN R, Trethlay, Boscastle 1919-1922, 1925-1928, 1931-1934, 1937-

HAIN EDWARD, St Ives 1895-1898
HAMLEY P, Westcott, Callington 1950-1953, 1955-1958, 1959-1962, 1968-1971, 1972-1975, 1976-1979
HANCOCK R G C, Helligan Barton,
St Mabyn 1990-1993
HARRIS A, Boscastle 1927-1930
HARRIS HEDLEY, Nancekuke, Illogan 1934-1937
HARRIS WILLIAM, Lanreath 1895-1898
HARVEY F I, Shepherds Fm, Newlyn East 1977-1980, 1982-1985, 1986-1989, 1990-1993
HAWKE GEORGE, Trehane, St Erme 1894-1897, 1900-1903, 1907-1910, 1914-1920
HAWKE WILLIAM, Treworgans, St Erme 1903-1906
HAWKE WILLIAM, Besoughan,
St Columb 1923-1926, 1930-1933
HAWKEN B, De Lank, St Breward 1892-1895
HAWKEY J GILES, Treharrock,
Wadebridge 1924-1927
HAWKEY J G, St Minver 1894-1897, 1900-1903
HAWKEY T, Goonhoskyn, St Enoder 1910-1913
HAWKEY T E, St Issey 1889-1892, 1895-1898
HAWKEY W E, Trescobeas Manor,
Falmouth 1921-1924
HEARD J E, West Petherwin, Launceston 1915-1921, 1924-1927, 1930-1933
HEARD N, Home Farm, Werrington 1925-1928, 1936-1939
HEARLE JOHN, St Clements 1894-1897
HEARLE JOSEPH T, St Antony, Helston 1888-1891, 1895-1898, 1903-1906, 1909-1912
HEARLE SAMUEL, Golden, Probus 1883-1886, 1890-1893, 1896-1899, 1903-
HEARLE WILLIAM, Trethowell, Kea,
Truro 1910-1913, 1919-1922, 1925-1928, 1932-1935, 1938-
HEARLE W L, Churchtown, Kea, Truro 1897-1900, 1903-1906
HELLYAR JAMES, Trenson, Mawgan 1893-1896
HELLYAR JOHN, Trenoon, St Columb 1911-1914, 1921- Deceased
HELLYAR R R, Gustivean, St Columb 1914-1920
HELLYAR WILLIAM, Rialton, St Columb 1910 — Deceased
HENDY IRA, Mawgan 1884-1887
HENDY JAMES, Trenoweth 1883-1886
HENDY JAMES, Truthan, St Ewe 1891-1894, 1899-1902, 1905-1908
HENDY Jas, Mawgan 1884-
HENDY JOHN, St Columb 1887-1890
HENWOOD P, Callington 1948-1951
HEXT A C, Tredethy, St Mabyn, Bodmin 1928-1931
HEXT F I, Tredethy, St Mabyn, Bodmin 1888-1891
HEXT GEORGE, Cowbridge, Lostwithiel 1902-1905
HICKS H W, Greenbank, Grampound Road 1961-1964, 1965-1968, 1969-1972, 1973-1976
HICKS WALTER, St Austell 1885-1888
HIGGINS G R, Hr Trenant, Dobwalls 1991-1994
HODGE HARRY, Trevarrick Hall,
St Austell 1926-1929
HODGE L, Six Acres, Burlawn,
Wadebridge 1973-1976, 1977-1980, 1981-1984, 1985-1988, 1989-1992
HODGE P, Pengelly Fm, Burlawn,
Wadebridge 1990-1993
HOLMAN J, St Michael Caerhayes 1885-1888
HOLMAN Jas, Govily, Tregony 1928-1931, 1935-1938

HOLMAN M H, Truthan, Truro	1890-1893, 1896-1899, 1902-1905, 1908-1911, 1914-1920, 1923-1926
HOLMAN N, Chyverton Hse, Zelah, Truro	1962-1965, 1966-1969
HOOPER W S, Blackwater, Scorrier	1899-1902
HORE S T, Higher Bore St, Bodmin	1933-1936, 1939-
HORE T, Laninval, Bodmin	1910-1912, 1921-1924, 1927-1930
HOSKEN E H, Cartuther Barton, Liskeard	1911-1914, 1922-1925, 1929-1932
HOSKEN SAMUEL, Hayle	1893-1896
HOSKEN W J, Pulsack, Hayle	1895-1898, 1901-1904, 1907-1910, 1913-1919
HOSKIN J, Trethurffe, Ladock	1938-1948
HOSKING A B, 18 Donnington Road, Penzance	1937-1940, 1950-1953, 1954-1957, 1958-1961, 1962-1965, 1966-1969, 1970-1973, 1974-1977
HOSKING E, Rosevidney, Ludgvan	1911-1914
HOSKING J, Trewinnard, St Erth	1909-1912
HOSKING JOHN, Penrose, Sennen	1915-1921
HOSKING N, Inchmahome, Sennen, Penzance	1977-1980, 1981-1984, 1985-1988
HOSKING S, Home Farm, St Anthony, Torpoint	1933-1936
HOSKING STANLEY, Penrose, Porthcurnow	1933-1936
HOSKING WILLIAM LORY, Fentongollan, Merther	1924-1927
HOSKING W J S, Fentongollan, Tresillian	1978-1981, 1982-1985, 1986-1989, 1990-1993
HOTTEN WILLIAM, Cornelly	1883-1886
HOWELL F B, Ethy, St Winnow	1890-1893
HOYLE J, Glebe Cott, St Austell	1932-1935
JACKA E, Lanner, Helston	1908-1911
JACKA E SPENCER, Truthall Manor, Helston	1957-1960, 1961-1964, 1965-1968, 1969-1972, 1973-1976
JACKA S, Sithney, Helston	1947-1950, 1953-1956
JACKA W G, Ninnis, Germoe, Marazion	1938-
JAMES H, Trethosa, St Stephens, St Austell	1960-1963
JAMES J G, Barteliver, Grampound Road	1902-1905, 1908-1911, 1914-1920, 1923-1926, 1929-1932
JAMES R, Barteliver, Grampound Road	1951-1954, 1955-1958, 1959-1962, 1963-1966, 1967-1970, 1971-1974
JAMES W, Grey Gables, Creed, Grampound Rd	1970-1973, 1974-1977, 1978-1981, 1982-1985, 1986-1989
JAMES WILLIAM, Barteliver, Grampound Rd	1881-1884, 1889-1892, 1898-1901, 1905-1908
JAMES W G, Trevarthian, Newlyn	1893-1896
JARVIS R C, Chywoone, Paul, Penzance	1928-1931
JEFFERY W, Lostwithiel	1948-1951
JEFFREE FRANK, Camborne	1939-
JENKIN H, Llanfair, St Austell	1927-1930, 1933 — Deceased
JENKIN S C, Biscovey, Par	1931-1934, 1937-1940, 1950-1953
JENNING M, Adelaide Rd, Redruth	1930-1933
JENNINGS A, Truro	1882-1885
JENNINGS J, Restormel, Lostwithiel	1906-1909
JOBSON Maj T H, Trewidden Frm, Penzance	1957-1960
JOHNS B, Treringey, Crantock	1921 — Deceased

JOHNS W A, Chy-an-Gwel, Kenwyn, Truro	1947-1950, 1953-1956, 1961-1964, 1965-1968, 1969-1972, 1973-1976, 1977-1980
JOHNS W REED, Treeza, Porthleven, Helston	1939-
JOHNS W T, Treringey, Crantock	1909-
JOHNSTONE Miss E A, Trewithen, Grampound	1953-1956, 1958-1961
JOHNSTONE Capt G H, Tregoose, Grampound	1922-1925
JOSE JOHN, Mellingey, Perranarworthal	1891-1894
JULYAN RICHARD, Creed	1886-1889, 1893-1896
KELLOW D C, Woodlands, Sweetshouse, Bodmin	1989-1992
KELLY C, Delamead, Brentor, Tavistock	1950-1953, 1954-1957, 1958-1961, 1962-1965, 1966-1969
KENDALL F, Gelasina, Portmelon, Mevagissey	1904-1907, 1911-1914, 1922-1925
KEY HART, St Breock	1887-1890, 1894-1897
KEY JOHN, Pawton, St Breock	1893-1896
KEY J, Perlees, Wadebridge	1906-1909
KITTOW D E, Patrieda, Callington	1927-1930, 1933-1936, 1939-
KITTOW H, Stourscombe, Launceston	1925-1928
KITTOW J BASIL, Patrieda, Callington	1911-1914
KIVELL C, Stratton	1927-1930, 1933-1936
KIVELL W, Launceston	1928-1931, 1934-1937, 1948-1951, 1954-1957, 1959-1961, 1963-1966, 1969-1972, 1973-1976
KNIGHT J H, Prideaux, Par Station	1891-1894
KNIGHT WILLIAM, Gorran	1897-1900
KNOWLES F P, Santola, St Mabyn	1948-1951
KNOWLES H, Treviades, Constantine	1937-1940, 1951-1954
KNOWLES J, Penhale, St Tudy	1931-1934, 1937-1940, 1950-1953
KNOWLES W A, Carwythick, Constantine	1911-1914, 1921-1924, 1927-1930
KNUCKEY B W, Green Lawns, Perranarworthal	1926-1929, 1932-1935, 1938-1948, 1953-1956, 1959-1962, 1963-1966, 1967-1970, 1971-1974
LAITY G T, Treworrick, St Ewe	1939-1949, 1952-1955
LAITY H, Gare, Lamorran, Probus	1925-1928
LAITY H, Rostheren, Marazion	1937-1940, 1950-1953, 1956-1959, 1960-1963, 1965-1968, 1969-1972, 1973-1976
LAITY H, 4 Frobisher Terrace, Falmouth	1950-1953
LAITY HARVEY, Trefusis, Falmouth	1936-1939
LAITY HERBERT, Bosistow, St Levan	1913-1919, 1923-1926, 1929-1932, 1935-1938
LAITY J, Polkinhorn, Gwinear	1933-1936
LAITY JACK, Bosfranken, St Buryan	1933-1936, 1947-1950, 1953-1956, 1957-1960
LAITY JOHN, Goldsithney	1891-1894
LAITY T, St Austell	1884-
LAMBRICK JAMES, St Keverne	1897-1900
LAMBRICK JOSEPH, St Keverne	1893-1896
LANGDON G T, Hill Hse, Uplands Cres, Truro	1948-1951, 1956-1959, 1961-1964, 1965-1968, 1969-1972
LANYON A C, Coswarth, St Columb	1926-1929, 1934-1937
LANYON EDWIN, Gorran	1891-1894
LANYON W, Cargoll, Newlyn East	1920-1923, 1927-1930
LAVERTON A, Truro	1890-1893
LAWRY F M, Varfell, Long Rock, Penzance	1957-1960, 1963-1966, 1967-1970
LAWRY JOHN, Gorran	1896-1899, 1904-1907

LAWRY R, Tregarne, Mawnan, Falmouth	1955-1958, 1962-1965, 1971-1974, 1975-1978, 1977-1980, 1981-1984
LAWRY S, Nancekuke, Illogan	1911-1914, 1919-1922
LAWRY T, Tregarne, Mawnan	1910-1912, 1915-1921
LAWRY, T P, Calamansack, Falmouth	1922-1925
LAWRY WILLIAM, Treveor, Gorran	1884-1887, 1894-1900, 1907-1910
LELLO H, Trevassick, Hayle	1971-1974, 1975-1978, 1979-1982, 1983-1986, 1987-1990, 1991-1994
LESBIREL J, Havett, Liskeard	1895-1898, 1901-1904, 1908-1911, 1920-1923
LETHBRIDGE J C, Tregear, Launceston	1882-1885
LEWARNE W, St Ingenger, Bodmin	1892- Deceased
LIDDELL HARRY, Bodmin	1903-1906
LITTLETON W J, Glenmore, Trelights	1926-1929, 1932-1935, 1938-1948, 1951-1954, 1955-1958, 1959-1962, 1964-1967
LOBB GEORGE, Lawhitton	1881-1884, 1885-1888, 1891-1894, 1897-1900, 1903-1906, 1910-1913
LOBB R J, Norton, Bodmin	1959-1962
LORD WILLIAM, Alma, Truro	1885-1888, 1892-1895, 1898-1901, 1904-1907, 1910- Deceased
LOVERING WILLIAM, The Grove, St Austell	1891-1894, 1899-1902, 1905-1908
LUGG W J, Parnoweth, Newlyn East	1921-1924
LUKES THOMAS, St Austell	1884-1887, 1894-1897
LUKEY J, Truro	1902-1905, 1908-1911, 1914-1920
LUTEY A J, Trenithon Fm, Summercourt	1990-1993
LUTEY H T, Carfury, New Mill, Penzance	1964-1967, 1968-1971, 1972-1975, 1976-1979, 1980-1983, 1984-1987
LUTEY J H, Winsfield, Summercourt	1967-1970, 1971-1974, 1975-1978, 1979-1982, 1983-1986, 1987-1990, 1991-1994
MABBOTT HOWELL, Penzance	1911-1914, 1919-1922, 1927-1930
MACLEAN W, Lostwithiel	1893-1896
MAGOR JOHN, Pennans, Creed	1882-1885, 1896-1899, 1902-1905, 1912-1915
MAGOR JOHN, Tregew, Feock	1888-1891
MAGOR RICHARD, Hillside, Truro	1911-1914
MANN A, Penwarne, Mawnan Smith	1954-1957
MANN H, Penwarne, Falmouth	1904-1907, 1910-1913, 1920-1923, 1926-1929, 1932- Deceased
MARTIN F, Treluswell Mount, Penryn	1947-1950
MARTIN G L, Tregartha, Liskeard	1960-1963, 1966-1969, 1970-1973, 1974-1977, 1978-1981, 1982-1985, 1986-1989, 1990-1993
MARTIN H, Carnsiddea, Stithians	1927-1930
MARTIN JOHN, Wadebridge	1898-1900, 1903- Deceased
MARTIN J KING, St Enoder	1899- Deceased
MARTIN N, Trewince, Grampound Road	1894-1897, 1900-1903
MARTIN R J, Crantock, Newquay	1893-1896
MARTIN R J, Trewince, Grampound Road	1927-1930
MARTIN R L, Kenwyn, Truro	1934-1937
MARTIN W H P, Truro	1888-1891, 1894- Deceased
MARTYN HENRY, Newquay	1908-1911
MARTYN J K, St Enoder	1881-1884
MASON GEORGE, Tretheake, Tregony	1927-1930, 1934- Deceased
MATTHEWS R, Marazion	1887-1890, 1893-1896
MATTHEWS T R, Trezelah, Gulval, Penzance	1966-1969, 1970-1973, 1974-1977, 1978-1981, 1982-1985, 1986-1989, 1990-1993
MATTHEWS W G, Trezelah, Gulval, Penzance	1932-1935, 1952-1955, 1956-1959, 1960-1963
MAY EDWIN, Truro	1904-1907, 1911-1914
MAY JOHN, Suffenten, St Teath	1890-1893, 1896-1898

MAY JOHN, Mawgan	1898-1901
MAY J T, St Columb	1949-1952, 1954-1957, 1958-1961, 1962-1965, 1967-1970
MAY T P, Marislea, Newquay	1937-1940, 1951-1954, 1955-1958, 1959-1962
MAY W H, Tregellas, Grampound Road	1952-1955
MEAD JOHN, Falmouth	1904-1907
MENHENICK J, Burniere, Wadebridge	1889-1892
MENHENICK J C, Wadebridge	1949-1952, 1955-1958
MENHENICK PERCY, Wadebridge	1905-1908, 1920-1923
MENHENICK WILLIAM, Hendra, St Kew	1899-1902
MENHENICK W E, Hendra, St Kew	1923-1926
MENHENITT J H, Tredustan, Wadebridge	1957-1960, 1961-1964, 1965-1968, 1969-1972, 1973-1976, 1977-1980, 1981-1984, 1985-1988
MENHENITT P J, Wadebridge	1930-1933, 1936- Deceased
MENHENNICK CHARLES, Lower Amble, St Kew	1888-1891, 1896-1899
MENHINICK P J C, Burniere, Wadebridge	1967-1970, 1971-1974, 1975-1978, 1979-1982, 1983-1986, 1987-1990
MENHINICK W E, Trenoon, Mawgan	1931-1934
MICHELL D, Coinagehall Street, Helston	1965-1968
MICHELL S, Lelant	1908-1911
MICHELL W T, Ladock	1891-1894, 1897-1900
MILLETT R N E, Marazion	1890-1893, 1897-1902
MILLS W G, Tolfrey, Par	1901- Deceased
MITCHELL A R, Marazion	1894-1897, 1898-1901
MITCHELL E, Trevella, Crantock, Newquay	1971-1974
MITCHELL EDWARD, Lelant	1906-1909
MITCHELL F W, Redruth	1884-1887
MITCHELL SAMSON, Beersheba, Lelant	1896-1898, 1902-1905
MITCHELL W, Tywardreath	1903-1906
MITCHELL WILLIAM, Trelowthas, Probus	1910-1913, 1919-1922, 1935-1938
MITCHELL WILLIAM E, Penzance	1915-1921, 1928-1931
MOLESWORTH ST AUBYN J, Pencarrow, Bodmin	1988-1991
MOORE J GWENNAP, Probus	1897-1900
MORCOM J G, Bargolva, Liskeard	1950-1953
MORKHAM H, Sawarne, St Martin in Meneage	1933-1936, 1939-
MORSHEAD V H, Tregadduk, Blisland	1930-1933, 1936-1939
MORTIMER G M, St Germans	1906-1909
MUCKLOW EDWARD, Bennetts, Holsworthy	1891-1894
MUTTON B N, Burlerrow, St Mabyn, Bodmin	1963-1966, 1967-1970, 1971-1974, 1975-1978, 1979-1982
MUTTON C, Burlerrow, St Mabyn, Bodmin	1981-1984, 1985-1988, 1989-1992
MUTTON N, Gwavas, St Tudy	1947-1950, 1953-1956, 1957-1960
NANCEKIVELL D W, Heatham, Kilkhampton, Bude	1991-1994
NEWTON H V, Pulstring, Camborne	1886-1889
NICHOLAS R, Menagissey, Mt Hawke, Scorrier	1915-1921
NICHOLLS A B, Trelan, St Keverne, Helston	1980-1983, 1984-1987, 1988-1991
NICHOLLS A J J, Rialton Barton, St Columb	1911-1913

NICHOLLS A M, Trelan, St Keverne, Helston	1951-1954, 1956-1959, 1961-1964, 1965-1968, 1969-1972, 1973-1976
NICHOLLS D A, Penvose, St Gluvias, Penryn	1952-1955
NICHOLLS J, St Columb	1882-1885
NICHOLLS L R, College Rd, Camelford	1950-1953
NICHOLLS W, Race, Camborne	1910-1913
NICHOLLS W J, Hallaze, St Austell	1930-1933
NICHOLLS W J C, Halvose, Helston	1947-1950
NICKELL GEORGE, Slades House, Wadebridge	1900-1903, 1906-1909
NICKELL HERBERT, Slades House, Wadebridge	1902-1905, 1912-1915, 1919-1922
NIGHTINGALE G, Pendoggett, Bodmin	1962-1965, 1966-1969, 1970-1973, 1975-1978, 1980-1983, 1984-1987, 1988-1991
NORTHEY G A, Gwarnick, Truro	1906-1909
NORTHEY L M, Gwarnick, Truro	1930-1933
OLDS J P, Menadarva, Camborne	1951-1954, 1963-1966, 1967-1970, 1971-1974, 1975-1978, 1977-1980, 1981-1984, 1985-1988
OLIVER G, Trehane, Probus	1889-1892, 1895-1898
OLIVEY REVD E, St Day, Scorrier	1883-1886
OLLIVER J, Cott, St Neot	1948-1951, 1954-1957, 1958-1961, 1962-1965, 1966-1969, 1970-1973, 1974-1977, 1978-1981, 1982-1985
OLVER J, Westnorth, Duloe	1904-1907
OLVER J C, Woodland Valley, Ladock	1921-1924
OLVER RICHARD, Trescowe, St Mabyn	1883-1886, 1889-1892
OLVER THOMAS, Truro	1884-1887
PAIGE RICHARD, St Germans	1891-1894
PAIGE W E, Treboul, St Germans	1901-1904, 1912-1915, 1919-1922
PAINTER RICHARD, Carclew, Mylor	1892-1895, 1898-1901
PARNALL J, Trecarne, Camelford	1910-1912
PARNELL JOHN, Trecarn, Advent	1891-1894
PARSONS BRENDON, Launceston	1896-1899
PARTRIDGE W R, Treyarnon Farm, St Merryn	1964-1967
PASCOE H, Trenear Road, Penzance	1934-1937
PASCOE J H, Trevilvas, Grampound Road	1912-1915, 1920-1923, 1926-1929
PASCOE N, Dugdale Bros, Penzance	1929-1932
PAULL A, Trelonk, Ruan	1904-1907
PAULL JOHN, Trelonk, Ruanlanihorne	1886-1889, 1893-1896
PAYNTER Col J H, Boskenna, St Buryan	1919-1922
PEARCE B, Trelissick, Hayle	1922-1925
PEARCE B W, Tremenhere, Ludgvan	1912-1915, 1920-1923
PEARCE B W, Carbis Bay, St Ives	1934-1937
PEARCE J B, Fair View, Callestick, Truro	1952-1955
PEARCE O, Tregatillian House, St Columb	1990-1993
PEARCE T H, Market Street, Falmouth	1955-1958
PEARSE H E L, Woodland, Callington	1908-1911, 1914-1920
PEASE WILLIAM, Castle, Lostwithiel	1894-1897
PENDER W R T, Budock Vean	1884-1887, 1890-1893
PENGELLY R, Trewithick, Trelights	1968-1971, 1972-1975, 1976-1979, 1980-1983, 1984-1987
PENNA W F, Treloweth, Pool, Redruth	1929-1932, 1938-1948, 1951-1954, 1956-1959, 1960-1963, 1964-1967, 1968-1971, 1972-1975
PENPRASE E, Trevingey House, Redruth	1926-1929

PENROSE WILLIAM, Helston	1896-1897
PENROSE W B, Trequean, Breage, Helston	1951-1954, 1955-1958, 1961-1964, 1965-1968, 1969-1972, 1973-1976, 1977-1980, 1981-1984, 1985-1988, 1989-1992
PERRY S, Lawhitton Barton, Launceston	1939-
PETER-HOBLYN W G, St Mabyn	1899-1901
PHILLIPS C, Clapper House, Wadebridge	1921-1924, 1932-1935, 1938-1948, 1951-1954
PHILLIPS J E, Bodieve, Wadebridge	1922-1925, 1929-1932, 1935-1938
PHILLIPS J L, Pentewan	1909-1912, 1915-1921, 1926-1929, 1932-1935
PHILLIPS P J, Pencorse, Summercourt	1960-1963
PHILLIPS WILLIAM, Court Place, Egloshayle	1896-1899, 1903-1906, 1910-1913, 1919-1922, 1925-1928
POLLARD CHARLES, St Kew	1881-1884
POLLARD R G, Bodieve, Wadebridge	1890-1893
POOLE Major, Cotswold House, Fowey	1914-1920
POOLEY Jas, Rosewarne, Hayle	1920-1923, 1928-1931
POOLEY J A, Blackwater	1927-1930, 1933-1936
POLWHELE A C, Polwhele, Truro	1920-1923, 1928-1931
PURCELL NOEL E, The Mount, Par	1921-1924
RABY GEORGE, Menheniot	1883-1886
RABY GRIGG, St Germans	1895-1898
RAIL WILLIAM, Calamansack, Constantine	1896-1899, 1912-1915
RATCLIFFE J V, Argal House, Falmouth	1954-1957
RAWLING JOHN, Oldwit, South Petherwin	1900-1903
READ R S, St Ives	1898-1901, 1909-1912
REED C, Chynoweth, Crantock, Newquay	1953-1956, 1957-1960, 1961-1964, 1966-1969
REED T C, Carines, Cubert	1924-1927
RESKELLY RICHARD, Chydaw, Kenwyn	1899-1902, 1908-1911
RICH THOMAS, Polsue, St Erme	1899-1902
RICHARDS C J, Penleigh, Rose-an-Grouse, Hayle	1962-1965, 1966-1969, 1970-1973, 1974-1977, 1978-1981, 1982-1985, 1986-1989, 1990-1993
RICHARDS C P, Splattenridden, Hayle	1981-1984, 1985-1988, 1989-1992
RICHARDS F, Lidcutt, Bodmin	1934-1937
RICHARDS F, The Mayoralty, Bodmin	1947-1950, 1953-1956
RICHARDS H M, Killivose, Truro	1989-1992
RICHARDS J, Lelant Downs, Hayle	1989-1992
RICHARDS J G, Gwills, Cury X Lns, Helston	1931-1934, 1937-1940
RICHARDS R J B, Goonhoskyn, Summercourt	1976-1979, 1982-1985, 1986-1989
RICHARDS WILLIAM, Godolphin, Hayle	1896-1899
RICHARDS W B, Killivose, St Allen	1925-1928, 1931-1934
RICHARDS W P, Killivose, Truro	1979-1982, 1983-1986, 1987-1990, 1991-1994
RICHARDS W R, Killivose, Truro	1947-1950, 1956-1959, 1960-1963, 1965-1968, 1969-1972, 1973-1976
RICKARD TOM, Wadebridge	1895-1898, 1901-1904, 1914-1920
RICKEARD S, Trenance, Newlyn	1890-1893, 1902-1905
ROACH PAUL, Beersheba Farm, Lelant	1926-1929, 1932-1935, 1938-
ROACH W, Trewidden, Penzance	1912-1915, 1922-1925
ROBERTS D C, Fernside, Praze, Camborne	1968-1971, 1972-1975, 1976-1979, 1980-1983, 1984-1987, 1988-1991
ROBERTS M, Farley Terrace, Truro	1952-1955, 1956-1959, 1960-1963, 1964-1967
ROBERTS P, St Levan	1881-1884
ROBERTS S G, Tregidgeo, Grampound Road	1929-1932, 1935-1938

ROBERTS W H, Padstow	1882-1885
ROBERTS W M, Wilton, St Germans	1909-1912, 1915-1921
ROBINS JOHN, Roche	1887-1890
ROBINS JOHN, Trescowe, St Mabyn	1905-1908, 1913-1919
ROGERS H MONTAGUE, Nansloe, Helston	1903-1906, 1910-1913, 1919-1922
ROGERS POWYS, Burncoose, Gwennap	1898-1901
ROSEWARNE H, Clifton Gardens, Truro	1915-1921, 1924-1927
ROSEWARNE HERBERT A, Coswinsawson, Gwinear	1895-1898, 1901-1904, 1907-1910, 1919-1922
ROSEWARNE JOHN, Phillack	1885-1888
ROSEWARNE J N, Ponsmere Road, Perranporth	1938-1948, 1954-1957, 1962-1965, 1966-1969
ROSKRUGE E P, St Keverne	1888-1891, 1899-1902
ROSKRUGE E P, Nantellan, Grampound	1925-1928, 1935-1938
ROSKRUGE T F, Tehidy Barton, Camborne	1892-1895, 1898-1901, 1904-1907, 1910-1913, 1919-1922, 1927-1930
ROW H W, Boconnoc, Lostwithiel	1929-1932, 1935-1938
ROWE F, Eddystone, Rumford, St Columb	1954-1957
ROWE FRANK, Colan, St Columb	1897-1900, 1904-1907, 1909-1912
ROWE F W, Trevego, Lostwithiel	1899-1902, 1905-1908, 1912-1915
ROWE HARRY, Treeza House, Porthleven	1914-1920, 1924-1927
ROWE HUGO, Lr Kergilliack, Falmouth	1921-1924, 1928-1931, 1934-1937
ROWE JOSEPH, Nancolleth, Newlyn East	1890-1893
ROWE Jos H, Nancolleth, Newlyn East	1911-1914
ROWE WILLIAM, Trega, Sithney	1894-1897
ROWE W, Plemming, Long Rock	1948-1951, 1954-1957, 1958-1961, 1963-1966
ROWE W E, Bodbrane, St Keyne, Liskeard	1932-1935
ROWS C, Tresprisson, Hayle	1901-1904, 1907-1910, 1913-1919, 1922-1925, 1928-1931
ROWSE H, Berrangoose, Probus	1912-1915, 1919-1922
ROWSE JOSEPH, Lancarffe, Bodmin	1895-1897
ROWSE WILLIAM, Bodmin	1897-1898
RUNDLE THOMAS, Colan	1907-1910
RUNDLE W, Dowstall, Mylor	1901-1904
SALMON W, Trewinnick, Rumford, Wadebridge	1973-1976, 1977-1980, 1981-1984, 1985-1988, 1989-1992
SAMBELL B, Restormel, Lostwithiel	1884-1887
SANDFORD L C P, St Minver	1897-1899
SAUNDRY J, St Levan	1887-1890
SCOBLE J M, Langarth, Kenwyn	1931-1934, 1937-1940, 1950-1953, 1955-1958, 1959-1962, 1963-1966, 1967-1970, 1971-1974
SEARLE J G, Calthrop, Truro	1923-1926, 1929-1932, 1935-1938
SEARLE WILLIAM, Trenithan, Probus	1884-1887, 1896-1899
SHORT W J, Bosent, Liskeard	1934-1937
SHUKER F W, Scorrier	1885-1888, 1897-1900, 1904-1907
SIMS W A, Sinns Barton, Redruth	1907-1910, 1913-1919, 1922-1925
SIMS W H, Sinns Barton, Redruth	1901-1904
SIMMONS M J, Greenlane Fm, Mawla, Redruth	1985-1988, 1989-1992
SIMPSON D H S, 12 St Marys Terr, Penzance	1964-1967, 1968-1971, 1972-1975
SLEEP W H, St Germans	1883-1886
SMEETH A J, Whalesboro Fm, Marhamchurch	1937-1940

SMEETH E J, Westland, Marhamchurch,
Bude — 1951-1954, 1955-1958, 1959-1962, 1963-1966, 1967-1970
SMITH F, Dunheved Road, Launceston — 1952-1955, 1956-1959, 1962-1965
SMITH G E STANLEY, Treliske, Truro — 1912-1915
SMITH H F, Ventonwyn, Grampound — 1903-1906
SMITH J, Parknoweth, Newlyn East — 1938-1948, 1951-1954
SMITH W J, Trengwethin, St Tudy — 1949-1952, 1956-1959, 1960-1963, 1964-1967
SNELL B, Wayton, Hatt — 1882-1885
SOBEY H, Pensipple, Liskeard — 1905-1908
SOBEY J, Trevartha, Liskeard — 1957-1960, 1961-1964, 1965-1968, 1969-1972, 1973-1976
SOBEY W F, Menheniot — 1905-1908, 1912-1915
ST AUBYN Hon PIERS, Chy-an-Eglos, Marazion — 1910-1913
ST AUBYN J MOLESWORTH, Tetcott, Holsworthy — 1939-
STANBURY J, Lady Cross Fm, Yeolmbridge — 1988-1991
STANBURY M, Holsworthy — 1930-1933, 1936-1939
STANIER J B, Duchy Offices, Liskeard — 1957-1960, 1964-1967, 1968-1971, 1972-1975, 1976-1979
STANIER T M, Trethellan, West Newquay — 1953-1956
STEPHENS JOHN, Treverbyn, Probus — 1886-1889
STEPHENS N, St Tudy — 1902-1905
STEPHENS R F, St Austell — 1881-1884
STEPHENS T F, Tremearne, Breage, Helston — 1937-1940, 1950-1953, 1955-1958
STEPHENS WESLEY, Hendra, St Kew — 1886-1889
STEPHENS W T, Whitley, Liskeard — 1936-1939
STEVENS J, Glebe, Sancreed — 1930-1933, 1936-1939
STOCKER F, Tregenna, St Ewe — 1887-1890
STRONG C, Elm Bank, Fowey — 1958-1961, 1963-1966

TAYLOR C, Home Farm, Tehidy, Camborne — 1953-1956, 1957-1960, 1962-1965, 1966-1969, 1970-1973
TEAGLE J C, Trevithick, St Columb — 1952-1955, 1956-1959
THOMAS EDWIN, St Coose, Truro — 1930-1933
THOMAS HENRY, Tolgarrick, Truro — 1888-1891, 1894-1897, 1900-1903
THOMAS JAMES, St Winnow — 1890-1893
THOMAS JOSEPH, Marazion — 1882-1885, 1893- Deceased
THOMAS JOSIAH, Tywardreath — 1885-1888
THOMAS R L, Katheledron, Camborne — 1914-1920
THOMAS T, Great Prideaux, Par — 1947-1950
THOMAS W, Tregenna, St Austell — 1915-1921
THOMPSON G W, Tregony — 1906-1909
TINNEY W, Nancewrath, Truro — 1933-1936, 1950-1953
TIPPETT A, Treludderow, Newlyn East — 1933-1936
TOM W H, St Columb — 1949-1952
TRANT BROOKING, Trethawle, Liskeard — 1897-1900, 1903-1906
TRANT RICHARD B, Tremabe, Liskeard — 1895-1898, 1901-1904, 1907-1910, 1913-1919, 1922-1925, 1928-1931
TREBILCOCK R R, St Columb — 1909-1912, 1915-1921
TREFFRY J DE CRESSY, Penarwyn, Par Stn — 1886-1889, 1892-1895, 1899-1902, 1909-1912
TREGARTHEN W B C, Collurian, Penzance — 1935-1938
TREGARTHEN W B C, Lelant — 1949-1952

TRELOAR P Q, Godolphin Hse, Breage	1920- Deceased
TRELOAR W J, Mankea, St Gluvias	1955-1958
TREMAINE H, Newlyn	1881-1884
TREMAINE HARRY, Trerice, Newlyn	1892-1895, 1902-1905, 1908-1911
TREMAINE WILLIAM, Polsue, Philleigh	1883-1886
TRENEER WILLIAM, Newlyn	1884-1887
TRENOUTH G, Trevone Farm, Padstow	1970-1973, 1974-1977, 1978-1981, 1982-1985, 1986-1989, 1990-1993
TRETHEWAY W J, Trencreek, Liskeard	1925-1928
TREVENEN WILLIAM, Helston	1889-1892
TREWHELLA W J, Helston	1949-1952
TREWIN J C, Trepool, Ruan Minor, Helston	1958-1961, 1962-1965, 1966-1969, 1970-1973, 1974-1977, 1978-1981
TREWIN K J, Higher Manaton, Callington	1974-1977, 1978-1981, 1982-1985, 1986-1989
TROUNCE W H, Trevilveth, Veryan	1886-1889, 1893-1896, 1900- Deceased
TRUDGIAN THOMAS, Tretheake, Veryan	1899-1902, 1905-1908, 1912-1915
TRUSCOTT H H, Brennal, St Stephens	1899-1902, 1907-1910, 1915-1921
TRUSCOTT H S G, St Stephens, St Austell	1950-1953
TRUSCOTT R, Wadebridge	1904-1907
TRUSCOTT S, Brennal, St Stephens	1888-1891
TUCKER CYRIL G, Molenick, St Germans	1914-1920
TUCKER E, Molenick, St Germans	1882-1885, 1892-1895, 1902-1905
TUCKER J J, Patheda, Menheniot	1904-1907, 1910-1913
TUCKER JOHN C, Patheda, Menheniot	1886-1889, 1892-1895, 1900- Deceased
TUCKER W H, Argyll House, Camelford	1952-1955
TUCKER W H A, Trezare, Fowey	1915-1921
TUCKETT A, Lanteglos, Fowey	1912-1915
TUCKETT W H, Penquite, St Sampson	1909-1912, 1921-1924
TYACKE J, Merthen, Constantine	1882-1885
TYZZER C N, Tregidgeo Farm, Grampound	1938-1948, 1951-1954
VENNING J, Sedgmoor House, St Austell	1937-
VIVIAN J, St Kew	1882-1885
VOSPER E, Old Treworgey, Liskeard	1928-1931, 1936-1939
VOWLER GUILLEM, Penheale, Egloskerry	1890-1893, 1896-1899
WARD J, Camelford	1949-1952, 1955-1958, 1964-1967
WARNE F J, The Old Rectory, Tregony	1976-1979, 1980-1983, 1984-1987, 1988-1991
WARNE JOHN, Tregonhayne, Tregony	1897-1900, 1906-1909, 1913-1919, 1922-1925, 1928- Deceased
WARNE J C, Tregonhayne, Tregony	1925-1928, 1931-1934, 1939-1949, 1952-1955, 1956-1959, 1960-1963, 1964-1967, 1968-1971, 1972-1975
WARNE T, Trevisquite, St Mabyn	1910-1913, 1919-1922, 1927-1930, 1934-1937
WARNE TOM, Trevisquite, St Mabyn	1931-1934
WARREN J P, Merthen, Gweek	1926-1929, 1934-1937, 1948-1951
WARRICK J, Treworder, Burlawn, Wadebridge	1985-1988, 1989-1992
WARRICK S, 54 Lemon Street Truro	1911-1914, 1923-1926
WEBBER T, Falmouth	1881-1884, 1888-1891
WEBSTER A M, The Coombe, Liskeard	1915-1921
WEVILL W, Lewannick, Launceston	1882-1885, 1892-1895
WEST JAMES, St Breock	1894-1897
WEST P C, Penwine, St Mabyn, Bodmin	1968-1971, 1972-1975, 1976-1979, 1980-1983, 1984-1987, 1988-1991
WHITE Major, Trewellard, St Just	1901-1904

WHITFORD C E, St Columb	1909-1912
WHITFORD WILLIAM, Truro	1883-1886, 1892-1895, 1901-1904
WICKET W, Redruth	1881-1884
WICKETT S, Tregweath, Redruth	1929-1932, 1935-1938
WICKETT T, 11 Trewirgie Rd, Redruth	1911-1914, 1921-1924
WICKETT WILLIAM, Redruth	1891-1894, 1900-1903
WILLEY J J H, Predannick Manor, Mullion	1967-1970, 1971-1974, 1975-1978, 1979-1982
WILLIAMS A, Elm House, Ponsanooth	1932-1935, 1938-
WILLIAMS A M, Werrington Park, Launceston	1921-1924
WILLIAMS D P, Tregonwell, Manaccan, Helston	1977-1980, 1981-1984, 1985-1988, 1989-1992
WILLIAMS F B, Fowey	1883-1886
WILLIAMS F JULIAN, Caerhayes Castle, Gorran	1961-1964
WILLIAMS GILES, Barn, Ruanlanihorne	1882-1885, 1889-1892
WILLIAMS Maj GODFREY, Perranarworthal	1933-1936
WILLIAMS G P, Scorrier	1922-1925, 1931-1934, 1937-
WILLIAMS HENRY, Lanreath	1881-1884
WILLIAMS J, Kergilliack, Penryn	1929-1932
WILLIAMS JOHN, Padstow	1885-1888
WILLIAMS JOHN, Ruanlanihorne	1886-1889
WILLIAMS JOHN, Scorrier	1889-1892
WILLIAMS J H, Devoran	1949-1952
WILLIAMS M P, Lanarth, St Keverne	1926-1929
WILLIAMS P H, Mullion	1908-1911, 1914-1920
WILLIAMS Brig STEPHEN, Scorrier	1947-1950
WILLIAMS T, Grouan House, St Day	1958-1961, 1963-1966
WILLIAMS T H, Royal Hotel, Falmouth	1906-1909, 1913-1919, 1923-1926
WILLIAMS T H, St Austell	1915-1921, 1924-1927
WILLIAMS WILLIAM, Porthmissen, Padstow	1898-1900
WILLIAMS Maj W B, Porthmissen, Padstow	1924-1927
WILLOUGHBY Mr, Illogan	1889-1892
WILLYAMS Maj E N, Carnanton, St Columb	1928-1931
WOODLEY F O, 51 Tregolls Road, Truro	1948-1951, 1962-1965, 1972-1975, 1976-1979
WOODLEY J M, Degembris, Newlyn East	1914-1920, 1923-1926, 1929-1932, 1935-1938
WOOLCOCK W H, Par	1886-1889

ENTRIES IN LIVESTOCK AND OTHER AGRICULTURAL CLASSES SINCE 1857

		Horses	Cattle	Sheep	Pigs	Goats	Horse Shoeing	Dairy Produce	Dairy Competitions	Milking	Judging by Points
1857	Truro	59	67	85	21						
1858	Camborne	65	91	128	24						
1859	St Austell	70	100	182	23						
1860	Penzance	90	150	200	27						
1861	Truro	Bath & West at Truro									
1862	Liskeard	42	98	120	20						
1863	Truro	70	137	110	25						
1864	Saltash	60	100	140	34						
1865	Falmouth	57	88	95	19						
1866		No Show due to Cattle Plague									
1867	Launceston	No Cattle due to Cattle Plague, other figures not known									
1868	Falmouth	Bath & West at Falmouth									
1869	Penzance	36	90	118	26						
1870	Launceston	31	105	145	26						
1871	Truro	89	135	88	56						
1872	Bodmin	152	144	100	37						
1873	Penryn	82	141	106	42						
1874	St Austell	134	129	88	38						
1875	Truro	200	147	103	45						
1876	Liskeard	124	148	88	44						
1877	Camborne	154	140	97	48						
1878	Saltash	144	148	97	44						
1879	Falmouth	149	126	61	24						
1880	Lostwithiel	193	169	31	57						
1881	Redruth	160	140	120	23						
1882	Launceston	188	164	53	35						
1883	Truro	176	188	81	34						
1884	Bodmin	186	174	79	24						

Entries in Livestock and Other Agricultural Classes since 1857 (cond.)

		Horses	Cattle	Sheep	Pigs	Goats	Horse Shoeing	Dairy Produce	Dairy Competitions	Milking	Judging by Points
1885	Penzance	134	132	53	23						
1886	St Austell	184	194	59	29						
1887	Camborne	182	145	57	16						
1888	Newquay	195	154	52	26						
1889	Helston	237	131	45	25						
1890	Truro	194	131	73	17						
1891	Par	146	144	46	15						
1892	Redruth	208	141	61	18						
1893	Liskeard	208	179	68	10			111			
1894	Falmouth	228	189	48	—		23				
1895	Wadebridge	205	213	56	21						
1896	St Ives	195	206	70	30		27				
1897	Lostwithiel	162	176	55	24						
1898	Penzance	113	219	47	25						
1899	Launceston	148	158	64	27		35	64			
1900	Truro	185	184	55	27		27	72			
1901	Bodmin	177	108	62	29		43	79			
1902	Camborne	147	140	62	32		46	108			
1903	St Austell	183	137	36	40		54	98			
1904	Falmouth	139	156	72	39		38	111			
1905	Newquay	143	167	42	46		39	152			
1906	Redruth	146	201	48	43		36	157			
1907	Liskeard	152	193	66	55		25	165			
1908	Helston	145	220	53	45		32	184			
1909	St Columb	149	181	49	46		35	184			
1910	St Ives	118	189	49	42		45	178			
1911	St Austell	176	173	31	53		32	174			
1912	Penzance	158	170	39	39		31				

215

Entries in Livestock and Other Agricultural Classes since 1857 (cond.)

		Horses	Cattle	Sheep	Pigs	Goats	Horse Shoeing	Dairy Produce	Dairy Competitions	Milking	Judging by Points
1913		Bath & West at Truro									
1914	Fowey	186	196	38	33		20	82	100		
1915	Camborne	185	142	30	47		13	69	83		
1916-17-18	No Shows										
1919	Truro	194	139	29	35		35	111	30		
1920	Callington	140	144	34	44		19	59	35		
1921	Falmouth	182	204	59	82		31	74	37		
1922	Newquay	171	191	42	61		51	59	48		
1923	Camborne	194	214	38	68		41	138	72	12	100
1924	Wadebridge	228	217	36	88		29	77	96	9	158
1925	Helston	251	301	51	99		39	90	103	19	290
1926	Launceston	206	240	87	98		25	101	65	5	413
1927	Truro	269	361	70	137		46	172	119	28	281
1928	Bodmin	235	284	66	72		25	78	98	26	316
1929	Penzance	244	311	73	84		17	108	119	20	371
1930	Liskeard	231	277	92	81		30	101	103	20	385
1931	St Columb	238	249	71	90		26	98	113	10	342
1932	Penryn	211	244	64	78		21	75	110	7	360
1933	St Austell	216	257	69	86		29	115	138	7	352
1934	Camborne	223	304	86	109		32	113	108	7	320
1935	Newquay	224	280	71	112		21	79	101	11	304
1936	St Ives	255	265	71	127		26	91	112	12	439
1937	Wadebridge	250	293	78	115		23	80	109	20	340
1938	Helston	266	324	108	130		35	84	115	20	491
1939	Bude	226	290	105	131	35	31	81	120	31	684
1947	Truro	257	289	63	82			44		25	1407
1948	Bodmin	284	402	95	152			45		23	810
1949	Falmouth	290	438	148	138	61	14	45		36	

Entries in Livestock and Other Agricultural Classes since 1857 *(cond.)*

		Horses	Cattle	Sheep	Pigs	Goats	Horse Shoeing	Dairy Produce	Dairy Competitions	Milking	Judging by Points
1950	Callington	343	454	131	224	10	18	19			850
1951	Newquay	304	453	135	204		16	65			348
1952	Redruth	239	360	96	152			39			194
1953	Launceston	256	447	114	189		5	34			172
1954	St Austell	239	390	117	216			25			215
1955	Bath & West at Launceston										
1956	Helston	270	355	88	163		8	46			165
1957	Wadebridge	345	389	166	233			25			195
1958	Truro	418	391	126	254			25			177
1959	Liskeard	434	389	144	212			27			181
1960	Wadebridge	358	368	109	252						186
1961	Wadebridge	425	361	118	213						170
1962	Wadebridge	402	343	149	214						213
1963	Wadebridge	358	333	136							147
1964	Wadebridge	417	329	134	210	79					200
1965	Wadebridge	424	380	126	202	68					144
1966	Wadebridge	540	425	140	172	72					137
1967	Wadebridge	571	401	159	174						158
1968	Wadebridge	565	333	161	170						120
1969	Wadebridge	482	346	131	186						99
1970	Wadebridge	527	243	111	169						99
1971	Wadebridge	546	293	125	115						48
1972	Wadebridge	656	326	132	121						60
1973	Wadebridge	722	349	135							50
1974	Wadebridge	944	318	127							53
1975	Wadebridge	904	397	113							48
1976	Wadebridge	990	473	138							60
1977	Wadebridge	967	588	125							

Entries in Livestock and Other Agricultural Classes Since 1857 (cond.)

		Horses	Cattle	Sheep	Pigs	Goats	Horse Shoeing	Dairy Produce	Dairy Competitions	Milking	Judging by Points
1978	Wadebridge	934	474	126							51
1979	Wadebridge	983	447	127							51
1980	Wadebridge	1111	514	160							61
1981	Wadebridge	1019	551	160							39
1982	Wadebridge	916	529	163							48
1983	Wadebridge	919	506	165		138					45
1984	Wadebridge	980	518	177		170					42
1985	Wadebridge	1000	464	180		203					51
1986	Wadebridge	1067	470	190		234					51
1987	Wadebridge	1263	485	209		223					39
1988	Wadebridge	1176	462	306		169					18
1989	Wadebridge	1419	570	494		204					24
1990	Wadebridge	1359	656	522		214					42
1991	Wadebridge	1321	660	598		197					33
1992	Wadebridge	1338	594	640		230					39

RESULTS — 1793 to 1992

These tables of results cover the Cattle, Sheep, Pig and Horse Sections of the Shows staged since 1794 and also include the results of various other competitions, when staged, from 1793. Compiled from the records held by the Association, newspaper reports and other documents, the tables are virtually complete and show the development of the various classes etc. over the 200 year period. With the introduction of breed classes from 1839, and the resulting large number of individual classes, space only permits the listing of breed championships and other major awards.

It should be remembered that many of these prizes were only offered from time to time, and not as a regular feature of the show, until relatively recently.

When two results for a class in one year are shown, the second result refers to a prize awarded at the second show staged by the Society, as listed in another section.

CATTLE SECTION RESULTS — 1794 to 1992

	BEST BULL	BEST BULL, property of a farmer of this county getting his livelihood solely by farming/rack renting	BEST BULL, not exceeding 2 years
	Exhibitor	*Exhibitor*	*Exhibitor*
1794	John Wevil, Lewannick		
1795	Rev H H Tremayne, Heligan		
1796	Paul Upton Oke, St Tudy		
1797	John Rogers, Quethiock		
1802	Thomas Julian, St Austell		Alexander Menhennick, St Mabyn
1803	John Rogers, Holwood		Thomas Key, St Breock
1804	Philip Rashleigh		Philip Rashleigh
1805	Robert Lovel Gwatkin, Killiow		William Gately
1805	Francis Enys		
1806	John Penhalow Peters, Creegmurrion, Philleigh	Nicholas Minhinnick, St Mabyn	
1806	William Osborne, St Hilary		
1807	Philip Rashleigh	Thomas Key, St Breock	
1807	John Shearm, Kilkhampton		
1808	Lord Falmouth, Tregothnan	P Oke, St Tudy	

	Exhibitor	Exhibitor	Exhibitor
1808	A Paul Camborne		
1809	Rev H H Tremayne Heligan		
1810	Philip Rashleigh	Mr Menhinnick	
1810	W Drew St Columb		
1811	No result known		
1811	John Shearm Kilkhampton		
1812	W Lane Liskeard	Nicholas Rundle Luxulyan	
1812	Lord Falmouth Tregothnan		
1813	R Dingle Ruan Lanyhorne	John Thomas St Wenn	
1814	Sir William Call, Bt	John Hawken Roach	
1814	Joseph Hendy Gunwalloe		
1815	Alexander Burrell St Erth	Robert Clemow St Enoder	
1815	John Shearm Kilkhampton		
1816	No result known		
1816	Adam Thomson Cardinham		
1817	Sir Christopher Hawkins Bt, Trewithen	Richard Dingle Ruanlanyhorne	
1817	Mr Foote Northhill		
1818	John Cardell Jnr Lower St Columb	John Sobey Luxulyan	
1819	F Hearle Rodd Trebartha		
1820	J Thomas St Wenn	J Cardall Lower St Columb	
1821	William Rashleigh Menabilly	John Thomas St Wenn	
1822	George Turner Upcott, Devon	John Buller Morval	
1825	F Hearle Rodd Trebartha	G Bulmer Newlyn	
	BEST BRED BULL Free of all England	BEST PAIR OF WORKING OXEN OR STEERS	BEST PAIR OF FED OXEN OR STEERS Worth most per cwt
1828	John Cardell Jnr Lower St Columb	Sir Christopher Hawkins Trewithen	Thomas Daniel Feock
1829	Mr Bullmore Newlyn	W Cardell Probus	The Execs of Sir Christopher Hawkins, Trewithen

	Exhibitor	Exhibitor	Exhibitor
1830	J H Tremayne St Ewe	E W W Pendarves Camborne	J H Tremayne St Ewe
1831	W Cardell Probus	John Hawkins Probus	Mr Polkinghorne St Austell
1832	M Doble Probus	Mr Cardell Lower St Columb	John Hawkins Probus
	BEST BULL To be worked in the County of Cornwall for 12 months subsequent to the meeting		BEST PAIR OF FED OXEN OR STEERS Any Breed
1833	Joseph Hicks St Columb	W Cardell Probus	J Hawkins Trewithen
1834	J H Tremayne Heligan	R Stevens Crantock	William Gatley Probus
1835	M Doble Probus	G Simmons Trevellas	James Hendy Trethurffe
1836	William Hodge Perranzabuloe	G Bullmore Newlyn	G W F Gregor Trewarthenick
1837	William Tremain Newlyn	G W F Gregor Trewarthenick	J H Tremayne Heligan
1838	Thomas Julian Creed	George Miners Cornelly	John Hawkins Trewithen
1939		W Cardell Probus	G W F Gregor Cornelly
1840		R Stephens Crantock	M A Doble Probus
1841		W Tremaine Newlyn	J Hendy Trethurffe
1842		W Pearce Truro	C H T Hawkins Trewithen
1843		W Treffry Ruan	C H T Hawkins Trewithen
1844		W Dingle Ruan	J Hendy Trethurffe
1845		M A Doble Probus	C H T Hawkins Trewithen
1846		T Julyan Creed	J Kendall Probus
1847		Mr Downing Ladock	R Cardell St Columb Minor
1848		R Cardell St Columb Minor	
1849		T Julyan Creed	

	BEST MADE COW having had 2 or more calves	BEST MADE HEIFER having had 1 calf	HEIFER, under 3 years old	HEIFER, not exceeding 4 years
	Exhibitor	*Exhibitor*	*Exhibitor*	*Exhibitor*
1799	John Tregaskis Tywardreath	John Hicks St Columb		
1802	John Slyman St Mabyn	John Slyman St Mabyn		
1804			Philip Rashleigh	
1805			William Gately	
1812				William Burrows Bodmin
1813				Lord Falmouth Tregothnan
1814				William Hicks St Wenn
1815				Mr Commins Bodmin

	BEST FED COW OR HEIFER Worth most per cwt	BEST BRED MILCH COW OR HEIFER
1828	Matthew Doble Probus	William Tremain Newlyn

		BEST BRED MILCH COW OR HEIFER Not rearing a buss
1829	R Harding St Michael Penkivell	W Cardell Probus
1830	J H Tremayne St Ewe	J H Tremayne St Ewe
1831	R Johns Gerrans	Matthew Doble Probus
1832	George Bosustow Tywardreath	W Gatley Probus

	BEST FED COW OR HEIFER Any Breed	
1833	J H Tremayne Heligan	J H Tremayne Heligan

	BEST FED COW OR HEIFER Any Breed *(cond)*	BEST BRED DAIRY COW OR HEIFER	BEST BRED DAIRY COW OR HEIFER belonging to Tenants occupying farms not exceeding £150 per year, getting their livelihood entirely by farming	BEST 3 YR OLD HEIFER IN MILCH OR WITH CALF
1834	J H Tremayne Heligan	Mr Cardell Probus		
1835	James Hendy Trethurffe	J H Tremayne Heligan		
1836	R Doble Philleigh	W Nattle Cadson Bury	W Hawke Ladock	
1837	J H Tremayne Heligan	George Bullmore Newlyn	Thomas Julyan Creed	Thomas Julyan Creed
1838	George Bosustow Tywardreath	George Bullmore Newlyn	P A Grieve Probus	Thomas Julyan Creed
1839	J H Tremayne Heligan	Thomas Julyan Creed	M Hotton Probus	
1840	J H Tremayne Heligan		F James Probus	
1841	R Lanyon Stithians		M Hotton Probus	
1842	Magor, Davey & Co Redruth		P A Grieve Probus	
1843	J H Tremayne Heligan		M Hotton Probus	
1844	J Thomas Hendra		P Grieve Probus	
1845	John Silvester Helston		E Kendall Probus	
1846	Mr Julyan Creed		J Hotton Probus	
1847	J D Gilbert Trelissick		David Rooke Cornelly	
1848			Charles Farley Kenwyn	
1849			E Kendall Probus	
1850			Withheld for want of merit	
1851			R V White Merther	
1852			Richard Tiddy St Clement	
1853			Charles Bice St Enoder	
1854			Charles Farley Kenwyn	
1855			William Kessel St Clement	
1856			William Lucus Kenwyn	

DURHAM CATTLE

	BEST DURHAM BULL	BEST DURHAM DAIRY COW OR HEIFER
	Exhibitor	Exhibitor
1839	James Dark Cornelly	
1840	J P Peters Philleigh	James Hendy Trethurffe
1841	J Davies Cornelly	G G Bullmore Newlyn
1842	James Hendy Trethurffe	James Hendy Trethurffe
1843	W Harvey St Erth	James Hendy Trethurffe

DEVON CATTLE

	BEST DEVON BULL	BEST DEVON DAIRY COW OR HEIFER	CHAMPION DEVON
1839	J H Tremayne Heligan		
1840	W Cardell Probus	W Tremaine Newlyn	
1841	S Anstey Tywardreath	J H Tremayne Heligan	
1842	Thomas Julyan Creed	M A Doble Probus	
1843	J H Tremayne Heligan	J H Tremayne Heligan	
1844	Thomas Julyan Creed	Thomas Julyan Creed	
1845	W Tremaine Newlyn	M A Doble Probus	
1846	Thomas Julyan Creed	R Arthur Creed	
		BEST DEVON DAIRY COW	
1847	R Davey Redruth	Thomas Julyan Creed	
1848	J H Tremayne Heligan	Richard Tremain	
1849	J D Gilbert Trelissick	Elizabeth James Lanivet	
1850	Mr Anstey Menabilly	Thomas Julyan Creed	
1851	Thomas Julyan Creed	Thomas Julyan Creed	
1852	Charles Davis St Enoder	Samuel Anstey Tywardreath	
1853	John Tremain Trerice, Newlyn	Samuel Anstey Menabilly	

1854	Roger Hawken St Minver	James Tremain Newlyn	
1881	Viscount Falmouth Tregothnan 'Sir Michael'		Viscount Falmouth Tregothnan 'Sir Michael'

BEST DEVON COW OR HEIFER

1888	Richard Bickle Bradstone Hall, Tavistock	Sir William Williams Bt Heanton, Barnstaple
1889	Richard Bickle Bradstone Hall, Tavistock	Alfred C Skinner Pounds Farm, Bishops Lydeard
1891	J C Williams Caerhayes Castle, St Austell	Alfred C Skinner Pounds Farm, Bishops Lydeard
1895	Mr Mucklow Holsworthy	
1900	John Button Bearock, Michaelstow	W Brent Clampit, Callington 'Hetty'
1901	William Trestain Lostwithiel	T Dyer Liskeard
1902		H E L Pearse Linkinhorne
1904	Abraham Trible Halsdon Barton, Holsworthy 'Joker'	H L Pearse Exwell, Callington
1907	W Stanbury & Sons Launceston	T Dyer Liskeard
1908	J Button Bearoak, Michaelstow 'Royalist'	
1909	W E Menhinick Hendra, St Kew 'Clampit Gay Boy'	
1910	W Hicks St Austell	J E & N Heard Launceston
1911	G Hicks Lower Menadue, Luxulyan 'Pound Forager'	H E L Pearse Exwell, Callington 'Exwell Linkinhorne 2nd'
1912	W Brent Callington 'Ford Plumper'	W Brent Callington
1914	J E Heard South Petherwin 'Overton Gentleman II'	Messrs Oliver & Crapp St Cleer, Liskeard 'Trewarrick Rose XV'
1915	W Brent Clampit, Callington	Messrs Oliver & Crapp St Cleer, Liskeard 'Stockleigh Nominator'
1919	Oliver & Crapp Treworrick, St Cleer	W G Brent Clampit, Callington
1922	J C Warne Grampound Road, Truro	
1923		N Heard Home Farm, Werrington 'Woodlands King'

1924			A Trible & Sons Halsdon Barton, Holsworthy 'Goldcoin 2nd'
1925			A Trible & Sons Halsdon Barton, Holsworthy
1926	A Trible & Sons Halsdon, Holsworthy	T Warne Trevisquite, St Mabyn	
1927	C Brent Clampit, Callington 'Pound Romper'	W Squance Beaworthy, Devon	
1928	C Brent Clampit, Callington 'Pound Romper'		C Brent Clampit, Callington 'Pound Romper'
1929			C Brent Clampit, Callington
1930			HRH The Prince of Wales Home Farm, Stokeclimsland
1931			C Brent Clampit, Callington
1935	J E Friend Stockleigh Pomeroy, Crediton 'Leigh Curly Coat'		
1937			A Trible & Sons Halsdon Barton, Holsworthy 'Stoke Rubicon'
1938			HM King George VI Home Farm, Stokeclimsland
1939			P H Hearn & Son Swaddledown, Bratton Clovelly 'Whitefield Ransom'
1947			F Stanbury Hammill, Launceston
1948			W B Nancekivell Pinkhill, Okehampton
1949			W Stanbury Carey Barton, Launceston 'Nutwell Esquire'
1950			C W Lewis Great Potheridge, Okehampton
1951			E J Alford West Illand, Congdon Shop
1952			Messrs C Brent & Son Clampit, Callington
1953			Messrs C Brent & Son Clampit, Callington 'Uggaton Highwayman 2nd'
1954			Lord Beaverbook Ilminster, Somerset

1956			P M Williams Burncoose, Redruth 'Burncoose Grouse'
1957			W B Nancekivell Pinkhill, Okehamton 'Pinkhill Dairymaid'
1958			W B Nancekivell Pinkhill, Okehampton 'Pinkhill Dairymaid'
1959			W B Nancekivell Pinkhill, Okehampton 'Pinkhill Dairymaid'
1960			W B Nancekivell Pinkhill, Okehampton
1961	S J Skinner Winkleigh, Devon	W R Yeo & Sons Netherby, Barnstaple	W R Yeo & Sons Netherby, Barnstaple
1962	W R Yeo & Sons Netherby, Barnstaple 'Roundswell Pride'	C Brent & Son Clampit, Callington 'Clampit Flirt 29th'	W R Yeo & Sons Netherby, Barnstaple 'Roundsell Pride'
1963	P M Williams OBE Burncoose, Redruth 'Burncoose Lira 2nd'	C Brent & Son Clampit, Callington 'Clampit Flirt 29th'	P M Williams OBE Burncoose, Redruth 'Burncoose Lira 2nd'
1964			W R Yeo & Sons Netherby, Barnstaple
1965	S J Skinner Winkleigh, Devon 'Burncoose No Use 5th'	W C May & Son Priorton, Crediton 'Priorton Showgirl'	S J Skinner Winkleigh, Devon 'Burncoose No use 5th'
1966	S J Skinner Winkleigh, Devon 'Burncoose No Use 5th'	C Brent & Son Clampit, Callington 'Clampit Norah 3rd'	S J Skinner Winkleigh, Devon 'Burncoose No Use 5th'
1967	W R Sanders Polyphant, Launceston 'Trerithick Defender'	W C May Priorton, Crediton 'Priorton Showgirl 54th'	W C May Priorton, Crediton 'Priorton Showgirl 54th'
1968	W R Sanders Polyphant, Launceston 'Clampit Groom'	W C May Priorton, Crediton	W R Sanders & Son Polyphant, Launceston 'Clampit Groom'
1969	W C May Priorton, Crediton 'Priorton Daimler 2nd'	W C May Priorton, Crediton 'Priorton Showgirl 56th'	W C May Priorton, Crediton 'Priorton Daimler 2nd'
1970	C Brent & Son Clampit, Callington 'Potheridge Big Ben'	W C May Priorton, Crediton 'Priorton Showgirl 56th'	C Brent & Son Clampit, Callington 'Potheridge Big Ben'
1971			C Brent & Son Clampit, Callington 'Potheridge Big Ben'
1972			C Brent & Son Clampit, Callington 'Potheridge Big Ben'
1973			W C May Priorton, Crediton
1974			Mrs A Matysiak Okehampton 'Lufton Fieldsman'

1975			Mrs A Matysiak Okehampton 'Lufton Fieldsman'
1976			W C May Priorton, Crediton
1977	J H Thomas Umberleigh, Devon 'Fairnington Prognession 13th'	W C May Priorton, Crediton 'Priorton Showgirl 126th'	W C May Priorton, Crediton 'Priorton Showgirl 126th'
1978	J H Thomas Umberleigh, Devon 'Fairnington Prognession 13th'	W C May Priorton, Crediton 'Priorton Showgirl 126th'	W C May Priorton, Crediton 'Priorton Showgirl 126th'
1979	Drake Brothers North Tawton, Devon 'Brightley Chancellor 47th'	W C May Priorton, Crediton 'Priorton Showgirl 126th'	Drake Brothers North Tawton, Devon 'Brightley Chancellor 47th'
1980	K Bidie Minehead, Somerset 'Heasley Pride'	W C May Priorton, Crediton 'Priorton Showgirl 182nd'	W C May Priorton, Crediton 'Priorton Showgirl 182nd'
1981	K Bidie Minehead, Somerset 'Heasley Pride'		K Bidie Minehead, Somerset 'Heasley Pride'
1982	Drake Brothers North Tawton, Devon 'Fishleigh Fieldsman 11th'	A G & B T Slee Bideford, Devon 'Halsbury Sandra 15th'	Drake Brothers North Tawton, Devon 'Fishleigh Fieldsman 11th'
1983	Drake Brothers North Tawton, Devon 'Fishleigh Fieldsman 11th'	A G & B T Slee Bideford, Devon 'Halsbury Marigold 12th'	Drake Brothers North Tawton, Devon 'Fishleigh Fieldsman 11th'
1984	K Bidie Minehead, Somerset 'Barton Brinkman'	P K James Broadbury, Okehampton Langworthy Madeleine 4th'	K Bidie Minehead, Somerset 'Barton Brinkman'
1985	Drake Brothers North Tawton, Devon 'Brightley Chancellor 53rd'	Mr & Mrs R K James Broadbury, Okehampton 'Langworthy Della 6th'	Drake Brothers North Tawton, Devon 'Brightley Chancellor 53rd'
1986	Rt Hon The Lord Clinton Okehampton 'Langworthy Rocket'	G Dart & Sons South Molton, Devon 'Champson Bribery 18th'	Rt Hon The Lord Clinton Okehampton 'Langworthy Rocket'
1987	Slee Partners Bideford, Devon 'Halsbury Harvester'	G Dart & Sons South Molton, Devon 'Champson Bribery 18th'	G Dart & Sons South Molton, Devon 'Champson Bribery 18th'
1988	G Dart & Sons South Molton, Devon 'Champson Security'	G Dart & Sons South Molton, Devon 'Champson Bribery 18th'	G Dart & Sons South Molton, Devon 'Champson Bribery 18th'
1989	Rt Hon The Lord Clinton Okehampton, Devon 'Langworthy Rocket 1st'	G Dart & Sons South Molton, Devon 'Champson Bribery 27th'	G Dart & Sons South Molton, Devon 'Champson Bribery 27th'
1990	W C May Priorton, Crediton 'Thorndale Baron 2nd'	G Dart & Sons South Molton, Devon 'Champson Bribery 27th'	G Dart & Sons South Molton, Devon 'Champson Bribery 27th'
1991	Slee Partners Bideford, Devon 'Halsbury Supreme 9th'	Slee Partners Bideford, Devon 'Halsbury Lily 39th'	Slee Partners Bideford, Devon 'Halsbury Supreme 9th'
1992			C Thornton Creeton, Grantham, Lincs 'Thorndale Baron'

SHORT HORN CATTLE
(Catalogued as Dairy Shorthorn from 1963)

	BEST SHORT HORN BULL	BEST SHORT HORN DAIRY COW OR HEIFER	CHAMPION SHORT HORN
	Exhibitor	*Exhibitor*	*Exhibitor*
1844	P Davies Probus		
1845	James Hendy Trethurffe	Lady Basset Tehidy	
1846	J Davies Cornelly	G G Bullmore Newlyn	
1847	Coryton Kemp Philleigh	James Hendy Trethurffe	
1848	G Stephens St Tudy	T H Tilley Tremough	
1849	Probus Farmers' Club Probus	M A Doble & Co Probus	
1850	M A Doble Probus	H Trethewy Probus	
1851	Probus Farmers' Club Probus	M A Doble & Co Probus	
1852	John Kendall & Co Probus	W R Pender Budock Vean	
1853	Probus Farmers' Club Probus	William Lemon Camborne	
1854	Probus Farmers' Club Probus	George Williams Trevince	
1887	W Chapman Trewithian House, Gerrans 'Earl of Oxford'		
1895	W J Hosken Hayle	W J Hosken Hayle	
1907		W J Hosken Hayle	
1908		W J Hosken Hayle	
1909		W J Hosken Hayle	
1919		W J Hosken Hayle	
1927	HRH The Prince of Wales KG Stokeclimsland	HRH The Prince of Wales KG Stokeclimsland	
1928		L V Garland Greenbank, Hayle 'Towan Nonpareil 3rd'	
1936		W Jeffery Priske, Mullion 'Priske Pauline 3rd'	
1937		T Lugg Hingey, Gunwalloe	T Lugg Hingey, Gunwalloe

1938		W Jeffery Priske, Mullion 'Priske Pauline 3rd'	W Jeffery Priske, Mullion 'Priske Pauline 3rd'
1939		H Stephens & Son Penwinnick, St Agnes	H Stephens & Son Penwinnick, St Agnes
1947	A V Flexman Royal Farm, Truro	W R Richards Killivose, Truro 'Church St Daisy V'	
1948	C S James Tredinnick, Grampound Rd	C S James Tredinnick, Grampound Rd	
1949	A V Flexman Royal Farm, Truro		
1950	A V Flexman Royal Farm, Truro	A V Flexman Royal Farm, Truro	
1951	C B Withers Pilning, Bristol	C S James Tredinnick, Grampound Road	
1952	S Hocking Sithney, Helston	James Brothers Grampound Road	
1953	W H Bond Trerulefoot, St Germans 'Beechwood Royal Foggathorpe 2nd Q'	C S James Tredinnick, Grampound Road 'Court Farm Beauty 31st'	
1954	James Brothers Grampound Road	James Brothers Grampound Road	
1956	S Hocking Sithney, Helston 'Snotterton Blenheim 2nd'	James Brothers Grampound Road 'Avenham Dorothy 7th'	
1957	A L Lerwill Brinsworthy, N Molton 'Effjay Ringletts Ringleader'	James Brothers Grampound Road 'Avenham Dorothy 7th'	
1958	A L Lerwill Brinsworthy, N Molton 'Effjay Ringletts Ringleader'	W L Stephenson & Son Ltd Blandford, Dorset 'Malahne Lorna Lilian 10th'	
1959	Michell & Andrew St Just Lane, Truro 'Roseland Bridesmaids Songster'	James Brothers Grampound Road	
1960	E J R Hutchings Crediton, Devon	James Brothers Grampound Road	
1961	G J Osborne & Son Bath	C S James Grampound Road	C S James Grampound Road
1962	James Brothers Grampound Road 'Hastoe Duke Magnum 3rd'	Mrs C Garnett Penhale, Wadebridge 'St Breock Bambi'	Mrs C Garnett Penhale, Wadebridge 'St Breock Bambi'
1963	James Brothers Grampound Road 'Hastoe Duke Magnum 3rd'	James Brothers Grampound Road 'Barteliver Dairymaid 91st'	James Brothers Grampound Road 'Barteliver Dairymaid 91st'
1964			James Brothers Grampound Road

1965	James Brothers Grampound Road 'Hastoe Duke Magnum 3rd'	James Brothers Grampound Road 'Barteliver Dairymaid 154th'	James Brothers Grampound Road 'Barteliver Dairymaid 154th
1966	E C Roose St Teath, Bodmin 'Mickle Reliance'	James Brothers Grampound Road 'Barteliver Dairymaid 145th'	E C Roose St Teath, Bodmin 'Mickle Reliance'
1967			E C Roose St Teath, Bodmin 'Rosehendra Joan'
1968			E C Roose St Teath, Bodmin 'Rosehendra Joan'
1969	William-Powlett Farms Ottery St Mary, Devon 'Calstone Teamster 2nd'	G T Withers Whimple, Exeter 'Gateshayes Wild Eyes'	William-Powlett Farms Ottery St Mary, Devon 'Calstone Teamster 2nd'
1970	E C Roose St Teath, Bodmin 'Orgreave Gentleman'	E C Roose St Teath, Bodmin 'Roosehendra Balloon 5th'	E C Roose St Teath, Bodmin 'Roosehendra Balloon 5th'
1971		G T Withers Whimple, Exeter 'Gateshayes Spring Daisy 3rd'	G T Withers Whimple, Exeter 'Gateshayes Spring Daisy 3rd'
1972		G T Withers Whimple, Exeter	G T Withers Whimple, Exeter

HEREFORD CATTLE

	BEST HEREFORD BULL *Exhibitor*	BEST HEREFORD COW OR HEIFER *Exhibitor*	CHAMPION HEREFORD *Exhibitor*
1859	John Sobey Trencreek, Menheniot	R Olver Trescowe, Washaway	
1860	George Lobb Lawhitton	John Sobey Trenceek, Menheniot	
1862	Richard Davey MP Polsue, Philleigh	Richard Davey MP Polsue, Philleigh	
1863	Thomas Olver Penhallow, Philleigh	Thomas Olver Penhallow, Philleigh	
1864	George Lobb Lawhitton	Thomas Olver Penhallow, Philleigh	
1865	Henry Esery Werrington, Launceston	Thomas Olver Penhallow, Philleigh	
1887	Wesley Grose Penpont, Wadebridge		
1965	Maj G E F North Rackenford, Devon 'Kingsfield 1 Star'	Miss R Anne Humbert Romsey, Hants 'Kiwi Hetty'	Maj G E F North Rackenford, Devon 'Kingsfield 1 Star'
1966	Maj G E F North Rackenford, Devon 'Tiverton 1 War Lord'	Maj N H Hambro Minehead, Somerset 'Westfarm Peach Flower 3rd'	Maj G E F North Rackenford, Devon 'Tiverton 1 War Lord'

1967	Maj G E F North Rackenford, Devon	Miss R Anne Humbert Romsey, Hants 'Kiwi 1 Eleanor'	Miss R A Humbert Romsey, Hants 'Kiwi 1 Eleanor'
1968	Maj G E F North Rackenford, Devon	Maj G E F North Rackenford, Devon	Maj G E F North Rackenford, Devon
1969	F G Hawke Hustyn, Wadebridge 'Eggbeer Artisan'	F G Hawke Hustyn, Wadebridge 'Nanscient Moorhen 3rd'	F G Hawke Hustyn, Wadebridge 'Eggbeer Artisan'
1970	Mr & Mrs Hambro Minehead, Somerset 'Rowington Dante'	Maj G E F North Rackenford, Tiverton 'Tiverton 1 Dowager 15th'	Mr & Mrs Hambro Minehead, Somerset 'Rowington Dante'
1971	Maj G E F North Rackenford, Devon 'Tiverton 1 Diamond'	Mr & Mrs N Hambro Minehead, Somerset 'Westfarm Contessa'	Maj G E F North Rackenford, Devon 'Tiverton 1 Diamond'
1972	B D Uglow Linkinhorne, Callington	J H Meinl Warminster, Wiltshire	J H Meinl Warminster, Wiltshire
1973	D Standerwick Bovey Tracey, Devon	J H Meinl Warminster, Wiltshire	D Standerwick Bovey Tracey, Devon
1974	Maj G E F North Rackenford, Devon 'Beaudesert of Freedom'	Cowley Farms Umberleigh, Devon 'Havenfield Pansy 10th'	Maj G E F North Rackenford, Devon 'Beaudesert of Freedom'
1975	Cowley Farms Umberley, Devon 'Cranford Senator'	D Standerwick Bovey Tracey, Devon 'Yawl 1 Festoon'	Cowley Farms Umberleigh, Devon 'Cranford Senator'
1976			D Standerwick Bovey Tracey, Devon
1977	Cowley Farms Umberleigh, Devon 'Cranford Bravado'	F S Hurndall-Waldron Moretonhampstead 'Quillet Dowager Gay'	Cowley Farms Umberleigh, Devon 'Cranford Bravado'
1978	D Standerwick Bovey Tracey, Devon 'Yawl 1 Machiavelli'	D Standerwick Bovey Tracey, Devon 'Yawl 1 Jane 7th'	D Standerwick Bovey Tracey, Devon 'Yawl 1 Machiavelli'
1979	D Standerwick Bovey Tracey, Devon 'Yawl 1 Nijinsky'	W Pascoe & Son Helston 'Tulsa Sunbeam 4th'	D Standerwick Bovey Tracey, Devon 'Yawl 1 Nijinsky'
1980	E J Sanders Altarnun, Launceston 'Tregrenna Goldfinger'	P M Prior-Wandesforde Timberscombe, Minehead 'Tryfan 1 Dowager 11th'	E J Sanders Altarnun, Launceston 'Tregrenna Goldfinger'
1981	W Pascoe & Son Helston 'Wenlock Butlin'	E J Sanders Altarnun, Launceston 'Barton 1 Beauty Songstress'	W Pascoe & Son Helston 'Wenlock Butlin'
1982	Mrs K M Jeffery Ilfracombe, Devon 'TBJ Nautilus'	W Pascoe & Son Helston 'Tulsa Sunbeam 4th'	Mrs K M Jeffery Ilfracombe, N Devon 'TBJ Nautilus'
1983	Mr & Mrs F G & R P Hawke Hustyn, Wadebridge 'Nanscient Chieftan'	Mr & Mrs E J Sanders Altarnun, Launceston 'The Barton 1 Phoebe 2nd'	Mr & Mrs E J Sanders Altarnun, Launceston 'The Barton 1 Phoebe 2nd'
1984	W Pascoe & Son Helston 'Wenlock Butlin'	W Pascoe & Son Helston 'Tulsa Sunbeam 4th'	W Pascoe & Son Helston 'Wenlock Butlin'

1985	E J Sanders Altarnun, Launceston 'Costhorpe 1 Trailblazer (p)'
1986	Mr & Mrs F G, Mr F J & Mr R P Hawke Hustyn, Wadebridge 'Westwood Argon'
1987	P Dingle Camborne 'Rosewarne Poll Trevithick'
1988	Mr & Mrs R W J Williams Helston 'Badlingham Broad Lad'
1989	Mr & Mrs R W J Williams Helston 'Badlingham Broad Lad'
1990	Mr & Mrs M E Ley & Son Torrington, Devon 'Durpley Mary 2nd'
1991	W Pascoe & Son Helston 'Smithston Equalizer'
1992	Mr & Mrs M E Ley & Son Torrington, Devon 'Longville Glen'

CHANNEL ISLAND CATTLE
BEST BULL

1872	Colonel Gilbert Bodmin
1873	T D Eva Camborne
1874	Mr Rendle Catel Farm, Guernsey
1876	Messrs N Saunders & Son Bodmin

JERSEY CATTLE

BEST JERSEY BULL	BEST JERSEY COW OR HEIFER	CHAMPION JERSEY
Exhibitor	*Exhibitor*	*Exhibitor*

1875	Messrs N Sanders & Son Bodmin
1877	Richard Hoskin Sancreed
1878	J Nicholls Lostwithiel

1879	Richard Hockin Gwithian	
1880	J Tremayne Heligan, St Austell	
1881	Richard Hockin Gwithian	
1882	Hon Mrs Tremayne Heligan, St. Austell	
1888	Colonel Fortescue Boconnoc	

BEST JERSEY COW OR HEIFER IN MILK

1909	R P Wheadon Ilminster	R P Wheadon Ilminster, Somerset	
1911	W J Rowlings Leedstown, Hayle 'Nimrod'	James E Spargo Parkengew, Penryn 'Flora'	
1915	James E Spargo Parkengew, Penryn 'Morning Star'		
1919	Mrs Rudd East Grinstead, Sussex 'Fire King'		
1920	F B Imbert-Terry Broadclyst, Devon 'Pro Dacchus'		
1921	J E Spargo Parkengew, Penryn 'Lucky Lad'		
1922	F B Imbert-Terry Broadclyst, Devon 'Blue Hayes Red Candy'	W J Rowlings Leedstown, Hayle 'Primrose 2nd'	
1927	Lt Col C F Miller DSO Great Trethew, Menheniot 'Wotton King of Clubs'	Lt Col C F Miller DSO Great Trethew, Menheniot 'Trethew Nerita'	Lt Col C F Miller DSO Great Trethew, Menheniot
1928	Lt Col C F Miller DSO Great Trethew, Menheniot	Miss Mable C S Williams Trannack House, Penzance 'Princess Amy'	
1929	R F Bolitho Ponsandane, Gulval 'Lingen Premier Valentine'	R F Bolitho Ponsandane, Gulval	
1930	Mrs Hayes-Sadler Ferring, Sussex 'Velveteens Oxford'	Lt Col C F Miller DSO GreatTrethew, Menheniot	Mrs Hayes-Sadler Ferring, Sussex 'Velveteens Oxford'
1931	Capt & Mrs F B Imbert-Terry Broadclyst, Devon	Lt Col C F Miller DSO Great Trethew, Menheniot 'Trethew Blue Belle'	Capt & Mrs F B Imbert-Terry Broadclyst, Devon
1932		Miss Mabel C S Williams Trannack House, Penzance	

1933	Miss K Curnow Treveor, St Michael Penkivel 'Lucky Hero'	W Hawkey St Elmo, Probus 'Owl's Dream'	
1934	W Hawkey Gainsborough House, Probus 'Green Hills Samaritan'	W J Kneebone Treskewes, Stithians 'Ashill Gazelle'	
1935		The Hon Lady Cook Porthallow, Looe 'Salsue'	
1936	Mrs E O Hawkey Gainsborough House, Probus 'Rochettes Royal Observer'	Mrs E O Hawkey Gainsborough House, Probus 'Rhyming Sultene'	
1937	Mrs E O Hawkey Gainsborough House, Probus 'Rochettes Royal Observer'	The Hon Lady Cook Porthallow, Looe	The Hon Lady Cook Porthallow, Looe
1938	Mrs E O Hawkey Gainsborough House, Probus 'Rochettes Royal Observer'	The Hon Lady Cook Porthallow, Looe	
1939	The Hon Lady Cook Talland Bay, Looe	Mrs E O Hawkey Gainsborough House, Probus	
1947		Mrs E O Hawkey Pencarrow, Probus	
1948		Mrs K G Truscott Treveor, Tresillian	
1949			Mrs E O Hawkey Pencarrow, Probus
1950		Maj A F West Ogbeare, North Tamerton	
1951		Sir Giles & Lady Sebright Millaton, Bridestowe	
1952		Mrs E O Hawkey Pencarrow, Probus	
1953		Maj A F West Ogbeare, North Tamerton 'Hutchings Val Leonie 2nd'	
1954		H S Benney & Sons Trevorlis, Breage	
1956		H E B Gundry Grange, Honiton 'Honiton Romance'	

1957		W E Eddy Buryas Bridge, Penzance 'Ecclesden Crystal G M O M'	
1958		H S Benney & Sons Trevorlis, Breage 'Roseboy's Mayqueen'	
1959		Mr & Mrs J A P Martin Chulmleigh, N Devon	
1960	W E Eddy Buryas Bridge, Penzance	H S Benney & Sons Trevorlis, Breage	
1961	Maj & Mrs P E Bisgood Chantry, Loddiswell 'Chesford Highfield Prince'	H S Benney & Sons Trevorlis, Breage	H S Benney & Sons Trevorlis, Breage
1962	Maj & Mrs P E Bisgood Chantry, Loddiswell	H S Benney & Sons Trevorlis, Breage	
1963	Maj & Mrs P E Bisgood Chantry, Loddiswell 'Aveton Orpheus'	W E Eddy Buryas Bridge, Penzance 'Chyanhall Sungleams Valonta'	
1964	Lower Coombe Farm Ltd East Allington, Totnes	H S Benney & Sons Trevorlis, Breage	
1965	H S Benney & Sons Trevorlis, Breage 'Trevorlis Dreaming Conqueror'	Lower Coombe Farm Ltd East Allington, Totnes 'Typro's Bramley Lady'	
1966	H S Benney & Sons Penhale-An-Drea, Breage 'Trevorlis Dreaming Conqueror'	Lower Coombe Farm Ltd East Allington, Totnes 'Typro's Bramley Lady'	
1967	H V Benney Penhale-An-Drea, Breage	H V Benney Penhale-An-Drea, Breage	
1968	H V Benney & Son Penhale-An-Drea, Breage	H F Huxtable & Sons Barnstaple	
1969	H V Benney Penhale-An-Drea, Breage 'Trevorlis Dreaming Conqueror'	H V Benney Penhale-An-Drea, Breage 'La Chasse Greek Rosey'	H V Benney Penhale-An-Drea, Breage 'Trevorlis Dreaming Conqueror'
1970	J E Taylor Loddiswell, Kingsbridge 'Chillaton Goliath'	H F Huxtable & Sons Barnstaple 'Sowden Oxford Design VHC'	H F Huxtable & Sons Barnstaple 'Sowden Oxford Design'
1971	Mrs J Ferguson Southampton 'Harfields Count Taurus'	J E Taylor Loddiswell, Kingsbridge 'Chillaton Saphire'	Mrs J Ferguson Southampton 'Harfields Count Taurus'
1972		H F Huxtable & Sons Barnstaple 'Sowden Dazzler's Enid'	H F Huxtable & Sons Barnstaple 'Sowden Dazzlers Enid'
1973		Mr & Mrs J K Morrish Wellington, Somerset	Mr & Mrs J K Morrish Wellington, Somerset

1974	H F Huxtable & Sons Barnstaple 'Sowden Prince'	H F Huxtable & Sons Barnstaple 'Sowden Dazzlers Girl 2nd'	H F Huxtable & Sons Barnstaple 'Sowden Dazzlers Girl 2nd'
1975	W R Richards Carnkie, Redruth 'Mistletoe Glorious Paradine'	Mr & Mrs W E Berryman St Austell 'Teifi Basil's Magic'	Mr & Mrs W E Berryman St Austell 'Teifi Basil's Magic'
1976	Mr & Mrs J E Taylor Loddiswell, Kingsbridge 'Chillaton Dainty Monarch'	Mr & Mrs J E Taylor Loddiswell, Kingsbridge	Mr & Mrs J E Taylor Loddiswell, Kingsbridge
1977	Harfields Estates Southampton 'Galinthia's Valentine Spark'	Mr & Mrs W E Berryman St Austell 'Teifi Basil's Magic'	Mr & Mrs W E Berryman St Austell 'Teifi Basil's Magic'
1978	Mr & Mrs J E Taylor Loddiswell, Kingsbridge 'Chilliton Snow Prince'	Mr & Mrs W E Berryman St Austell 'Teifi Basil's Magic'	Mr & Mrs W E Berryman St Austell 'Teifi Basil's Magic'
1979	J F W & J Colwell Trerice, St Newlyn East 'Quintrell Kenwyn'	Mr & Mrs J E Taylor Loddiswell, Kingsbridge 'Mayfield Dazzling Scottie'	Mr & Mrs J E Taylor Loddiswell, Kingsbridge 'Mayfield Dazzling Scottie'
1980		Mr & Mrs W E Berryman St Austell 'Crinnisbay Serenade'	Mr & Mrs W E Berryman St Austell 'Crinnsibay Serenade'
1981	Mr & Mrs J Colwell Trerice, St Newlyn East 'Windsor Fillpail Beautiful'	Mr & Mrs W E Berryman St Austell	Mr & Mrs W E Berryman St Austell
1982	R & H Daft S Treviddo, Menheniot 'Richan King Arthur'	Mr & Mrs W E Berryman St Austell 'Crinnisbay Serenade'	Mr & Mrs W E Berryman St Austell 'Crinnisbay Serenade'
1983	Mr & Mrs J Colwell Trerice, St Newlyn East 'Quintrell Marvellous Charmer'	H V Benney Trevorlis, Breage 'Trevorlis Harvesters Janet'	H V Benney Trevorlis, Breage 'Trevorlis Harvesters Janet'
1984	Mr & Mrs J Colwell Trerice, St Newlyn East 'Quintrell Marvellous Charmer'	H V Benney Trevorlis, Breage 'Trevorlis Harvesters Janet'	H V Benney Trevorlis, Breage 'Trevorlis Harvesters Janet'
1985	P W & P A Booth Yarner, Dartington 'Allwood Mimosas Reflection'	H V Benney Trevorlis, Breage 'Trevorlis Harvesters Janet'	H V Benney Trevorlis, Breage 'Trevorlis Harvesters Janet'
1986	P W & P A Booth Yarner, Dartington 'Allwood Mimosas Reflection'	Mr & Mrs W E Berryman St Austell 'Allwood Dazzler's Damask'	Mr & Mrs W E Berryman St Austell 'Allwood Dazzler's Damask'
1987	Mr & Mrs H V Benney Trevorlis, Breage 'Windsor Louise's Napoleon'	W E & E Berryman St Austell 'Allwood Dazzler's Damask'	W E & E Berryman St Austell 'Allwood Dazzler's Damask'

1988	P & A Booth Yarner, Dartington 'Allwood Mimosas Reflection'	P & A Booth Yarner, Dartington 'Dartington Dairy Dawn'	P & A Booth Yarner, Dartington 'Dartington Dairy Dawn'
1989	P & A Booth Yarner, Dartington 'Allwood Mimosas Reflection'	P & A Booth Yarner, Dartington 'Dartington Dairy Dawn'	P & A Booth Yarner, Dartington 'Dartington Dairy Dawn'
1990	Mr & Mrs H V Benney Penhale-An-Drea, Breage 'Trevorlis Victorious N'Dume'	W E & E Berryman St Austell 'Allwood Dazzler's Damask'	W E & E Berryman St Austell 'Allwood Dazzler's Damask'
1991	Mr & Mrs H V Benney Penhale-An-Drea, Breage 'Trevorlis Victorious N'Dume'	Mr & Mrs W E Berryman St Austell 'Allwood Dazzler's Damask'	Mr & Mrs W E Berryman St Austell 'Allwood Dazzler's Damask'
1992		W E & E Berryman St Austell 'Allwood Dazzler's Damask'	W E & E Berryman St Austell 'Allwood Dazzler's Damask'

GUERNSEY CATTLE

	BEST GUERNSEY BULL	BEST GUERNSEY COW OR HEIFER	CHAMPION GUERNSEY
	Exhibitor	*Exhibitor*	*Exhibitor*
1875	T D Eva Troon, Camborne		
1877	T D Eva Troon, Camborne		
1878	T D Eva Troon, Camborne		
1879	T D Eva Troon, Camborne		
1880	T D Eva Troon, Camborne		
1881	T D Eva Troon, Camborne		
1883	T D Eva Troon, Camborne		
1884	Nicholas Saunders Bodmin		
1885	Henry Roberts Trevedra, St Buryan		
1886	T D Eva Troon, Camborne		
1887	W T Richards Gweallavellan, Troon, Camborne		
1888	John C E Osborne Trevorian, Sancreed		
1889	John Pearce Tregonebras, Sancreed		

1890	John Pearce Alverton Farm, Penzance	
1891	John Pearce Alverton Farm, Penzance	
1892	Richard Pearce Tregonebras, Sancreed	
1894	John Rowe Newbridge	
1909		E St Aubyn Glynn, Bodmin
1910	T B Bolitho	Rev Canon S R Raffles-Flint Nansawsan, Ladock, Truro
1911	T R Bolitho Trengwainton, Penzance 'Good Friday'	G Blight Tregonning, Breage 'Trewince Fussie'
1912	Paul Q Christopher Gwithian, Hayle	Rev Canon S R Raffles-Flint Nansawsan, Ladock, Truro
1914	W Roach Bejowans, St Buryan 'Calartha King'	Rev Canon S R Raffles-Flint Nansawsan, Ladock, Truro 'Ladock Sweet Briar'
1915	Mrs W J Thomas Tremethick, Penzance 'Trewidden Brilliant'	A W Bailey Hawkins Stagenhoe Park, Welwyn, Herts 'Tempsford Beauty'
1919	G Blight Tregonning, Breage	Mrs R C Bainbridge Plympton, Devon
1921	G Blight Tregonning, Breage 'Tregonning Field Marshall'	
1923	S Christopher Engew, Gwithian, Hayle 'Lynchmere Lord Roberts 13th'	The Rt Hon Viscount Falmouth Tregothnan, Truro 'Tregothnan Mary'
1924	W Roach Trewidden, Penzance	W Roach Trewidden, Penzance
1925	George Blight & Son Tregonning, Breage	T R Bolitho Trengwainton
1926	George Blight & Son Tregonning, Breage	E Jenkins Netherleigh, Hayle
1927	George Blight & Son Tregonning, Breage	S Christopher Engew, Gwithian, Hayle
1928	W Roach Trewidden, Penzance	George Blight & Son Tregonning, Breage
1929	George Blight Tregonning, Breage 'Trestrayle Myrtle Boy'	George Blight Tregonning, Breage
1930	W Roach Trewidden, Penzance	W Roach Trewidden, Penzance
1931	W Roach Trewidden, Penzance 'Halwyn Majesty'	George Blight Tregonning, Breage

1932	E Gerrish Carrallack House, St Just 'Valeries Honour'	George Blight Tregonning, Breage
1933	E Gerrish Carrallack House, St Just 'Valeries Honour'	H Johns Wheal Davey, St Agnes 'Mithian Creamer'
1934	E Gerrish Carrallack House, St Just	J E Hoskin Boscarne, St Buryan 'Boscarne Bantam'
1935	E Gerrish Carrallack House, St Just 'Valeries Honour'	H Johns Wheal Davey, St Agnes
1936	W Penrose Trewavas, Breage 'Foundry Daylight'	J E Hoskin Boscarne, St Buryan
1937	The Lady Seaton Bosahan, Manaccan, Helston	George Blight Tregonning, Breage
1938	W Roach Trewidden, Penzance 'Fernhill Slogan 4th'	George Blight Tregonning, Breage
1939	The Lady Seaton Bosahan, Manaccan, Helston	George Blight Tregonning, Breage
1947	G F dee Shapland Green Farm, Claverham, Bristol	George Blight Tregonning, Breage
1948	H Johns Wheal Davey, St Agnes	P O Eddy Tregoose, Grampound Road
1949	P O Eddy Tregoose, Grampound Road 'Malverleys Roberts Lad 3rd'	G H & Miss E A Johnstone Trewithen, Grampound Road
1950	P O Eddy Tregoose, Grampound Road 'Malverleys Roberts Lad 3rd'	G H & Miss E A Johnstone Trewithen, Grampound Road
1951	H A Y Dyson Steyning, Sussex	Trewithen Pedigree Farms Trewithen, Grampound Road
1952	P O Eddy Probus	Trewithen Pedigree Farms Trewithen, Grampound Road
1953	Mr & Mrs L W B Smith Treliske, St Erme 'Whiteladies Majestic 24th'	R J Oates Trenewjack, Marazion 'Lockinge Duchess 27th'
1954	H A Y Dyson Steyning, Sussex	J W Eddy
1956	S H Chellew Polkinghorne, Gulval 'Malverleys Robert's Lad 13th'	J W Eddy Trewidden, Penzance 'Petwood Phoebe 3rd'
1957	S H Chellew Polkinghorne, Gulval	Capt W H C Daniel Stockleigh, Crediton

1958	S H Chellew Polkinghorne, Gulval	S E Leighton Twyford, Berks	
1959	S E Leighton Twyford, Berks	S E Leighton Twyford, Berks	
1960	R J Oates Trenewjack, Marazion	Mrs M Howell Ivybridge, Devon	
1961	R J Oates Trenewjack, Marazion	Capt W H C Daniel Stockleigh, Crediton	
1962	I Rowe Boscean, St Just 'Glovers Tip Geraniums King 4th'	Mrs M Howell Ivybridge, Devon 'Lukesland Gentian'	
1963	J H Eddy Rosewarrick, Lanivet 'Trevalgan Programmes Robert'	G Blight Jnr Trelowarren, Mawgan 'Buttercup 20th of Tregonning Farm'	
1964	J H Eddy Rosewarrick, Lanivet	Miss B G Jenkin Perranwell Station, Truro	
1965	C P Richards & Son Ltd Splattenridden, Hayle 'Melchbourne Supreme'	A J Oliver Steer Villa, Marazion 'Hillshorn Kildare'	
1966	C P Richards & Son Ltd Splattenridden, Hayle 'Melchbourne Supreme'	Mr & Mrs L W Heard Titson, Marhamchurch 'Cherrytop Graceful'	Mr & Mrs L W Heard Titson, Marhamchurch 'Cherrytop Graceful'
1967	C P Richards & Son Ltd Splattenridden, Hayle 'Melchbourne Supreme'	L Heard Titson, Marhamchurch 'Cherrytop Graceful'	
1968	C Edyvean Breezeleigh, Roche 'Rosewarrick Programmes Prosperous'	Mr & Mrs D Coryn Wheal Boys, Redruth 'Reens Queenie 34th'	
1969	Mr & Mrs W L Hollow Towednack, St Ives 'Irroy 5th of Browning'	H C S Guinness Newbury, Berks 'Brookhill Fame's Grace'	
1970	Mr & Mrs W L Hollow Towednack, St Ives 'Irroy 5th of Browning'	J T Jelbert Carnkie, Helston 'Reens Queenie 55th'	Mr & Mrs W L Hollow Towednack, St Ives 'Irroy 5th of Browning'
1971	Mr & Mrs W L Hollow Towednack, St Ives 'Irroy 5th of Browning'	M O Eddy Treloweth, St Erth 'Treloweth Rachel 35th'	Mr & Mrs W L Hollow Towednack, St Ives 'Irroy 5th of Browning'
1972	H Worth Luxulyan	Mr & Mrs C J Waters Kirthen, Hayle 'Reens Crocus 18th'	Mr & Mrs C J Waters Kirthen, Hayle 'Reens Crocus 18th'
1973	M Eddy Treloweth, St Erth	C J & E Waters Kirthen, Hayle 'Reens Crocus 18th'	C J & E Waters Kirthen, Hayle 'Reens Crocus 18th'
1974	Mr & Mrs W L Hollow St Ives 'Ross of Stokechurch'	H C S Guinness Newbury, Berks 'White Ladies Golden Ruby'	H C S Guinness Newbury, Berks 'White Ladies Golden Ruby'
1975	D W J Hawken Landrake, Saltash 'Chytodden Prix de Montreux 2nd'	H C S Guinness Newbury, Berks 'White Ladies Ruby 3rd'	D W J Hawken Landrake, Saltash 'Chytodden Prix de Montreux 2nd'

| 1976 | | H C S Guinness
Newbury, Berks | D W J Hawken
Landrake, Saltash |
|---|---|---|---|
| 1977 | Mr & Mrs J H Menhennitt
Tredustan, Wadebridge
'Leybrad Supreme 7th' | A B Thomas (C J & E A
Waters)
Hayle
'Kirthen Helen's Bardia' | A B Thomas (C J & E A
Waters)
Hayle
'Kirthen Helen's Bardia) |
| 1978 | Mr & Mrs J H Menhennitt
Tredustan, Wadebridge
'Leybrad Supreme 7th' | Mr & Mrs T K Emm
Dorchester, Dorset
'Fisherton Fuschia' | Mr & Mrs T K Emm
Dorchester, Dorset
'Fisherton Fuchsia' |
| 1979 | | W G Matthews & Son
Penzance
'Javelin's Rosetoo of
Holbatch' | W G Matthews & Son
Penzance
'Javelin's Rosetoo of
Holbatch' |
| 1980 | Mr & Mrs M C Bone
Penzance
'Kelsmor Meadowsweet
Montrose 5th' | H C S Guinness
Newbury, Berks
'White Ladies Gillian
6th' | H C S Guinness
Newbury, Berks
'White Ladies Gillian
6th' |
| 1981 | Mr & Mrs M C Bone
Penzance
'Kelsmore Meadowsweet
Montrose 5th' | H C S Guinness
Newbury, Berks
'White Ladies Ruby
23rd' | H C S Guinness
Newbury, Berks
'White Ladies Ruby
23rd' |
| 1982 | A B Thomas (C J & E A
Waters)
Hayle
'Kirthen Helen's Prince' | A B Thomas (C J & E A
Waters)
Hayle
'Kirthen Crocus Lena' | A B Thomas (C J & E A
Waters)
Hayle
'Kirthen Crocus Lena' |
| 1983 | I J Rowe
Penzance
'White Ladies Ruby's
Maybourne' | Mr & Mrs M Grose
Manaccan, Helston
'Trewarnevas Gillian 2nd' | Mr & Mrs M Grose
Manaccan, Helston
'Trewarnevas Gillian 2nd' |
| 1984 | I J Rowe
Penzance
'White Ladies Ruby's
Maybourne' | Sir Howard Guinness
Newbury, Berks
'White Ladies Design
4th' | Sir Howard Guinness
Newbury, Berks
'White Ladies Design
4th' |
| 1985 | I J Rowe
Penzance
'White Ladies Ruby's
Maybourne' | Mr & Mrs M Grose
Manaccan, Helston
'Trewarnevas Happy
Maid 3rd' | Mr & Mrs M Grose
Manaccan, Helston
'Trewarnevas Happy
Maid 3rd' |
| 1986 | I J Rowe
Penzance
'White Ladies Ruby's
Maybourne' | Mr & Mrs M Grose
Manaccan
'Trewarnevas Primrose
Wallflower 2nd' | I J Rowe
Penzance
'White Ladies Ruby's
Maybourne' |
| 1987 | I J Rowe
Penzance
'White Ladies Ruby's
Maybourne' | A B Thomas (C J & E A
Waters)
Hayle
'Kirthen Janet's Primrose' | A B Thomas (C J & E A
Waters)
Hayle
'Kirthen Janet's Primrose' |
| 1988 | A B Thomas (C J & E A
Waters)
Hayle
'Kirthen Crocus
Governor' | A B Thomas (C J & E A
Waters)
Hayle
'Kirthen Crocus Tassel' | A B Thomas (C J & E A
Waters)
Hayle
'Kirthen Crocus Tassel' |
| 1989 | I J Rowe
Penzance
'Kenilworth Supreme' | Mr & Mrs M C Bone
Madron, Penzance
'Lanyon Meadowsweet
2nd' | Mr & Mrs M C Bone
Madron, Penzance
'Lanyon Meadow Sweet
2nd' |

1990	I J Rowe Penzance 'Kenilworth Supreme'	Lt Col & Mrs H C Watson Wincanton, Somerset 'Shalford Justine 2nd'	Lt Col & Mrs H C Watson Wincanton, Somerset 'Shalford Justine 2nd'
1991		Lt Col & Mrs H C Watson Wincanton, Somerset 'Clifford Marigold 10th'	Lt Col & Mrs H C Watson Wincanton, Somerset 'Clifford Marigold 10th'
1992		Sir Howard Guinness Sherborne, Dorset 'White Ladies Dryad 58th'	Sir Howard Guinness Sherborne, Dorset 'White Ladies Dryad 58th'

SOUTH DEVON CATTLE

	BEST SOUTH DEVON BULL	BEST SOUTH DEVON COW OR HEIFER	CHAMPION SOUTH DEVON
	Exhibitor	*Exhibitor*	*Exhibitor*
1892	John S Ford Jnr Luson, Holbeton, Ivybridge	John S Ford Hall Tors, Yealmpton	
		BEST SOUTH DEVON COW	
1910	H Hawken & Son Kingsbridge	Ben Luscombe Langston, Kingsbridge	
1911	J Leach Carwen, Lanreath 'New Year's Gift'	Ben Luscombe Langston, Kingsbridge 'Fidget 5th'	
1921	B Luscombe Bowden, Yealmton 'Bowden Strawberry Boy'		
1923	Capt J T Coryton Pentillie Castle, St Mellion 'Mothecombe Milkman'		
1924	Capt J T Coryton Pentillie Castle, St Mellion 'Mothecombe Milkman'		
1925	Capt J T Coryton Pentillie Castle, St Mellion		
		BEST SOUTH DEVON COW OR HEIFER	
1926	S Every & Son Tinnell, Landulph, Hatt	Henry Chaffe Harestone, Brixton, Devon	
1927	S Every & Son Tinnell, Landulph, Hatt 'Lixton Councillor 5th'	R W Chaffe Revelstoke, Devon	
1928	S Every & Son Tinnell, Landulph, Hatt 'Lixton Councillor 5th'		
1929			Messrs S Every & Son Tinnell, Landulph
1930			J Every Landulph
1931		Henry Chaffe Brixton, Devon	J Wakeham Rowden, Newton Ferrers 'Cadet'

1932		Henry Chaffe Hareston, Brixton	Henry Chaffe Hareston, Brixton
1933			W Louch Lanlawren, Polperro 'Pamflete New Fashion'
1934			W Louch Lanlawren, Polperro 'Pamflete New Fashion'
1935			G C Maddever Looe Down, Liskeard 'Milkman'
1936			J Wakeman Rowden, Newton Ferrers 'Pamflete New Fashion'
1937			J Wakeham & Son Rowden, Newton Ferrers 'Keynedon Sir William'
1938			J P Cundy & Sons Estover Farms, Crownhill 'Pamflete Buck'
1939			W Pearce Ashcombe Barton, Dawlish
1947		G Wills Haccombe, Newton Abbot	Brig Stephen Williams Scorrier Home Farm, Scorrier 'East Farm No 58'
1948	E V Bunday Rydon, Newton Abbot	E V Bunday Rydon, Newton Abbot	E V Bunday Rydon, Newton Abbot
1949			G Nicholls Tregasso, Manaccan
1950	J Olliver Wenmouth, St Neot 'Trevellas Gay Boy'	J T Dennis & Son Winsor, Yealmpton, Devon	J Olliver Wenmouth, St Neot 'Trevellas Gay Boy'
1951	S F Thomas Lancellan, Mevagissey	J A Irish Edmeston, Modbury, Devon	S F Thomas Lancellan, Mevagissey
1952	J Olliver Wenmouth, St Neot	G Eustice Gwinear	J Olliver Wenmouth, St Neot
1953	J A Irish Modbury, Devon 'Edmeston General 25th'	Cecil E Harvey Yealmpton, Devon 'Yealmpton Melva 2nd'	J A Irish Modbury, Devon 'Edmeston General 25th'
1954	J A Irish Modbury, Devon	J T Dennis & Son	J A Irish Modbury, Devon
1956	J E Sobey Liskeard 'Edmeston Derrick 16th'	J A Irish Modbury, Devon 'Edmeston Hilda 19th'	J E Sobey Liskeard 'Edmeston Derrick 16th'
1957	Sir Charles Hanson Bt Fowey 'Lawhyre Matchless'	J A Irish Modbury, Devon 'Edmeston Hilda 19th'	Sir Charles Hanson Bt Fowey 'Lawhyre Matchless'
1958	Lello Brothers Ltd Hayle 'Edmeston Derrick 16th'	J A Irish Modbury, Devon 'Edmeston Hilda 19th'	Lello Brothers Ltd Hayle 'Edmeston Derrick 16th'

1959	John G Tozer Lifton, Devon 'Dunterton 61st'	J A Irish Modbury, Devon 'Edmeston Hilda 19th'	John G Tozer Lifton, Devon 'Dunterton 61st'
1960	E J Sanders Higher Clicker, Liskeard 'Edmeston Flash 30th'	Cecil Harvey Yealmpton, Devon	E J Sanders Higher Clicker, Liskeard
1961	E J Sanders Higher Clicker, Liskeard 'Edmeston Flash 30th'	G Nicholls Manaccan, Helston	E J Sanders Higher Clicker, Liskeard 'Edmeston Flash 30th'
1962	E J Sanders Higher Clicker, Liskeard 'Edmeston Flash 20th'	Cecil E Harvey Yealmpton, Plymouth 'Yealmpton Marigold'	E J Sanders Higher Clicker, Liskeard 'Edmeston Flash 20th'
1963	E J Sanders Higher Clicker, Liskeard 'Edmeston Flash 30th'	The Rt Hon The Earl of Devon Powderham Castle, Exeter 'Powderham Sapphire 12th'	E J Sanders Higher Clicker, Liskeard 'Edmeston Flash 30th'
1964			J B Wakeham Yealmpton, Devon
1965	J C Warne & Son Tregony, Truro 'Wishworthy Flash 12th'	G Trenouth Trevone, Padstow 'Treworder Pansy 1st'	J C Warne & Son Tregony, Truro 'Wishworthy Flash 12th'
1966	J C Warne & Son Tregony, Truro 'Wishworthy Flash 12th'	R K L Rundle Kestle Mill, Newquay 'Trevean Wendy 13th'	J C Warne & Son Tregony, Truro 'Wishworthy Flash 12th'
1967	E W Camp Modbury, Devon 'Edmeston Magnet 3rd'	R K L Rundle Kestle Mill, Newquay	E W Camp Modbury, Devon 'Edmeston Magnet 3rd'
1968	G N Lancaster Wishworthy, Launceston 'Edmeston Magnet 3rd'	W G Rowe Pollinick, S Petherwin	G Lancaster Wishworthy, Launceston 'Edmeston Magnet 3rd'
1969	J Rossiter Kingsbridge, Devon 'Edmeston Express 1st'	J C Warne & Son Tregony, Truro 'Tregonhayne Gwenie 4th'	J Rossiter Kingsbridge, Devon 'Edmeston Express 1st'
1970	S R Lancaster Tavistock, Devon 'Wishworthy Flash 19th'	F B Thomas Tregerrick, Gorran 'Tregerrick Hyacinth'	S R Lancaster Tavistock, Devon 'Wishworthy Flash 19th'
1971	H Tully Brixham, Devon 'Trewinney 83rd'	H Tully Brixham, Devon 'Waddeton Hilda 7th'	H Tully Brixham, Devon 'Trewinney 83rd'
1972	H Tully & Sons Brixham, Devon		H Tully & Sons Brixham, Devon
1973	H Tully & Sons Brixham, Devon 'Edmeston Romany 8th'		H Tully & Sons Brixham, Devon 'Edmeston Romany 8th'
1974		H Tully Brixham, Devon 'Waddeton Hilda 7th'	H Tully Brixham, Devon 'Waddeton Hilda 7th'
1976	F J Warne & Son Tregony, Truro	Lord Courtney Powderham Castle, Exeter	Lord Courtney Powderham Castle, Exeter
1977	H Tully & Sons Brixham, Devon 'Edmeston Romany 27th'	H Tully & Sons Brixham, Devon 'Waddeton Ann'	H Tully & Sons Brixham, Devon 'Waddeton Ann'

1978	F B & F C Thomas Tregerrick, Gorran 'Tregerrick Tom 2nd'	H Tully & Sons Brixham, Devon 'Waddeton Ann'	H Tully & Sons Brixham, Devon 'Waddeton Ann'
1979	Lord Courtenay Powderham Castle, Exeter 'Powderham Gem 9th'	H Tully & Sons Brixham, Devon 'Waddeton Dolly 13th'	Lord Courtenay Powderham Castle, Exeter 'Powderham Gem 9th'
1980	Lord Courtenay Powderham Castle, Exeter 'Powderham Gem 9th'	F B & F C Thomas Tregerrick, Gorran 'Tregerrick Angela 18th'	Lord Courtenay Powderham Castle, Exeter 'Powderham Gem 9th'
1981	R R B Harvey & A R Lee Ivybridge, Devon 'Sexton Marquis 3rd'	F B & F C Thomas Tregerrick, Gorran	R R B Harvey & A R Lee Ivybridge, Devon 'Sexton Marquis 3rd'
1982	R R B Harvey & A R Lee Ivybridge, Devon 'Sexton Marquis 3rd'	F J Warne & Son Tregony, Truro 'Tregonhayne Model'	R R B Harvey & A R Lee Ivybridge, Devon 'Sexton Marquis 3rd'
1983	F B & F C Thomas St Austell 'Tregerrick Magnet 2nd'	W W Williams & Son Mawgan, Helston 'Roskymer Cowslip 3rd'	F B & F C Thomas St Austell 'Tregerrick Magnet 2nd'
1984	W W Williams & Sons Mawgan, Helston 'Grove Ferdinand'	R J Eustice Newquay 'Trevowah Dorothy 6th'	R J Eustice Newquay 'Trevowah Dorothy 6th'
1985	R R B Harvey Ivybridge, Devon 'Sexton Jester 18th'	R J Eustice & Sons Newquay 'Trevowah Dorothy 6th'	R R B Harvey Ivybridge, Devon 'Sexton Jester 18th'
1986	D Eustice Hayle 'Grove Antony'	D J Thomas Treguddick, Launceston 'Eastcott Polly 10th'	D Eustice Hayle 'Grove Antony'
1987	W W Williams & Sons Mawgan, Helston 'Roskymer Poldark'	W W Williams & Sons Mawgan, Helston 'Roskymer Dimples 23rd'	W W Williams & Sons Mawgan, Helston 'Roskymer Poldark'
1988	D Eustice Hayle 'Grove Marlborough'	J A Ward Devichoys, Perranar- worthal 'Brookhay Cowslip 5th'	D Eustice Hayle 'Grove Marlborough'
1989	D Eustice Hayle 'Grove Marlborough'	H Lello Hayle 'Trevassack Aster 23rd'	D Eustice Hayle 'Grove Marlborough'
1990	D Eustice Hayle 'Grove Marlborough'	H Lello Hayle 'Trevassack Pride 95th'	D Eustice Hayle 'Grove Marlborough'
1991	W W Williams & Sons Helston 'Roskymer Challenger 3rd'	J A Ward Devichoys, Perranar- worthal 'Brookhay Cowslip 7th'	W W Williams & Sons Helston 'Roskymer Challenger 3rd'
1992	H Lello Hayle 'Crokers Favourite'	H Lello Hayle 'Trevassack Pride 95th'	H Lello Hayle 'Crokers Favourite'

BRITISH FRIESIAN/HOLSTEIN FRIESIAN CATTLE

	BEST BRITISH FRIESIAN BULL	BEST BRITISH FRIESIAN COW OR HEIFER	CHAMPION BRITISH FRIESIAN
	Exhibitor	*Exhibitor*	*Exhibitor*
1922	G P Williams Scorrier		
1923	W G Berridge Castlezens, Tregony		
1924	Mrs J Putnam Farringdon House, Exeter		
1925	B Pearce Trelissick, Hayle		
1926	B Pearce Trelissick, Hayle		
1927	B Pearce Trelissick, Hayle	T Mansfield Winterbourne, Glos	B Pearce Trelissick, Hayle 'Iken Pel Beatty 11th'
1928	B Pearce Trelissick, Hayle	E Hosking Jnr Pulsack, Hayle	
1929	E Hosking, Jnr Pulsack, Hayle	E Hosking Jnr Pulsack, Hayle	
1947	T A Barker Perranwell Station	W H May Tregellas, Grampound Road	T A Barker Perranwell Station 'Treverras Conqueror'
1949			R E Pomeroy Treverras, Truro
1950			W H May Tregellas, Grampound 'Ragdale Bertha 12th'
1951			R E Pomeroy Treverras, Truro
1952	Messrs W J Osborne & Sons Tolverne Barton, Philleigh	Messrs W J Osborne & Sons Tolverne Barton, Philleigh	
1953	Messrs W J Osborne & Sons Tolverne Barton,Philleigh	Messrs W J Osborne & Sons Tolverne Barton, Philleigh	
1954	Messrs W J Osborne & Sons Tolverne Barton, Philleigh	Mr & Mrs Rossowiecki Looe	
1956		W B Penrose Trequean, Breage 'Trequean Beatrice 2nd'	
1957	J Viggers & Sons Bere Abbot, Devon 'Collins Adema Romeo'	Maj C Wheaton-Smith Winsham, Chard 'Rurik Minerva'	
1958	Mr & Mrs M I N Strachan Whorridge, Cullompton 'Bowerchalke Pathfinder'	J Simmons & Son Greenlane, Mawla 'Crellow Tareen 2nd'	
1959	H R Graham-Vivian Manaccan, Helston	Maj C Wheaton-Smith Winsham, Chard	
1960	S C Tibbetts Longlands, Kingskerswell	Maj C Wheaton-Smith Winsham, Chard 'Ruik Penelope II'	

1961	S C Tibbetts
Longlands,
Kingskerswell	S J K Hammett
Tregonna,
Little Petherick	
1962	A T Axford
Uppacott, Kilkhampton
'Uppacott Nobleman'	F G Summerhayes
Upottery, Honiton
'Ringwell Night 4th'	
1963	A M Nicholls & Son
Trelan, St Keverne
'Kingburton Omar 5th'	F G Summerhayes
Upottery, Honiton
'Haccombe Allpail 2nd'	
1965	Mark May
Trewethert, Port Isaac
Trewethert Welbred'	G B Smale
South Hellescott, North Petherwin
'Hurle Dauntless Skybird'	
1966	Mr & Mrs F R Cameron
Trewollock, Gorran
'Trewollock Nanadema'	D J Smale
Glebe, North Petherwin
'Ashwater Gerald's Poppy 2nd'	
1967	G Brown
Tresmarrow, Launceston	D J Smale
Glebe, North Petherwin	
1968	C P Richards & Son Ltd
Splattenridden, Hayle
'Mawle Tarsus'	D J Smale
Glebe, North Petherwin	
1969	H W Upham
Langaller, Bovey Tracey
'Langaller Eve's Renown'	D J Smale
Glebe, North Petherwin
'Glebeston Fancymaid 9th'	
1970	H W Upham
Langaller, Bovey Tracey
'Langaller Eve's Renown'	D J Smale
Glebe, North Petherwin
'Walscombe Landcon Rebecca'	D J Smale
Glebe, North Petherwin
'Walscombe Landcon Rebecca'
1971	D W Batten & Co
Treweese, Quethiock
'Grove Patriot'	G B Smale
South Hellescott,
N Petherwin
'Glebewin Violet 7th'	G B Smale
South Hellescott,
N Petherwin
'Glebewin Violet 7th'
1972			W R Pearce
Crediton
1973	C G Horswell
Scrawson, Golberdon
'Trehill Jack'	G B Smale
South Hellescott,
N Petherwin
'Glebewin Herald Snowflake 9th'	G B Smale
South Hellescott,
N Petherwin
'Glebewin Herald Snowflake 9th'
1974	Harvey Brothers
Shepherds, Newlyn East
'Grove Sensation'	G B Smale
South Hellescott,
N Petherwin
'Glebewin Herald Snowflake 9th'	G B Smale
South Hellescott,
N Petherwin
'Glebewin Herald Snowflake 9th'
1975	S Hoskin & Son &
Francis Trewin Farms
Tregilgas, Gorran
'Gradefield Hiawatha'	R Stafford-Smith
Little Dartmouth,
Dartmouth
'Shopland Edleet Ruth 6th'	R Stafford-Smith
Little Dartmouth,
Dartmouth
'Shopland Edleet Ruth 6th'
1976	Mr & Mrs S E Coryn
Sparnock, Kea, Truro	Whitsbury Farm & Stud Ltd
Fordingbridge, Hants	Whitsbury Farm & Stud Ltd
Fordingbridge, Hants

1977	L C & L Fuller & Son Timberscombe, Minehead 'Wilowna Hawk'	J M & R H Harding Compton Abbas, Shaftesbury 'Bowerchalke Tilly 45th'	J M & R H Harding Compton Abbas, Shaftesbury 'Bowerchalke Tilly 45th'
1978	G Fuller, D Martin & R Earley Timberscombe, Minehead 'Wilowna Hawk'	G B Smale South Hellescott, N Petherwin 'Glebewin Baby 6th'	G B Smale South Hellescott, N Petherwin 'Glebewin Baby 6th'
1979	L C & L Fuller & Son, D R Martin & R J Earley Timberscombe, Minehead 'Wilowna Hawk'	R L & J M Harding Ltd Long Crichel, Wimborne 'Crichel Marliene 5th'	R M & J L Harding Ltd Long Crichel, Wimborne 'Crichel Marliene 5th'
1980	Mr & Mrs E Comley Penscawn, Summercourt 'Summercourt Sovereign 2nd'	Harvey Brothers Shepherds, Newlyn East 'Shepherds Queen Madge 4th'	Harvey Brothers Shepherds, Newlyn East 'Shepherds Queen Madge 4th'
1981	Mr & Mrs E Comley Penscawn, Summercourt 'Summercourt Sovereign 2nd'	Winfrith Farms Dorchester 'Crichel Alline 177th'	Winfrith Farms Dorchester 'Crichel Alline 177th'
1982	Mr & Mrs E Comley Penscawn, Summercourt 'Summercourt Sovereign 2nd'	W Gubb & Sons Brayford, Barnstaple 'Glebelands Alice'	W Gubb & Sons Brayford, Barnstaple 'Glebelands Alice'
1983	R H P Redwood (Farms) Ltd Holcombe Rogus, Wellington 'Holcombe Rogus Jubilee Boy'	C N Hart Eastcourt, Malmesbury 'Barton Hoo Nugget Meg'	C N Hart Eastcourt, Malmesbury 'Barton Hoo Nugget Meg'
1984	M J Simmons Greenlane, Mawla 'Mawle Royal Bracken'	F I & M I Harvey Shepherds, Newlyn East 'Shepherds Vaakje'	F I & M I Harvey Shepherds, Newlyn East 'Shepherds Vaakje'
1985	E & G Comley Penscawn, Summercourt 'Moffaview Alabama'	D J & L A Smith Pilehayes, Woodbury Salterton 'Shopland Ambassador Viola'	D J & L A Smith Pilehayes, Woodbury Salterton 'Shopland Ambassador Viola'
1986	G B Smale & Sons S Hellescott, N Petherwin 'Glebewin Babylon'	D J & L A Smith Pilehayes, Woodbury Salterton 'Oakroyal Edleet Ruth ET'	D J & L A Smith Pilehayes, Woodbury Salterton 'Oakroyal Edleet Ruth ET'
1987	D J & L A Smith Pilehayes, Woodbury Salterton 'Oakroyal Gadget'	D J & L A Smith Pilehayes, Woodbury Salterton 'Oakroyal Seabreeze 2nd'	D J & L A Smith Pilehayes, Woodbury Salterton 'Oakroyal Seabreeze 2nd'
1988	F E Thomas & Son Condurrow, Camborne 'Conrow Homeguardsman'	D J & L A Smith Pilehayes, Woodbury 'Silverpark Pamela 2nd RM VG'	D J & L A Smith Pilehayes, Woodbury 'Silverpark Pamela 2nd RM VG'

	BEST HOLSTEIN FRIESIAN BULL	BEST HOLSTEIN FRIESIAN COW OR HEIFER	CHAMPION HOLSTEIN FRIESIAN
1989	G B Smale & Sons S Hellescott, N Petherwin 'Glebewin Sam'	W H Ley Sutcombe, Holsworthy 'Thuborough Annie 3rd'	W H Ley Sutcombe, Holsworthy 'Thuborough Annie 3rd'
1990	G B Smale & Sons S Hellescott, N. Petherwin 'Glebewin Sam'	W H Ley Sutcombe, Holsworthy 'Thuborough Spangle 10th H85'	W H Ley Sutcombe, Holsworthy 'Thuborough Spangle 10th H85'
1991	F E Thomas & Son Condurrow, Camborne 'Conrow Valiant'	W H Ley Sutcombe, Holsworthy 'Thuborough Keepsake 3rd H87'	W H Ley Sutcombe, Holsworthy 'Thuborough Keepsake 3rd H87'
1992	E Comley Penscawn, Summercourt 'Summercourt Trixie Warden'	W H Ley Sutcombe, Holsworthy 'Thuborough Flora 7th'	W H Ley Sutcombe, Holsworthy 'Thuborough Flora 7th'

ABERDEEN ANGUS CATTLE

	BEST ABERDEEN ANGUS BULL	BEST ABERDEEN ANGUS COW OR HEIFER	CHAMPION ABERDEEN ANGUS
	Exhibitor	*Exhibitor*	*Exhibitor*
1927	Rt Hon Viscount Falmouth Tregothnan		F H Turnbull Llantwit Major, Cardiff
1928	Port Eliot Estate St Germans 'Premier of Bellever'		Port Eliot Estate St Germans 'Premier of Bellever'
1929	Rt Hon Viscount Falmouth Tregothnan 'Vain Vicar'		Rt Hon Viscount Falmouth Tregothnan 'Vain Vicar'
1930	Rt Hon Viscount Falmouth Tregothnan		Rt Hon Viscount Falmouth Tregothnan
1931	W B Richards Killivose, Truro		W B Richards Killivose, Truro
1932	W B Richards Killivose, Truro		W B Richards Killivose, Truro
1933			W B Richards Killivose, Truro
1934	W B Richards Killivose, Truro		W B Richards Killivose, Truro
1935	W B Richards Killivose, Truro	T Salmon Fore Street, St Columb 'Pride 19th of Soilerrie'	T Salmon Fore Street, St Columb 'Pride 19th of Soilerrie'
1936	W B Richards Killivose, Truro	W B Richards Killivose, Truro	W B Richards Killivose, Truro
1937			W B Richards Killivose, Truro
1938	Rt Hon Viscount Falmouth Tregothnan	Rt Hon Viscount Falmouth Tregothnan	Rt Hon Viscount Falmouth Tregothnan
1939		Rt Hon Viscount Falmouth Tregothnan	Rt Hon Viscount Falmouth Tregothnan

AYRSHIRE CATTLE

	BEST AYRSHIRE BULL	BEST AYRSHIRE COW OR HEIFER	CHAMPION AYRSHIRE
	Exhibitor	*Exhibitor*	*Exhibitor*
1947	W C Pegley Treviades, Constantine	The Rt Hon The Earl of Portsmouth Farleigh Wallop, Basingstoke	
1948	W C Pegley Treviades, Constantine		
1949	W C Pegley Treviades, Constantine	S J Woolway Fosgrove, Shoreditch, Taunton	
1950	Capt E J Edwards-Heathcote Timberscombe, Minehead	Capt E J Edwards-Heathcote Timberscombe, Minehead	
1951		Lt Col & Mrs F Blake Tokenbury, Pensilva, Liskeard	
1952		S W Curnow Helston	
1953		Capt E J Edwards-Heathcote Timberscombe, Minehead 'Chilworth Rio'	
1954	Lady Coleraine Bocaddon, Lanreath	Capt E J Edwards-Heathcote Timberscombe, Minehead	
1956	W J Treloar Menkea, Ponsanooth 'Upwoods True Form'	W G Bradley Trelyn, St Agnes 'Raglan Honey'	
1957	Lady Coleraine Bocaddon, Lanreath 'Bocaddon Winston'	Sqd Ldr F R Offord MBE Puddington, Tiverton	
1958	W G Bradley Trelyn, St Agnes 'Raglan Full Cry'	J C E Copland Templecombe, Somerset	
1959	Lady Coleraine Bocaddon, Lanreath 'Bocaddon Economist'	T P Williams & Son Manaccan, Helston	
1960	K F Stanbury Downland, Hr Clovelly 'Downland Early Harvest'	Col C T Mitford-Slade DL JP Montys Court, Taunton	
1961	K F Stanbury Downland, Hr Clovelly 'Downland Early Harvest'	T Jewell Caralsa, Zelah	
1962	Mr & Mrs P Young Upton St Leonards, Glos 'Rowde Neptune'	Col C T Mitford-Slade DL JP Montys Court, Taunton 'Fenhill Rosemary 5th'	
1963	T Jewell Caralsa, Zelah 'Caralsa Eversure'	Mr & Mrs J C Jewell Lanteague, Zelah 'Copthall Willow'	
1964		Mr & Mrs J C Jewell Lanteague, Zelah 'Wellhouse Banana 26th'	

1965	M G Gregory Crosspark, Holsworthy 'Willstock Willie'	Mr & Mrs J C Jewell Perranporth Dairies 'Wellhouse Banana 26th'
1966	M G & J M Gregory Crosspark, Holsworthy 'Crosspark Silver Boy'	M G & J M Gregory Farthings Fm, Holsworthy 'Crosspark Emerald'
1967	Mr & Mrs J C Jewell Perranporth Dairies 'Zellamar Sportsmaster'	Mr & Mrs J C Jewell Lanteague, Zelah 'Wellhouse Banana 26th'
1968	Mr & Mrs J C Jewell Perranporth Dairies 'Zellamar Sportsmaster'	
1969	D & P Coryn Whealboys, Redruth 'Iveshead North Shield'	Mr & Mrs J C Jewell Ropers Walk Fm, Mount Hawke 'Zellamar Banana'
1975		J & E M Weaver Lostwithiel 'Tennacott Tonga'
1976		J & E M Weaver Lostwithiel
1977		N D Simpson St Mabyn, Bodmin 'Fleminghill Dolly 32nd'
1978		P M Bickford-Smith Trevarno, Helston 'Trevarno Verbenna'
1979		J Weaver Lostwithiel 'Croftside Molly 25th'
1980		T P Williams & Son Manaccan, Helston 'Tregonwell White Mist'
1981		T P Williams & Son Manaccan, Helston 'Tickle'
1982		T P Williams & Son Manaccan, Helston 'Hartley Tickle 2nd'
1983		T P Williams & Son Manaccan, Helston 'Tregonwell Donnalyn'
1984		T P Williams & Son Manaccan, Helston 'Tregonwell Annabella'
1985		T P Williams & Son Manaccan, Helston 'Tregonwell Annabella'
1986		W R C Christophers Tregony, Truro 'Rosehill Amber VG'
1987		W R C Christophers Tregony, Truro 'Rosehill Amber'

1988	W R C Christophers Probus, Truro 'Rosehill Amber EX'
1989	W R C Christophers Probus, Truro 'Rosehill Jean'
1990	W R C Christophers Probus, Truro 'Rosehill Jean'
1991	D P Williams & Son Manaccan, Helston 'Tregonwell Lady Lockett'
1992	G A & S K Creeper St Clether, Launceston 'Cuthill Towers Zara 3rd VG'

RED POLL CATTLE

	BEST RED POLL BULL	BEST RED POLL COW OR HEIFER	CHAMPION RED POLL
	Exhibitor	*Exhibitor*	*Exhibitor*
1953			G W & A J Sadlow Shepton Mallet, Somerset 'Kirton Catseye'
1954		J S Blewett Trewhella, Goldsithney	
1956	J S Blewett Trewhella, Goldsithney 'Honest Rota'	J S Blewett Trewhella, Goldsithney 'Hailsham Lucy'	
1957	C C Wickett & Son 'Wenhaston Peter 2nd'	E E Trump Broadclyst, Exeter	
1959		I J Hayes Bideford	
1960		Col M H Aird South Pool, Kingsbridge	
1961		Mr & Mrs P Young Cheltenham, Glos	
1962		Mr & Mrs P Young Cheltenham, Glos 'Bromborough Ting'	

BRITISH CHAROLAIS CATTLE

	BEST BRITISH CHAROLAIS BULL	BEST BRITISH CHAROLAIS COW OR HEIFER	CHAMPION BRITISH CHAROLAIS
	Exhibitor	*Exhibitor*	*Exhibitor*
1979	R De Pass Pewsey, Wiltshire 'Tavy Nelson'	R De Pass Pewsey, Wiltshire 'Tavy Immaculate'	R De Pass Pewsey, Wiltshire 'Tavy Immaculate'

1980	Mr & Mrs P J Stephens Lanreath, Looe 'Cullum Prefect'	R De Pass Pewsey, Wiltshire 'Tavy Lilac'	R De Pass Pewsey, Wiltshire 'Tavy Lilac'
1981	R De Pass Pewsey, Wiltshire 'Mount Pleasant Professor'	D H Lightfoot Duloe, Liskeard 'De Crespigny Pommade'	R De Pass Pewsey, Wiltshire 'Mount Pleasant Professor'
1982	Mr & Mrs J J B & Miss A Pascoe Helston 'Whitchurch Simeon'	W T Northmore & Son Yelverton, Devon 'Loveton Noblesse'	Mr & Mrs J J B & Miss A Pascoe Helston 'Whitchurch Simeon'
1983	W T Northmore & Son Yelverton, Devon 'Cross Lanes Taurus'	Mr & Mrs C R White Helston 'Furrydance Sophie'	W T Northmore & Son Yelverton, Devon 'Cross Lanes Taurus'
1984	W T Northmore & Son Yelverton, Devon 'Loveton Tavistock'	Mr & Mrs J J B & Miss A Pascoe Helston 'Pengwedna Twiggy'	W T Northmore & Son Yelverton, Devon 'Loveton Tavistock'
1985	C J Hutchings & Sons Wiveliscombe, Somerset 'Pacha'	P G Throssell St Mabyn, Bodmin 'Hawkslow Olwen'	C J Hutchings & Sons Wiveliscombe, Somerset 'Pacha'
1986	C J Hutchings & Son Wiveliscombe, Somerset 'Rosherne Pacha'	Miss A Pascoe Helston 'Pengwedna Viola'	C J Hutchings & Sons Wiveliscombe, Somerset 'Rosherne Pacha'
1987	W T Northmore & Son & J G Martin Yelverton 'Ragdale Bonus'	Mr & Mrs P J Stephens Looe 'Cockerington Alice'	Mr & Mrs P J Stephens Looe 'Cockerington Alice'
1988	Mr & Mrs P J Stephens Looe 'Woodsaws Conrad'	Miss A Pascoe Helston 'Pengwedna Alice'	Miss A Pascoe Helston 'Pengwedna Alice'
1989	Mrs & Miss A E Pascoe Helston 'Farleycopse Capulet'	G W J Blewett Boscastle 'Gredington Campanula'	Mrs & Miss A E Pascoe Helston 'Farleycopse Capulet'
1990	W T Northmore & Son & J G Martin Yelverton 'Tattenhall Cosmonaut'	G W J Blewett Boscastle 'Cullom Beaujolais'	W T Northmore & Son & J G Martin Yelverton, Devon 'Tattenhall Cosmonaut'
1991	W T Northmore & Son & J G Martin Yelverton 'Tattenhall Cosmonaut'	Mrs F Evans Taunton, Somerset 'Claybory Annabella'	W T Northmore & Son & J G Martin Yelverton 'Tattenhall Cosmonaut'
1992	W B Gubbin & Sons Coads Green, Launceston 'Rawlings Esprit'	Mrs & Miss A E Pascoe Helston 'Pengwedna Alice'	W B Gubbin & Sons Coads Green, Launceston 'Rawlings Esprit'

BRITISH LIMOUSIN CATTLE

	BEST BRITISH LIMOUSIN BULL	BEST BRITISH LIMOUSIN COW OR HEIFER	CHAMPION BRITISH LIMOUSIN
	Exhibitor	*Exhibitor*	*Exhibitor*
1988	F & C Walters Exeter, Devon 'Columbjohn Cavalier'	M F G S Clarke Buckfastleigh, Devon 'Rainette'	M F G S Clarke Buckfastleigh, Devon 'Rainette'

1989	F & C Walters Exeter, Devon 'Columbjohn Drake'	F & C Walters Exeter, Devon 'Columbjohn Amber'	F & C Walters Exeter, Devon 'Columbjohn Amber'
1990	F & C Walters Exeter, Devon 'Columbjohn Elephant'	E A Down & Son Crediton, Devon 'Creedy Trula'	E A Down & Son Crediton, Devon 'Creedy Trula'
1991	F & C Walters Exeter, Devon 'Columbjohn Fantastic'	E W Quick & Sons Crediton, Devon 'Loosebeare Dainty'	E W Quick & Sons Crediton, Devon 'Loosebeare Dainty'
1992	F & C Walters Exeter, Devon 'Columbjohn Drake'	J H Neale & Son Launceston 'Nealford Flojo'	F & C Walters Exeter, Devon 'Columbjohn Drake'

BRITISH SIMMENTAL CATTLE

	BEST BRITISH SIMMENTAL BULL	BEST BRITISH SIMMENTAL COW OR HEIFER	CHAMPION BRITISH SIMMENTAL
	Exhibitor	*Exhibitor*	*Exhibitor*
1989	R H Widdicombe Ashprington, Totnes 'Starling Volunteer'	M J Skinner Winkleigh, Devon 'Cannee Ticky'	M J Skinner Winkleigh, Devon 'Cannee Ticky'
1990	R H Widdicombe Ashprington, Totnes 'Starling Wagner'	Miss B E Goddard Winkleigh, Devon 'Reveley Lotti 8th RM'	Miss B E Goddard Winkleigh, Devon 'Revelex Lotti 8th RM'
1991	P J Johnston Helston 'Lesceave Cliff Accolade'	P B Borlase Hertford, Herts 'Sacombe Patricia'	P B Borlase Hertford, Herts 'Sacombe Patricia'
1992	W C May Priorton, Crediton 'Priorton Brilliant'	B D Uglow Linkinhorne, Callington 'Solway Celeste'	B D Uglow Linkinhorne, Callington 'Solway Celeste'

ANY OTHER PURE BREED CATTLE

CHAMPION ANY OTHER PURE BREED CATTLE

	Exhibitor	*Breed*
1991	E Haste Beaworthy, Devon 'Stoneleigh Gemma'	British Belgian Blue

CHAMPION ANY OTHER PURE BREED CATTLE
OF CONTINENTAL ORIGIN

1992	E Haste Beaworthy, Devon 'Stoneleigh Devonian Dusk'	British Belgian Blue

CHAMPION ANY OTHER PURE BREED CATTLE
OF BRITISH NATIVE ORIGIN

1992	F J Warne Tregony, Truro 'Balmyle Tulip 7th'	Beef Shorthorn

STEERS AND CROSS BRED HEIFERS (Cross Bred Heifer added in 1989)
CHAMPION STEER

	Exhibitor	Breed
1965	W R Cook Bude 'Creathorne Good Boy'	
1966	Maj & The Hon Mrs D C Russell Dorchester, Dorset	
1967	HRH The Prince of Wales Home Farm, Stokeclimsland	Devon
1968	C Jasper Altarnun, Launceston	
1969	Dennis Brothers Okehampton	
1970	C Jasper Altarnun, Launceston 'Tresmeake Favorite'	Hereford X
1971	C Jasper Altarnun, Launceston 'Tresmeake Pride'	Devon
1972	H Tully Brixham, Devon 'Waddeton Don'	
1973	C Jasper Altarnun, Launceston 'Tresmeake Pride'	Devon
1974	H Tully Brixham, Devon 'Waddeton Nevermindhim'	
1975	J J B Pascoe Nancegollan, Helston	
1976	H Tully & Sons Brixham, Devon	
1977	Miss A E Pascoe Helston 'Rocky'	
1978	Miss A E Pascoe Helston 'Jasper'	Charolais X
1979	Miss A Martin Liskeard 'Rupert'	
1980	N J Hawkey St Wenn, Bodmin 'Biscuits'	
1981	N J Hawkey St Wenn, Bodmin 'Joey'	
1982	L E Thomas Liskeard 'Albert'	
1983	R A Bersey & Son Torpoint	

1984	J A Eustice & Sons St Ervan, Wadebridge	
1985	D A J Hodge Tavistock, Devon 'Sam'	Limousin X South Devon
1986	R Thomas Liskeard 'Rambo'	Limousin X
1987	Mrs A M Gynn Liskeard 'Oliver'	Limousin X South Devon
1988	A J Lobb St Austell 'Jim The Lim'	South Devon X Limousin

CHAMPION STEER OR CROSS BRED HEIFER

1989	W A Hutchings Holsworthy, Devon	Belgian Blue X Limousin
1990	R R Stephens Liskeard 'Rocket'	Limousin X
1991	R & C Anstey Quality Livestock Bruton, Somerset 'Just Jimmy'	Limousin X Friesian X Limousin
1992	L E Thomas & Son Liskeard	South Devon X Limousin

MAJOR CHAMPIONSHIP AWARDS

BEST BULL IN YARD

	Exhibitor	Breed
1875	Thomas Julyan Creed	Devon
1877	Mr Hawken St Breward	Hereford 'King Koffee', bred by Wesley Grose, St Kew
1878	Viscount Falmouth Tregothnan	Devon 'Romaney Rye'
1879	Mr Lutley Brockhampton, Worcs	Hereford
1880	Viscount Falmouth Tregothnan	Devon 'Sir Michael'
1881	Viscount Falmouth Tregothnan	Devon 'Sir Michael'
1882	Mr Whitford Trehane, St Erme	Short Horn 'Miriam's Heir'
1883	William Chapman Trewithian, Gerrans	Short Horn 'Earl of Oxford'
1884	William Chapman Trewithian, Gerrans	Short Horn 'Earl of Oxford'
1885	William Perry Alder, Lewdown	Devon 'Benedict'
1886	Alfred C Skinner Pounds Farm, Bishops Lydeard	Devon 'General Gordon'
1893	T F Roskruge Trenethick, Helston	Short Horn 'Robin Hood'

BEST BULL IN THE YARD, 3 YEARS OLD AND ABOVE—The Par Challenge Cup

	Exhibitor	Breed
1893	J H Arkwright Hampton Court, Leominster	Hereford 'Good Cross'
1894	J H Arkwright Hampton Court, Leominster	Hereford 'Good Cross'
1895	Mr Mucklow Holsworthy	Devon
1898	Mr Mucklow Holsworthy	Devon 'Magna Carta'
1900	The Hon E W Portman Hestercombe, Taunton	Devon 'Duke of Pound'
1901	J C Williams Caerhayes Castle, St Austell	Devon
1903	J C Williams Caerhayes Castle, St Austell	Devon 'Muca'

BEST COW OR HEIFER IN YARD

	Exhibitor	Breed
1875	Messrs Hosken & Son Hayle	Short Horn
1877	Messrs Hosken & Son Hayle	Short Horn 'Alexandria'
1878	Mrs Langdon Northmolton	Devon 'Cherry 10th'
1879	Messrs Hosken & Son	Short Horn 'Carnation'
1880	Messrs Hosken & Son Hayle	Short Horn 'Rose of Oxford'
1881	Messrs Hosken & Son Hayle	Short Horn 'Rose of Oxford 4th'
1882	Messrs Hosken & Son Hayle	Short Horn 'Rose of Oxford 4th'
1883	Messrs Hosken & Son Hayle	Short Horn 'Gertrude 5th'
1885	Messrs Hosken & Son Hayle	Short Horn 'Alexandria'
1886	F J Gough Bordesley Hall, Reddich	Hereford 'Mabella'
1893	T F Roskruge Trenethick, Helston	Short Horn 'Ruth 58A'

BEST COW OF ANY BREED, FOR DAIRY PURPOSES—
Silver Cup, value £5, given by C Childs Esq

	Exhibitor	Breed
1863	Messrs Tedder & Pridham Truro	Jersey

BEST THREE MILCH COWS OF ANY BREED—
Special Prize given by the President, John Tremayne Esq

	Exhibitor	Breed
1900	T Manuell Trevoroe, Probus	'Lizzie', 'Ruth 2nd' & 'Ruby'

BEST ANIMAL OF ANY BREED—£10 given by Viscount Falmouth

	Exhibitor	Breed
1870	Messrs Hosken & Son
Hayle	Short Horn Heifer

CATTLE SUPREME CHAMPION—The King George VI Perpetual Challenge Cup

	Exhibitor	Breed
1966	S J Skinner
Winkleigh, Devon	Devon 'Burncoose No Use 5th'
1967	W C May
Crediton, Devon	Devon 'Priorton Showgirl 34th'
1968	G N Lancaster
Wishworthy, Launceston	South Devon 'Edmeston Magnet 3rd'
1969	W C May
Crediton, Devon	Devon 'Priorton Showgirl 56th'
1970	Mr & Mrs Hambro
Minehead, Somerset	Hereford 'Rowington Dante'
1971	Maj G E F North
Tiverton, Devon	Hereford 'Tiverton 1 Diamond'
1972	H Tully
Brixham, Devon	South Devon
1973	H Tully
Brixham, Devon	South Devon 'Edmeston Roman 8th'
1974	H Tully
Brixham, Devon	South Devon 'Waddeton Hilda 7th'
1975	R Stafford-Smith
Dartmouth, Devon	British Friesian 'Shopland Edleet Ruth 6th'
1976	Whitsbury Farm & Stud Ltd
Fordingbridge, Hants	British Friesian
1977	Mr & Mrs W E Berryman
St Austell	Jersey 'Teifi Basil's Magic'
1978	G B Smale
North Petherwin, Launceston	British Friesian 'Glebewin Baby 6th'
1979	R L & J M Harding Ltd
Wimborne, Dorset	British Friesian 'Crichel Favourite 130th'
1980	Rt Hon the Lord Courtenay
Powderham Castle, Exeter	South Devon 'Powderham Gem 9th'
1981	W Pascoe & Son
Helston	Hereford
1982	Mr & Mrs W E Berryman
St Austell	Jersey 'Crinnisbay Serenade'
1983	T P Williams & Son
Manaccan, Helston	Ayrshire 'Tregonwell Donnalyn'
1984	T P Williams & Son
Manaccan, Helston	Ayrshire 'Tregonwell Annabella'
1985	H V Benney
Helston	Jersey 'Trevorlis Harvesters Janet'
1986	W R C Christophers
Tregony, Truro	Ayrshire 'Rosehill Amber VG'
1987	W W Williams & Son
Helston	South Devon 'Roskymer Poldark'

1988	D Eustice Hayle	South Devon 'Grove Marlborough'
1989	Mrs & Miss A E Pascoe Helston	British Charolais 'Farleycopse Capulet'
1990	D Eustice Hayle	South Devon 'Grove Marlborough'
1991	W H Ley Holsworthy, Devon	Holstein Friesian 'Thuborough Keepsake 3rd H87'
1992	W B Gubbin & Sons Coads Green, Launceston	British Charolais 'Rawlings Esprit'

CATTLE — BEST EXHIBIT, OWNED & BRED BY EXHIBITOR—
The William Kivell Perpetual Memorial Trophy

	Exhibitor	*Breed*
1978	G B Smale North Petherwin	British Friesian 'Glebewin Baby 6th'
1979	R L & J M Harding Ltd Wimborne, Dorset	British Friesian 'Crichel Favourite 130th'
1980	Rt Hon the Lord Courtenay Powderham Castle, Exeter	South Devon 'Powderham Gem 9th'
1981	Mr & Mrs N D Simpson St Mabyn, Bodmin	British Friesian 'Kyles Fairy'
1982	Mr & Mrs W E Berryman St Austell	Jersey 'Crinnisbay Serenade'
1983	T P Williams & Son Manaccan, Helston	Ayrshire 'Tregonwell Donnalyn'
1984	T P Williams & Son Manaccan, Helston	Ayrshire 'Tregonwell Annabella'
1985	H V Benney Helston	Jersey 'Trevorlis Harvesters Janet'
1986	G Dart & Sons South Molton, Devon	Devon 'Champson Bribery 18th'
1987	W W Williams & Son Helston	South Devon 'Roskymer Poldark'
1988	P & A Booth Dartington, Devon	Jersey 'Dartington Dairy Dawn'
1989	Mrs & Miss A E Pascoe Helston	British Charolais 'Pengwedna Alice'
1990	W R C Christophers Probus, Truro	Ayrshire 'Rosehill Jean'
1991	W H Ley Holsworthy, Devon	Holstein Friesian 'Thuborough Keepsake 3rd H87'
1992	C Thornton Creeton, Grantham, Lincs	Devon 'Thorndale Baron'

BEST BULL OF A RECOGNISED DAIRY BREED —
From 1926 The Cornwall Milk Recording Society Perpetual Challenge Cup

	Exhibitor	*Breed*
1924	Mrs J Putnam Farringdon House, Exeter	
1925	B Pearce Trelissick, Hayle	

1926	B Pearce, Trelissick, Hayle	British Friesian
1927	B Pearce, Trelissick, Hayle	British Friesian 'Iken Pel Beatty II'
1928	B Pearce, Trelissick, Hayle	British Friesian 'Iken Pel Beatty II'
1929	Capt F B Imbert-Terry MC, Broadclyst, Exeter	Jersey 'Blye Hayes Piper'
1931	W Roach, Trewidden, Penzance	Guernsey
1932	W Roach, Trewidden, Penzance	Guernsey
1933	W J Kitto, Tregeagle, Probus	Guernsey
1934	E Gerrish, Carrallack House, St Just	Guernsey
1935	J E Hoskin, Boscarne, St Buryan	Guernsey
1936	J E Hoskin, Boscarne, St Buryan	Guernsey
1937	E Gerrish, Carrallack House, St Just	Guernsey
1938	G H Johnstone, Trewithen, Grampound Road	Guernsey
1939	The Lady Seaton, Bosahan, Manaccan	Guernsey
1947	W H May, Tregellas, Grampound Road	British Friesian
1948	Antron Model Farm, Mabe, Penryn	Guernsey
1949	J V Ratcliffe, Argal, Falmouth	Jersey
1950	S W Curnow, Pollard Farm, Helston	Ayrshire
1951	S W Curnow, Pollard Farm, Helston	Ayrshire
1953	Mr & Mrs L W B Smith, Treliske, St Erme	Guernsey 'Whiteladies Majestic 24th'
1954	R J Cardell, St Keverne	
1956	Trewithen Pedigree Farms, Trewithen, Grampound Road	Guernsey 'Trewithen Renown'
1962	J Simmons & Son, Mawla, Redruth	British Friesian 'Whitsbury Brenman'
1963	J Simmons & Son, Mawla, Redruth	British Friesian 'Whitsbury Brenman'
1965	Mr & Mrs F R Cameron, Trewollock, Gorran	British Friesian 'Trewollock Nanadema'
1966	Mr & Mrs F R Cameron, Trewollock, Gorran	British Friesian 'Trewollock Regal Prince'

1967	G Brown, Tresmarrow, Launceston	British Friesian 'Tresmarrow Sea Rover'
1968	C J Trewin & Son, Ruan Minor, Helston	British Friesian
1969	C J Trewin & Son, Ruan Minor, Helston	British Friesian 'Blackadon Donald'
1970	C J Trewin & Son, Ruan Minor, Helston	British Friesian 'Blackadon Donald'
1971	D W Batten & Co, Treweese, Quethiock	British Friesian 'Grove Patriot'
1972	Francis Trewin Farms, Ruan Minor, Helston	British Friesian 'Blackadon Donald'
1973	H H Rowe, Godcott, North Petherwin	British Friesian
1974	H H Rowe, Godcott, North Petherwin	British Friesian 'Romax Pirate'
1975	S Hoskin & Son & Francis Trewin Farms, Tregilgas, Gorran	British Friesian 'Gradefield Hiawatha'
1976	S Hoskin & Son & J C Trewin & Son, Tregilgas, Gorran	British Friesian
1977	S Hoskin & Son, Tregilgas, Gorran	British Friesian 'Gradefield Hiawatha'
1978	H H Rowe, Godcott, North Petherwin	British Friesian 'Romax Julian 2nd'
1979	E & G Comley, Penscawn, Summercourt	British Friesian 'Ullswater Royal Standard'
1980	H H Rowe, Godcott, North Petherwin	British Friesian
1981	H H Rowe, Godcott, North Petherwin	British Friesian
1982	H H Rowe, Godcott, North Petherwin	British Friesian 'Romax Brigadoon'
1983	J C P Hawken, Poltreworgey, Port Isaac	British Friesian 'Poltreworgey Monopoly'
1984	J C P Hawken, Poltreworgey, Port Isaac	British Friesian 'Poltreworgey Monopoly'
1985	F I & M I Harvey, Shepherds, Newlyn East	British Friesian 'Hunday Baronet'
1986	F I & M I Harvey, Shepherds, Newlyn East	British Friesian 'Hunday Baronet ET'
1987	F E Thomas & Son, Condurrow, Camborne	British Friesian 'Conrow Homeguardsman'
1988	Richard Brothers, Tregaminion, St Keverne	British Friesian 'Pralla Sir Gallahad'
1989	J C P Hawken, Poltreworgey, Port Isaac	British Friesian 'Poltreworgey Bert'
1990	J C P Hawken, Poltreworgey, Port Isaac	Holstein Friesian 'Poltreworgey Bert'
1991	F E Thomas & Son, Condurrow, Camborne	Holstein Friesian 'Conrow Valiant'
1992	E Comley, Penscawn, Summercourt	Holstein Friesian 'Summercourt Trixie Warden'

BEST DAIRY COW OF ANY BREED

	Exhibitor	Breed
1907	J H Lesbirel Liskeard	
1926	E Jenkins Netherleigh, Hayle	
1939	The Hon Lady Cook Talland Bay, Looe	Jersey
1947	Trewithen Pedigree Farms Trewithen, Grampound Road	Guernsey
1948	Mrs K G Truscott Treveor, Tresillian	Jersey
1949	A M Bennett Hr Kehelland, Camborne	British Friesian
1950	J H Gee Pentire, Wadebridge	Jersey
1951	W J Osborne & Sons Tolverne Barton, Philleigh	British Friesian
1953	Trewithen Pedigree Herds Trewithen, Grampound Road	Guernsey
1954	Mr & Mrs Rossowieki Looe	British Friesian
1956	G Blight Tregonning, Breage	Guernsey 'Tregonning Buttercup 92nd'
1962	W B Penrose Trequean, Breage	British Friesian 'Trequean Beatrice 2nd'
1963	S J K Hammett Tregonna, Little Petherick	British Friesian 'Tregonna Breeze 2nd'
1965	Mr & Mrs F R Cameron Trewollock, Gorran	British Friesian 'Hamdon Mavis 12th'
1966	Mr & Mrs F R Cameron Trewollock, Gorran	British Friesian 'Hamdon Mavis 12th'
1967	H T Rowe & Sons Godcott, North Petherwin	British Friesian
1968	C J Trewin & Son Ruan Minor, Helston	British Friesian
1969	M May Trewethart, Port Isaac	British Friesian 'Bristock Dauntless Twinstar'
1970	H T Rowe & Sons Godcott, North Petherwin	British Friesian 'Collacombe Tippett 2nd'
1971	C J Trewin & Son Ruan Minor, Helston	British Friesian 'Summerhayes Stella 2nd'
1972	Francis Trewin Farms Ruan Minor, Helston	British Friesian
1973	Francis Trewin Farms Ruan Minor, Helston	British Friesian
1974	Francis Trewin Farms Trethvas, The Lizard	British Friesian 'Summerhayes Stella 2nd'
1975	H H Rowe Godcott, North Petherwin	British Friesian 'Edenford Mona'

1976	M H Drover & E J Pool
Trevillick, Tintagel	Ayrshire
1977	A B Thomas (C J & E A Waters)
Townshend, Hayle	Guernsey 'Kirthen Helen's Bardia'
1978	G B Smale
S Hellescott, N Petherwin	British Friesian 'Glebein Baby 6th'
1979	J C P Hawken
Poltreworgey, Port Isaac	British Friesian 'Dalesend Vievers Moll'
1980	Mr & Mrs E Comley
Penscawn, Summercourt	British Friesian 'Warkton Fairy 108th'
1981	J C P Hawken
Poltreworgey, Port Isaac	British Friesian 'Poltreworgey Stiletto 9th'
1982	M J Simmons
Mawla, Redruth	British Friesian 'Hungerford Bracket 38th'
1983	M J Simmons
Mawla, Redruth	British Friesian 'Hungerford Bracket 38th'
1984	M J Simmons
Mawla, Redruth	British Friesian 'Hungerford Bracket 38th'
1985	J C P Hawken
Poltreworgey, Port Isaac	British Friesian 'Poltreworgey Irene 5th'
1986	W R C Christophers
Tregony, Truro	Ayrshire 'Rosehill Amber'
1987	E & G Comley
Penscawn, Summercourt	British Friesian 'Pencorse Mandy 2nd'
1988	J C P Hawken
Poltreworgey, Port Isaac	British Friesian 'Poltreworgey Britannia EX'
1989	J C P Hawken
Poltreworgey, Port Isaac	Holstein Friesian 'Poltreworgey Lana 12th EX'
1990	F I & M I Harvey
Shepherds, Newlyn East	Holstein Friesian 'Shepherds Annie 19th RM VG'
1991	D P Williams & Son
Manaccan, Helston	Ayrshire 'Hartley Princess 4th'
1992	D P Williams & Son
Manaccan, Helston	Ayrshire 'Hartley Princess 4th EX Elite'

BEST COW OF A RECOGNISED DAIRY BREED, BRED BY OWNER/EXHIBITOR
— The Western Morning News Perpetual Challenge Cup

	Exhibitor	*Breed*
1926	Henry Chaffe
Harestone, Brixton	South Devon
1927	S Semmens
Trevarthian, Marazion	Guernsey
1928	Messrs G Blight & Son
Tregonning, Breage	Guernsey
1929	G Blight
Tregonning, Breage	Guernsey
1930	Mrs Hayes-Sadler
Seale Hayne Agri College	Jersey
1931	Henry Chaffe
Harestone, Brixton	South Devon
1932	G Blight
Tregonning, Breage	Guernsey

1933	H Johns Wheal Davey, St Agnes	Guernsey
1934	J E Hoskin Boscarne, St Buryan	Guernsey
1935	J Rossiter Cholwells, Kingsbridge	South Devon
1936	J Rossiter & Son Cholwells, Kingsbridge	South Devon
1937	J Hendy Alston, Holbeton	South Devon
1938	G Blight Tregonning, Breage	Guernsey
1939	G Blight Tregonning, Breage	Guernsey
1947	E W Fowler Stowey, Timberscombe, Minehead	Ayrshire
1948	P O Eddy Tregoose, Grampound Road	Guernsey
1949	G H & Miss E A Johnstone Trewithen, Grampound Road	Guernsey
1950	A V Flexman Royal Farm, Kenwyn, Truro	Short Horn
1951	Trewithen Pedigree Farms Trewithen, Grampound Road	Guernsey
1952	G Eustice Bezurrell, Gwinear	South Devon
1953	H A Y Dyson Steyning, Sussex	Guernsey 'Bramble 3rd of Payhay'
1954	H A Y Dyson Steyning, Sussex	Guernsey
1956	W G Bradley Trelyn, St Agnes	Ayrshire 'Raglan Honey'
1957	Maj C Wheaton-Smith Winsham, Chard	British Friesian
1958	H S Benney & Sons Trevorlis, Breage	Jersey
1959	Mr & Mrs J A P Martin Colleton Manor, Chulmleigh, Devon	Jersey
1960	Maj C Wheaton-Smith Winsham, Chard	British Friesian
1961	W B Penrose Trequean, Breage	British Friesian
1962	H S Benney & Sons Trevorlis, Breage	Jersey 'Trevorlis Jesters Muriel'
1963	S J K Hammett Tregonna, Little Petherick	British Friesian 'Hurle Zeph Daffodil'
1964	T Jewell Zelah	Ayrshire
1965	M G Gregory Farthings, Crosspark, Holsworthy	Ayrshire 'Crosspark Doreen'

1966	S J K Hammett Tregonna, Little Petherick	British Friesian 'Tregonna Breeze 2nd'
1967	C G Newman Woolminstone Farm, Crewkerne, Somerset	British Friesian
1968	M G & J M Gregory Farthings, Crosspark, Holsworthy	Ayrshire
1969	Mr & Mrs J C Jewell Ropers Walk Farm, Mount Hawke, Truro	Ayrshire 'Zellamar Banana'
1970	G B Smale South Hellescott, North Petherwin	British Friesian 'Glebewin Violet 7th'
1971	G B Smale Ropers Walk Farm, Mount Hawke, Truro	British Friesian 'Glebewin Violet 7th'
1972	Mr & Mrs J C Jewell Ropers Walk Farm, Mount Hawke, Truro	Ayrshire 'Zellamar Banana'
1973	J K Morrish Wellington, Somerset	Jersey
1974	G B Smale South Hellescott, North Petherwin	British Friesian 'Glebewin Herald Snowflake 9th'
1975	G B Smale South Hellescott, North Petherwin	British Friesian 'Glebewin Herald Snowflake 9th'
1976	H C S Guinness Newbury, Berks	Guernsey
1977	J M & R H Harding Compton Abbas, Shaftesbury, Dorset	British Friesian
1978	G B Smale South Hellescott, North Petherwin	British Friesian 'Glebewin Baby 6th'
1979	Harvey Brothers Shepherds, Newlyn East	British Friesian 'Shepherds Queen Madge 4th'
1980	Mr & Mrs W E Berryman London Apprentice, St Austell	Jersey 'Crinnisbay Serenade'
1981	Mr & Mrs N D Simpson Trethick, St Mabyn	British Friesian
1982	Mr & Mrs W E Berryman London Apprentice, St Austell	Jersey 'Crinnisbay Serenade'
1983	T P Williams & Son Manaccan, Helston	Ayrshire 'Tregonwell Donnalyn'
1984	T P Williams & Son Manaccan, Helston	Ayrshire 'Tregonwell Annabella'
1985	H V Benney Trevorlis, Breage	Jersey 'Trevorlis Harvesters Janet'
1986	T P Williams & Sons Manaccan, Helston	Ayrshire 'Tregonwell Annabella'

1987	E & G Comley Penscawn, Summercourt	British Friesian 'Summercourt Gay Trixie'
1988	D J & L A Smith Pilehayes, Woodbury Salterton, Exeter	British Friesian 'Oakroyal Susette 2 H88'
1989	W R C Christophers Trevorva, Probus	Ayrshire 'Rosehill Jean'
1990	W R C Christophers Trevorva, Probus	Ayrshire 'Rosehill Jean'
1991	W H Ley Thuborough, Sutcombe, Holsworthy	Holstein Friesian 'Thuborough Keepsake 3 H87'
1992	W R C & J E Christophers Trevorva, Probus	Ayrshire 'Rosehill Babycham'

BEST BEEF RECORDED BULL

	Exhibitor	Name of Animal	Breed
1990	W C May Crediton, Devon	'Priorton Wacko'	British Simmental
1991	G W J Blewett Boscastle	'Reddivallen France'	British Charolais
1992	W C May Crediton, Devon	'Priorton Brilliant'	British Simmental

BEST COW OF A RECOGNISED BEEF BREED, BRED BY THE OWNER/EXHIBITOR

	Exhibitor	Breed
1949	W R Cook Creathorne, Bude	Devon
1950	W R Cook Creathorne, Bude	Devon
1951	C Brent Clampit, Callington	Short Horn
1953	W B Nancekivell Pinkhill, Okehampton	Devon 'Pinkhill Cherry 5th'
1954	Mr & Mrs R L Leach	South Devon
1956	P M Williams Burncoose, Redruth	Devon 'Burncoose Narcissus 5th'
1957	J A Irish Edmeston, Modbury, S Devon	South Devon
1958	W B Nancekivell Pinkhill, Okehampton	Devon
1959	W B Nancekivell Pinkhill, Okehampton	Devon
1960	W B Nancekivell Pinkhill, Okehampton	Devon
1961	Cecil Brent & Son Clampit, Callington	Devon
1962	Cecil Brent & Son Clampit, Callington	Devon 'Clampit Flirt 29th'
1963	Cecil Brent & Son Clampit, Callington	Devon 'Clampit Flirt 29th'
1964	C E Harvey Higher Luxon, Yealmpton, Devon	South Devon

1965	W C May & Son, Priorton, Crediton	Devon 'Priorton Showgirl'
1966	W C May & Son, Priorton, Crediton	Devon 'Priorton Showgirl'
1967	W C May & Son, Priorton, Crediton	Devon
1968	W C May & Son, Priorton, Crediton	Devon
1969	W C May, Priorton, Crediton	Devon 'Priorton Showgirl 56th'
1970	W C May, Priorton, Crediton	Devon 'Priorton Showgirl 56th'
1971	H Tully, Brixham, Devon	South Devon 'Waddeton Hilda 7th'
1972	J Meinl, Warminster, Wilts	Hereford
1973	H Tully, Brixham, Devon	South Devon 'Waddeton Hilda 7th'
1974	H Tully, Brixham, Devon	South Devon 'Waddeton Hilda 7th'
1975	W C May, Priorton, Crediton	Devon 'Priorton Snowie 2nd'
1976	W C May, Priorton, Crediton	Devon
1977	H Tully & Sons, Brixham, Devon	South Devon 'Waddeton Ann'
1978	H Tully & Sons, Brixham, Devon	South Devon 'Waddeton Ann'
1979	R De Pass, Pewsey, Wilts	British Charolais 'Tavy Immaculate'
1980	F B & F C Thomas, Tregerrick, Gorran	South Devon 'Tregerrick Angela 18th'
1981	R De Pass, Pewsey, Wilts	British Charolais
1982	R J K Rundle & Sons, Kestle Mill, Newquay	South Devon 'Trevean Edna 5th'
1983	W W Williams & Sons, Mawgan, Helston	South Devon 'Roskymer Cowslip 3rd'
1984	P K James, Langworthy, Okehampton	Devon 'Langworthy Madeleine 4th'
1985	R J Eustice & Son, Trevowah, Crantock	South Devon 'Trevowah Dorothy 6th'
1986	G Dart & Sons, South Molton, Devon	Devon 'Champson Bribery 18th'
1987	G Dart & Sons, South Molton, Devon	Devon 'Champson Bribery 18th'
1988	Miss A Pascoe, Nancegollan, Helston	British Charolais 'Pengwedna Viola'
1989	Mrs & Miss A E Pascoe, Nancegollan, Helston	British Charolais 'Pengwedna Alice'
1990	Mrs & Miss A E Pascoe, Nancegollan, Helston	British Charolais 'Pengwedna Alice'
1991	Mrs & Miss A E Pascoe, Nancegollan, Helston	British Charolais 'Pengwedna Alice'
1992	Mrs & Miss A E Pascoe, Nancegollan, Helston	British Charolais 'Pengwedna Alice'

PRESIDENT'S TROPHY—
EXHIBITOR GAINING MOST POINTS IN THE CATTLE SECTION

Year	Exhibitor	Year	Exhibitor
1961	Mr & Mrs P Young, Cheltenham	1977	H Tully & Son, Brixham, Devon
1962	C E Harvey, Yealmpton, Devon	1978	Mr & Mrs J E Taylor, Kingsbridge, Devon
1963	James Bros, Grampound Road, Truro	1979	W C May, Crediton, Devon
1965	James Bros, Grampound Road, Truro	1980	Mr & Mrs W E Berryman, St Austell
1966	E C Roose, St Teath, Bodmin	1981	R R B Harvey, Ivybridge, Devon
1967	H V Benney, Breage, Helston	1982	T P Williams & Son, Helston
1968	H V Benney, Breage, Helston	1983	Winfrith Farms, Dorchester, Dorset
1969	E C Roose, St Teath, Bodmin	1984	W Pascoe & Son, Helston
1970	E C Roose, St Teath, Bodmin	1985	T P Williams & Son, Helston
1971	J E Taylor, Kingsbridge, Devon	1986	D J & L A Smith, Exeter, Devon
1972	J C Jewell, Ropers Walk, Mount Hawke, Truro	1987	T P Williams & Son, Helston
1973	H Tully, Brixham, Devon	1988	W R C Christophers, Helston
1974	H Tuly, Brixham, Devon	1989	W R C Christophers, Helston
1975	R R B Harvey, Ivybridge, Devon	1990	W R C Christophers, Helson
1976	D Standerwick, Dunley, Bovey Tracey, Devon	1991	W H Ley, Holsworthy, Devon
		1992	W R C & J E Christophers, Probus, Truro

SHEEP SECTION RESULTS — 1794 to 1992

	BEST RAM	BEST RAM, yeaned in County of Cornwall	BEST HOG RAM, yeaned in County of Cornwall	BEST RAM, property of a farmer of this county getting his livelihood solely by farming/rack renting
	Exhibitor	*Exhibitor*	*Exhibitor*	*Exhibitor*
1794	John Slyman St Mabyn			
1795	James Drew Probus			
1796	Anthony Hawkey St Issey			
1797	John P Peters Philleigh			
1799	John Roberts Newlyn	John Roberts Newlyn		
1802	John P Peters Philleigh	John Gummow Probus		
1803	John P Peters Philleigh	John P Peters Philleigh	Rev Robert Walker St Winnow	
1804	John P Peters Philleigh	Rev Robert Walker St Winnow		
1805	S Symons and J Roberts Lower St Columb	John Cardell	J Roberts	
1805	Francis Enys	Mr Sickler Gwinear	Nathaniel Roberts Manaccan	
1806	John P Peters Philleigh	John P Peters Philleigh	John Roberts Newlyn	Michael Mill St Columb
1806	S Plomer Manaccan	J Plomer Roscrow, Gluvias	Joseph Jacka Wendron	
1807	John P Peters Philleigh	John P Peters Philleigh	John Blight Truro	George Vercoe Newlyn
1807	John Shearm Kilkhampton	Mrs Judith Adams Moorwinstow	Mr Trood Poughill	
1808	S Symons Newlyn	John P Peters Philleigh	S Symons Newlyn	Mr R Cardell St Columb Minor
1808	Thomas Grylls Helston	William Plomer St Martins		
1809	J Roberts Newlyn	D Whitley Tregoney	John Cardell Lower St Columb	
1810	Mark Symons Newlyn	Lord Falmouth Tregothnan	John P Peters Philleigh	John Whitford Newlyn
1810	Alex Paul Jnr Camborne		W. Symons Newland	Joseph Hendy Gunwalloe
1811	Thomas Trood Poughill	John Shearm Kilkhampton	Henry Adams Morwinstow	Mrs Judith Adams Moorwinstow
1812	John Roberts Newlyn	Richard Lawer Newlyn	George Johns Cornelly	William Rowe Newlyn
1812	William Searle Newlyn		John Cardell St Columb	John Glasson Breage
1813	William Tremayne Newlyn	John P Peters Philleigh	Daniel Whitley Tregony	George Johns Cornelly

270

1812	William Searle
Newlyn		John Cardell
St Columb	John Glasson
Breage
1813	William Tremayne
Newlyn	John P Peters
Philleigh	Daniel Whitley
Tregony	George Johns
Cornelly
1814	John Turner
Cadbury	John Cardell
Lower St Columb	John Williams
Ruan-Lanyhorn	Thomas A Lee
Boconnoc
1814	James Busustow
Cury	John Plomer
Mylor	Richard Lawer
Newlyn	
1815	John Cardell
Lower St Columb	William Searle
Newlyn	John Roberts
Newlyn	Richard Lawer
Newlyn
1815	John Shearm
Kilkhampton		Thomas Trood
Morwinstow	
1816	John Plomer, Milor
& James Davis,
Probus		John Plomer
Milor	
1817	Mr Norway,
Wadebridge
& Mr Reynolds,
Thorverton	John Roberts
Newlyn	Philip Hawke
Newlyn	George Varcoe
Newlyn
1817	George Gater
Cadbury		Robert Hawke
Newlyn	
1818	John Cardell Jnr
Lower St Columb	John Buller
Morval	Robert Clemow
St Enoder	John Roberts
Newlyn
1820	George Gater
Cadbury	John Cardell
Lower St Columb	P Hawke
Newlyn	J Roberts
Newlyn
1821	John P Peters
Philleigh	Robert Hawke
Newlyn	John Symons
Newlyn	Joseph Cardell
Nanswhyden,
St Columb
1822	John P Peters
Philleigh	Robert Clemow
St Enoder	Stephen Darke
St Columb Minor	Samuel Hawke
St Eval
1825	John P Peters
Philleigh	Mr S Darke
St Columb	Mr R Clemow
St Enoder	Mr W Rowe
Newlyn
	BEST BRED RAM	BEST BRED HOG
RAM	BEST BRED 10
EWES, that have
reared their lambs
this season	BEST BRED 10
EWE HOGS
	Exhibitor	*Exhibitor*	*Exhibitor*	*Exhibitor*
1828	Mr Rowe
Newlyn	Sir Christopher
Hawkins
Trewithen	Sir Christopher
Hawkins
Trewithen	Sir Christopher
Hawkins
Trewithen
1829	A Dingle
Philleigh	William Rowe
Newlyn	W Cardell
Probus	W Cardell
Probus
1830	H Hicks
St Just	R Varcoe
Newlyn	Matthew Doble
Probus	M Doble
Probus
1831	J P Peters
Philleigh	A Dingle
Philleigh	John Hawkins
Probus	John Hawkins
Probus
1832	G Bullmore
Newlyn	G Bullmore
Newlyn	J Kendall
Probus	N Ball
St Stephens
	BEST BRED RAM,
to be worked in the
County of Cornwall
for 12 months
subsequent to the
meeting			
1833	G Bullmore
Newlyn	G Bullmore
Newlyn	J Hawkins
Trewithen	W Cardell
Probus

	BEST BRED RAM, to be worked in the County of Cornwall until the end of the year	BEST BRED HOG RAM, to be worked in the County of Cornwall until the end of the year		
	Exhibitor	*Exhibitor*		
1834	R Clemow St Enoder	W Tremaine Newlyn	M Doble Probus	George Bullmore Newlyn
1835	G Bullmore St Enoder	J Richards Probus	M Doble Probus	M Doble Probus
1836	William Hodge Perranzabuloe	G Bullmore Newlyn	G Bullmore Newlyn	M Doble Probus
1837	George Bullmore Newlyn	George Bullmore Newlyn	M Doble Probus	M Doble Probus
1838	George Bullmore Newlyn	William Hodge Perranzabuloe	George Bullmore Newlyn	William Tremain Newlyn

			BEST BRED 5 EWES, that have reared their lambs this season	BEST BRED 5 EWE HOGS
			Exhibitor	*Exhibitor*
1839	W Tremayne Newlyn	J H Tremayne Heligan	G Bullmore Newlyn	J Hawkins Probus
1840	George G Bullmore Newlyn	M A Doble Probus	G G Bullmore Newlyn	G G Bullmore Newlyn
1841	J H Tremayne Heligan	W Hodge Lambourne	G G Bullmore Newlyn	G G Bullmore Newlyn
1842	J Cardell Jnr Lower St Columb	J H Tremayne Heligan	M A Doble Probus	G G Bullmore Newlyn
1843	T Glanville St Columb	R Cock Craven	M A Doble Probus	Mr Tremaine Newlyn
1844	M A Doble Probus	M A Doble Probus	M A Doble Probus	W Tremaine Newlyn
1845	W Tremaine Newlyn	M A Doble Probus	M A Doble Probus	C H T Hawkins Trewithen
1846	Lake Blee St Enoder	J Tremain Trerice	M A Doble Probus	C H T Hawkins Trewithen
1847	Mrs Oliver Tolverne	John Tremain Newlyn		C H T Hawkins Trewithen
1848	Peter Davis	John Tremain Newlyn	C H T Hawkins Trewithen	C H T Hawkins Trewithen
1849	F Stocker Gorran	M A Doble Probus	M A Doble Probus	M A Doble Probus
1850	James Tremain Trevarton, Newlyn	J Tremain Newlyn	C H T Hawkins Trewithen	C H T Hawkins Trewithen
1851	William Clark St Ewe	James Tremain Newlyn		C H T Hawkins Trewithen

1852	William Drew Creed	James Tremain Trevarton, Newlyn	C H T Hawkins Trewithen	C H T Hawkins Trewithen
1853	James Tremain Trevarton, Newlyn	James Tremain Trevarton, Newlyn	C H T Hawkins Trewithen	C H T Hawkins Trewithen
1854	James Tremain Trevarton, Newlyn	James Tremain Trevarton Newlyn	C H T Hawkins Trewithen	C H T Hawkins Trewithen
	BEST RAM	BEST HOG RAM	BEST 5 OLD EWES	
1855	James Tremain Newlyn	James Tremain Newlyn	C H T Hawkins Trewithen	C H T Hawkins Trewithen
1856	James Tremain Newlyn	James Tremain Newlyn	C H T Hawkins Trewithen	C H T Hawkins Trewithen
1857	James Tremain Newlyn	James Tremain Newlyn	C H T Hawkins Trewithen	C H T Hawkins Trewithen
	BEST FAT SHEEP	BEST FAT WETHER SHEEP under 2 years of age	BEST FAT WETHER SHEEP under 3 years of age	BEST LOT OF HOG EWES not less than five
	Exhibitor	*Exhibitor*	*Exhibitor*	*Exhibitor*
1797	Mr Monk			
1799	Samuel King Egloshayle			
1802	John Roberts Newlyn		Rev John Molesworth	
1803	Thomas Trood Poughill	John P Peters Philleigh		
1804		Rev J Molesworth	John P Peters Philleigh	
1805		Rev J Molesworth	Rev C P Brune Padstow	
1805	John Roberts			
1806		Rev J Molesworth & William Norway	Rev John Molesworth	
1806			Mr Sickler Gwinear	
1807		William Norway Egloshayle	Rev John Molesworth	
1807			John Shearm Kilkhampton	
1810		Rev John Kempe St Mabyn	William Norway Egloshayle	
1812				William Kendall Padstow
1813		William Norway Egloshayle	William Norway Egloshayle	William Norway Egloshayle
1814		Francis Hearle Rodd Trebartha		
1815				Rev J Kempe St Mabyn
1820		J Cardell Lower St Columb	R Andrew St Mabyn	

	BEST 10 FAT SHEEP (Wethers or Ewes)	BEST 10 WETHER HOGS
	Exhibitor	*Exhibitor*
1828	Sir Christopher Hawkins Trewithen	Sir Christopher Hawkins Trewithen
1829	Matthew Doble Probus	Matthew Doble Probus
1830	J Kendall Probus	W Cardell Probus
1831	G Simmons St Erme	J Kendall Probus
1832	P A Cleeve Probus	John Hawkins Probus
1833	E Hoblyn Philleigh	T Symons Probus
1834	G Huddy Probus	J Hawkins Trewithen
1835	W Tremain Newlyn	J Hawkins Probus
1836		R Doble Philleigh
1837		John Hawkins Trewithen
1838		John Hawkins Trewithen

	BEST 5 WETHER HOGS
1839	A Doble Probus
1840	T Croggon Creed
1841	J Hawkins Trewithen
1842	M A Doble Probus
1843	C H T Hawkins Trewithen
1844	M A Doble Probus
1845	C H T Hawkins Trewithen
1846	C H T Hawkins Trewithen
1847	C H T Hawkins Trewithen
1848	M A Doble Probus
1849	M A Doble Probus

1850	C H T Hawkins Trewithen	
1851	C H T Hawkins Trewithen	
1852	M A Doble Probus	
1953	C H T Hawkins Trewithen	
1854	C H T Hawkins Trewithen	
1855	Richard Doble Probus	
1856	C H T Hawkins Trewithen	
1857	C H T Hawkins Trewithen	

SHEEP—PREMIUMS OFFERED FOR STOCK BELONGING TO TENANTS OCCUPYING FARMS NOT EXCEEDING £150 PER YEAR, GETTING THEIR LIVELIHOOD ENTIRELY BY FARMING

	BEST BRED 10 EWES, that have reared their lambs this season	BEST BRED 10 EWE HOGS	BEST 10 WETHER HOGS
	Exhibitor	*Exhibitor*	*Exhibitor*
1836	P A Grieve Jnr Probus	P A Grieve Jnr Probus	
1837	Mr Grieve Probus	Mr Grieve Probus	Mr Grieve Probus
1838	Mr Hotten Probus	Robert Trethewey Ladock	Robert Trethewey Ladock
	BEST BRED 5 EWES, that have reared their lambs this season	BEST BRED 5 EWE HOGS	BEST 5 WETHER HOGS
1839	M Hotton Probus	M Hotton Probus	Robert Trethewey Ladock
1840	M Hotton Probus	M Hotton Probus	R D Bone Jnr Ladock
1841	M Hotton Probus	M Hotton Probus	Robert Trethewey Ladock
1842	M Hotton Probus	Mrs E James Probus	M Hotton Probus
1843	R Trethewey Ladock	R Trethewey Ladock	M Martin Newlyn
1844	Edward Kendall Probus	Edward Kendall Probus	Edward Kendall Probus
1845	Edward Kendall Probus	Mark Martyn Newlyn	Edward Kendall Probus
1846	Edward Kendall Probus	Edward Kendall Probus	J Blackford St Just
1847	Edward Kendall Probus	P A Grieve Probus	Edward Kendall Probus

1848	Edward Kendall Probus	Edward Kendall Probus	Messrs W & J Wallis
1849	R H Earle Probus	Edward Kendall Probus	Edward Kendall Probus
1850	M Hotton Probus	Edward Kendall Probus	J Plummer Kenwyn
1851	Mrs T Grieve Probus	Edward Kendall Probus	J Plummer Kenwyn
1852	Edward Kendall Probus	John Plummer Kenwyn	John Plummer Kenwyn
1853	John Plummer Kenwyn	John Plummer Kenwyn	John Plummer Kenwyn
1854	Mrs T Grieve Probus	John Plummer Kenwyn	John Plummer Kenwyn

BEST 5 OLD EWES, that have reared their lambs this season

1855	John Plummer Kenwyn	John Plummer Kenwyn	John Plummer Kenwyn
1856	John Plummer Kenwyn		John Plummer Kenwyn

COTSWOLD SHEEP

BEST COTSWOLD RAM

Exhibitor

- 1858 Richard Davey MP
Redruth
- 1859 William Cossetine
St Veep
- 1862 John Speare
Stokeclimsland
- 1871 Mr Gatley

SOUTH DOWN SHEEP

BEST SOUTH DOWN RAM

Exhibitor

- 1858 Mr Armitage
Sithney
- 1859 Rt Hon Viscount Falmouth
Tregothnan
- 1860 Richard F Bolitho
Ponsandane, Penzance

DOWNS SHEEP

	BEST DOWNS SHEEP —MALE	BEST DOWNS SHEEP —FEMALE	CHAMPION DOWNS SHEEP
	Exhibitor	*Exhibitor*	*Exhibitor*
1900		W H Treweeke Chipping Norton	
1901		G Blight Tregonning, Breage	
1902		J Joyce Milverton	
1903	Edwin Jacka Helston	Edwin Jacka Helston	
1904		Viscount Falmouth Tregothnan	
1905	Edwin Jacka Helston	Viscount Falmouth Tregothnan	
1906	Edwin Jacka Helston	C R Rosewarne Gwinear	
1907		Viscount Falmouth Tregothnan	
1908		Viscount Falmouth Tregothnan	
1909		Viscount Falmouth Tregothnan	
1910		Viscount Falmouth Tregothnan	
1911		Viscount Falmouth Tregothnan	
1912		Viscount Falmouth Tregothnan	
1914		Viscount Falmouth Tregothnan	
1915		Viscount Falmouth Tregothnan	
1919		George Blight Breage	
1920	George Blight Breage	George Blight Breage	
1922	George Blight Breage	George Blight Breage	
1923		E Jacka Helston	
1924	E Jacka Helston	G Blight Breage	
1925		G Blight Breage	
1926		E Jacka Helston	

1927		Maj & Mrs Jervoise Basingstoke, Hants	G Blight & Son Tregonning, Breage
1928		S Jacka Helston	
1929		S Jacka Helston	
1930	S Jacka Helston	S Jacka Helston	
1931	S Jacka Helston	S Jacka Helston	
1932		S Tripp Philleigh	
1933		S Jacka Helston	
1934		G C Tory Blandford, Dorset	
1935	S Jacka Helston	G Tripp Philleigh	
1936		R E Pomeroy St Just Lane, Truro	
1937		R E Pomeroy St Just Lane, Truro	
1938	S Jacka Helston	S Jacka Helston	
1939	S Jacka Helston	S Jacka Helston	
1947	G Blight Breage		
1948	G Blight Breage		
1949	G Blight Tregonning, Breage		G Blight Tregonning, Breage
1950	W R Yeo Barnstaple, Devon		
1954	R W M Hocken Chard, Somerset		R W M Hocken Chard, Somerset
1957	E E Bluett Poundstock, Bude		

SOUTH DEVON SHEEP

	BEST SOUTH DEVON MALE	BEST SOUTH DEVON FEMALE	CHAMPION SOUTH DEVON SHEEP
	Exhibitor	*Exhibitor*	*Exhibitor*
1924	J R Hallett Brixton, Plymouth	W C Bice St Columb	
1925	J R Hallett Brixton, Plymouth		

1926	W Hawke Besoughan, Colan	
1927	W Hawke Besoughan, Colan	W Hawke Besoughan, Colan
1928	H Whitley Paignton, Devon	H Whitley Paignton, Devon
1929	W C Bice & Son Nanswhyden, St Columb	
1930	E A Stidston & Son Kingsbridge, Devon	E A Stidston & Son Kingsbridge, Devon
1931	W J Short Bosent, Liskeard	W J Short Bosent, Liskeard
1932		J N Grose Pennare, Gorran
1933	T R Brown Tremadart, Duloe	T R Brown Tremadart, Duloe
1939	A C Lanyon Coswarth, St Columb	A C Lanyon Coswarth, St Columb
1947		W H Tom Coswarth, St Columb
1948	W F R Bice Nanswhyden, St Columb	W F R Bice Nanswhyden, St Columb
1949	W J Osborne Philleigh, Truro	W F R Bice Nanswhyden, St Columb
1950	F O Woodley Trenance, Newlyn East	W F R Bice Nanswhyden, St Columb
1951	W F R Bice Nanswhyden, St Columb	F O Woodley Trenance, Newlyn East
1952	F J Rowe St Ewan	W F R Bice Nanswhyden, St Columb
1953	W F R Bice Nanswhyden, St Columb	P T Guy Penvose, Tregony
1954	W F R Bice Nanswhyden, St Columb	P T Guy Penvose, Tregony
1956	H H Mildren Trewall, Torpoint	F O Woodley Trenance, Newlyn East
1957	P T Guy Penvose, Tregony	J H Cumming & Son Bovey Tracey, Devon
1958	W Hawke & Son Colan, Newquay	F O Woodley Trenance, Newlyn East
1959	W Hawke & Son Colan, Newquay	R W Darke & Son Kingsbridge, Devon
1960	W Hawke & Son Colan, Newquay	F O Woodley Trenance, Newlyn East
1961	J F Julyan Penhallow, Ruanhighlanes	F O Woodley Trenance, Newlyn East
1962	A G Daniel Trescowe, Bodmin	J F Julyan Penhallow, Ruanhighlanes
1963	M & R Williams Presingoll, St Agnes	J F Julyan Penhallow, Ruanhighlanes

1964	F O Woodley Trenance, Newlyn East	J F Julyan Penhallow, Ruanhighlanes
1965	W Hawke & Son Besoughan, Colan	J F Julyan Penhallow, Ruanhighlanes
1966	W Hawke & Son Besoughan, Colan	J F Julyan Penhallow, Ruanhighlanes
1967	W Hawke & Son Besoughan, Colan	W Hawke & Son Besoughan, Colan
1968	W Hawke & Son Besoughan, Colan	W Hawke & Son Besoughan, Colan
1969	F O Woodley Trenance, Newlyn East	J F Julyan & Son Penhallow, Ruanhighlanes
1970	W Hawke & Son Besoughan, Colan	F R Cameron Trewollock, Gorran
1971	W Hawke & Son Besoughan, Colan	F R Cameron Trewollock, Gorran
1972	W T R Hawke & Son Trebudannon, Newquay	F O Woodley Newlyn East
1973	J F Julyan & Sons Penhallow, Ruanhighlanes	J F Julyan & Sons Penhallow, Ruanhighlanes
1974	J F Julyan & Sons Penhallow, Ruanhighlanes	Presingoll Farms Ltd St Agnes
1975	W T R Hawke & Son Tregudannon, Newquay	P J Woodley Trenance, Newlyn East

DARTMOOR SHEEP

	BEST DARTMOOR MALE	BEST DARTMOOR FEMALE	CHAMPION DARTMOOR
	Exhibitor	*Exhibitor*	*Exhibitor*
1926	R P Luce Tavistock, Devon		
1927	R P Luce Tavistock, Devon	H J Rich Lewdown, Devon	H J Rich Lewdown, Devon
1928	R P Luce Tavistock, Devon	H J Rich Lewdown, Devon	
1929	R P Luce Tavistock, Devon	R P Luce Tavistock, Devon	
1930		J M Cole & Sons Tavistock, Devon	
1931	J M Cole & Sons Tavistock, Devon	J M Cole & Sons Tavistock, Devon	
1932	R P Luce Tavistock, Devon	R P Luce Tavistock, Devon	
1933	J M Cole Tavistock		
1934	R P Luce Tavistock, Devon		

1937		J H Cole Tavistock, Devon	
1938	J H Cole Tavistock, Devon		
1939	J H Cole Tavistock, Devon	R P Luce Tavistock, Devon	J H Cole Tavistock, Devon
1948	J H Cole Tavistock, Devon	J H Cole Tavistock, Devon	J H Cole Tavistock, Devon
1949	R G Rogers Ivybridge, Devon	J H Cole Tavistock, Devon	
1950	J N Colton & Sons Buckfastleigh, Devon	J H Cole Tavistock, Devon	
1952		J H Cole Tavistock, Devon	
1953	R R Dawe Tavistock, Devon	J H Cole Tavistock, Devon	J H Cole Tavistock, Devon
1954			A R Ellis Slapton
1956	W W Dawe & Son Langford, Tavistock	M J Bowden Manaton, Newton Abbot	W W Dawe & Son Langford, Tavistock, Devon
1957	W J Sprague Chagford, Devon	J H Cole Tavistock, Devon	W J Sprague Chagford, Devon
1958	W J Sprague Chagford, Devon		W J Sprague Chagford, Devon
1959	J H Cole Tavistock, Devon		J H Cole Tavistock, Devon
1960	J H Cole Tavistock, Devon	J H Cole Tavistock, Devon	J H Cole Tavistock, Devon
1962	W J Sprague Chagford, Devon	C T Every CBE Chagford, Devon	W J Sprague Chagford, Devon
1963	J H Cole Tavistock, Devon	J H Cole Tavistock, Devon	J H Cole Tavistock, Devon
1964			W W Dawe & Son Tavistock, Devon
1965	W J Sprague Chagford, Devon	W J Sprague Chagford, Devon	W J Sprague Chagford, Devon
1966	J H Cole Tavistock, Devon	Miss O Luce Tavistock, Devon	J H Cole Tavistock, Devon
1967			J H Cole Tavistock, Devon
1968			W J Sprague Chagford, Devon
1969	J H Cole Tavistock, Devon	W J Sprague Chagford, Devon	J H Cole Tavistock, Devon
1971	W J Sprague Chagford, Devon	R T Alford & Son Tavistock, Devon	W J Sprague Chagford, Devon

DEVON LONG-WOOLLED SHEEP

	BEST DEVON LONG-WOOLLED MALE *Exhibitor*	BEST DEVON LONG-WOOLED FEMALE *Exhibitor*	CHAMPION DEVON LONG-WOOLED *Exhibitor*
1926	F White Williton, Somerset	F White Williton, Somerset	
1927	F White Torweston, Williton	F White Torweston, Williton	F White Torweston, Williton
1928	F White Williton, Somerset	F White Williton, Somerset	
1929	F White Williton, Somerset	F White Williton, Somerset	
1930	F White Williton, Somerset	F White Williton, Somerset	
1931	F White Williton, Somerset	F White Williton, Somerset	
1934	F White Williton, Somerset	F White Williton	
1935	M H Watts & Son Knowle, Okehampton	M H Watts & Son Knowle, Okehampton	
1937	R Lawrence Cullompton, Devon	R Lawrence Cullompton, Devon	
1938	R Lawrence Cullompton, Devon		
1939	M H Watts & Son Knowle, Okehampton	W M Snell Cullompton, Devon	M H Watts & Son Knowle, Okehampton
1948	W M Snell & Son Kentisbere, Devon		
1949	W M Snell & Sons Orway, Cullompton		W M Snell & Sons Orway, Cullompton
1950	J H Strout Pengelly, Linkinhorne		J H Strout Pengelly, Linkinhorne
1951			M H Watts & Son Knowle, Okehampton
1952			M H Watts & Son Knowle, Okehampton
1953	H J Medland Holsworthy, Devon	M H Watts Knowle, Okehampton	M H Watts Knowle, Okehampton, Devon
1954	J H Strout Pengelly, Linkinhorne		J H Strout Pengelly, Linkinhorne
1956	H J Medland Holsworthy, Devon		H J Medland Holsworthy, Devon
1958	A J Moore Washaway, Bodmin	K W Watts Okehampton, Devon	A J Moore Washaway, Bodmin
1959		K W Watts Okehampton, Devon	K W Watts Okehampton, Devon
1960			W S Lucas & Son Marhamchurch, Bude
1961	K W Watts Okehampton, Devon	K W Watts Okehampton, Devon	

1962	W R Cook Bude		W R Cook Bude
1963	A J Moore Holsworthy, Devon	K W Watts Okehampton, Devon	K W Watts Okehampton, Devon
1964			T W Strout Boyton, Launceston
1965		K W Watts Okehampton, Devon	K W Watts Okehampton, Devon
1966	W S Lucas & Son Bude	K W Watts Okehampton, Devon	W S Lucas & Son Bude
1967			K W Watts Knowle, Okehampton
1968			W S Lucas & Son Greendale, Marhamchurch
1969	W J Johns Alverdiscott, Bideford	K W Watts Okehampton, Devon	K W Watts Okehampton, Devon
1970		K W Watts Okehampton, Devon	K W Watts Okehampton, Devon
1971	K W Watts Okehampton, Devon		K W Watts Okehampton, Devon
1972		K W Watts Okehampton, Devon	K W Watts Okehampton, Devon
1973	Drake Bros North Tawton, Devon		Drake Bros North Tawton, Devon
1974	W J Johns Alverdiscott, Bideford	K W Watts Okehampton, Devon	K W Watts Okehampton, Devon
1975	Drake Bros North Tawton, Devon	Drake Bros North Tawton, Devon	Drake Bros North Tawton, Devon
1976			Drake Bros North Tawton, Devon

HAMPSHIRE & DORSET DOWN SHEEP

	BEST HAMPSHIRE & DORSET DOWN RAM	BEST HAMPSHIRE & DORSET DOWN FEMALE	CHAMPION HAMP- SHIRE & DORSET DOWN
	Exhibitor	*Exhibitor*	*Exhibitor*
1961	C Burrough Taunton, Somerset	C Burrough Taunton, Somerset	
1962		C Burrough Taunton, Somerset	
1963		W R Yeo & Sons Barnstaple, Devon	
1969	A T Guard Torrington, Devon	A T Guard Torrington, Devon	
1970	A T Guard Torrington, Devon	A T Guard Torrington, Devon	
1971	A T Guard Torrington, Devon		
1974	A T Guard Torrington, Devon	A T Guard Torrington, Devon	

1975	G W Winter Bideford, Devon	A T Guard Torrington, Devon	
1977	A T Guard Torrington, Devon	A T Guard Torrington, Devon	
1978	A T Guard Torrington, Devon		A T Guard Torrington, Devon
1979		A T Guard Torrington, Devon	D C & V G Kellow Delabole
1980	A T Guard Torrington, Devon	A T Guard Torrington, Devon	A T Guard Torrington, Devon
1981	Mrs M J Whitlock Torrington, Devon		Mrs M J Whitlock Torrington, Devon
1982	Mrs M J Whitlock Torrington, Devon		Mrs M J Whitlock Torrington, Devon
1983	A T Guard Torrington, Devon		A T Guard Torrington, Devon
1984	A T & I E Guard Torrington, Devon		A T & I E Guard Torrington, Devon
1985	D C & V G Kellow Delabole	D C & V G Kellow Delabole	D C & V G Kellow Delabole
1986	A T Guard Torrington, Devon	A T Guard Torrington, Devon	A T Guard Torrington, Devon
1987	A T Guard Torrington, Devon	D C & V G Kellow Delabole	D C & V G Kellow Delabole
1988	A T Guard Torrington, Devon	A T Guard Torrington, Devon	A T Guard Torrington, Devon

SUFFOLK or ANY OTHER DOWN BREED SHEEP

	BEST SUFFOLK or ANY OTHER DOWN BREED RAM	BEST SUFFOLK or ANY OTHER DOWN BREED FEMALE	
	Exhibitor	*Exhibitor*	
1962	G H Tripp Portscatho, Truro	G H Tripp Portscatho, Truro	
1963	G H Tripp Portscatho, Truro	G H Tripp Portscatho, Truro	
1965	G H Tripp Portscatho, Truro	G H Tripp Portscatho, Truro	
1966	J P W Julyan Tregony, Truro		
1970	R E Pomery St Just Lane, Truro	R E Pomery St Just Lane, Truro	
1971	R E Pomery St Just Lane, Truro		

DEVON CLOSEWOOL SHEEP

	BEST DEVON CLOSE- WOOL MALE	BEST DEVON CLOSE- WOOL FEMALE	CHAMPION DEVON CLOSEWOOL
	Exhibitor	*Exhibitor*	*Exhibitor*
1963	A L Lerwill North Molton, Devon	A L Lerwill North Molton, Devon	A L Lerwill North Molton, Devon

1964			A L Lerwill North Molton, Devon
1965	A L Lerwill North Molton, Devon	J I Reddaway & Son Okehampton, Devon	A L Lerwill North Molton, Devon
1966	J I Reddaway & Son Okehampton, Devon	A L Lerwill North Molton, Devon	J I Reddaway & Son Okehampton, Devon
1967			A L Lerwill North Molton, Devon
1968			Rt Hon Lord Clinton Huish, Okehampton
1969	A & E Bulled Barnstaple, Devon		A & E Bulled Barnstaple, Devon
1970	A & E Bulled Barnstaple, Devon	L W Kent-Smith & Son Barnstaple, Devon	L W Kent-Smith & Son Barnstaple, Devon
1971	L W Kent-Smith & Son Barnstaple, Devon	L W Kent-Smith & Son Barnstaple, Devon	L W Kent-Smith & Son Barnstaple, Devon
1972			L W Kent-Smith Barnstaple, Devon

SCOTCH BLACK-FACED SHEEP

	BEST SCOTCH BLACK-FACED RAM *Exhibitor*	BEST SCOTCH BLACK-FACED EWE *Exhibitor*	CHAMPION SCOTCH BLACK-FACED *Exhibitor*
1966	C W Abel Tavistock, Devon	Leslie R Huggins Okehampton, Devon	C W Abel Tavistock, Devon
1967	Leslie R Huggins Lydford, Okehampton	J Rowe Chagford, Devon	L R Huggins Lydford, Okehampton
1968	C W Abel Tavistock, Devon	J Jordan Chagford, Devon	C W Abel Tavistock, Devon
1969	C W Abel Tavistock, Devon	W J Jordan Chagford, Devon	W J Jordan Chagford, Devon
1970	J Jordan Chagford, Devon	J Jordan Chagford, Devon	J Jordan Chagford, Devon
1971	J Jordan Newton Abbot, Devon	J Jordan Newton Abbot, Devon	J Jordan Newton Abbot, Devon
1972			C W Abel Tavistock, Devon
1974	C W Abel Tavistock, Devon	J Rowe Chagford, Devon	C W Abel Tavistock, Devon
1975	J Rowe Chagford, Devon	J Rowe Chagford, Devon	J Rowe Chagford, Devon
1976			P Cornelius St Clether, Launceston
1977	J A T Hodge Okehampton, Devon	J A T Hodge Okehampton, Devon	J A T Hodge Okehampton, Devon
1978	J A T Hodge Okehampton, Devon	J A T Hodge Okehampton, Devon	J A T Hodge Okehampton, Devon
1979	P H Cornelius Launceston	J Jordan Chagford, Devon	P H Cornelius Launceston
1980	J A T Hodge Okehampton, Devon	J A T Hodge Okehampton, Devon	J A T Hodge Okehampton, Devon

1981	I G Mortimer Chagford, Devon	I G Mortimer Chagford, Devon	I G Mortimer Chagford, Devon
1982	I G Mortimer Chagford, Devon	I G Mortimer Chagford, Devon	I G Mortimer Chagford, Devon
1983	J Jordan Chagford, Devon	I G Mortimer Chagford, Devon	I G Mortimer Chagford, Devon
1984	P H Cornelius Camelford	J Jordan Chagford, Devon	P H Cornelius Camelford
1985	P H Cornelius Camelford	I G Mortimer Chagford, Devon	P H Cornelius Camelford
1986	I G Mortimer Chagford, Devon	I G Mortimer Chagford, Devon	I G Mortimer Chagford, Devon
1987	P H Cornelius Camelford	I G Mortimer Chagford, Devon	P H Cornelius Camelford
1988	I G Mortimer Chagford, Devon	P H Cornelius Camelford	I G Mortimer Chagford, Devon
1989	I G Mortimer Chagford, Devon	I G Mortimer Chagford, Devon	I G Mortimer Chagford, Devon
1990	W J Jordan Chagford, Devon	I G Mortimer Chagford, Devon	W J Jordan Chagford, Devon
1991	W J Jordan Chagford, Devon	I G Mortimer Chagford, Devon	W J Jordan Chagford, Devon
1992	W J Jordan Chagford, Devon	W J Jordan Chagford, Devon	W J Jordan Chagford, Devon

DORSET HORN & POLL DORSET SHEEP

	BEST DORSET HORN & POLL DORSET RAM	BEST DORSET HORN & POLL DORSET EWE	CHAMPION DORSET HORN & POLL DORSET
	Exhibitor	*Exhibitor*	*Exhibitor*
1973	P Kingdon & Sons Newlyn East, Newquay	O G Masters & Son Delabole	
1974	P Kingdon & Sons Newlyn East, Newquay		
1975	P Kingdon & Sons Newlyn East, Newquay		P Kingdon & Sons Newlyn East, Newquay
1976			P Kingdon & Sons Newlyn East, Newquay
1977	O G Masters & Son Delabole		P Kingdon & Sons Newlyn East, Newquay
1978	P Kingdon & Sons Newlyn East, Newquay		P Kingdon & Sons Newlyn East, Newquay
1979	P Kingdon & Sons Newlyn East, Newquay		P Kingdon & Sons Newlyn East, Newquay
1980	P Kingdon & Sons Newlyn East, Newquay	P Kingdon & Sons Newlyn East, Newquay	P Kingdon & Sons Newlyn East, Newquay

1981	P Kingdon & Sons Newlyn East, Newquay	P Kingdon & Sons Newlyn East, Newquay	P Kingdon & Sons Newlyn East, Newquay
1982	P Kingdon & Sons Newlyn East, Newquay	P Kingdon & Sons Newlyn East, Newquay	P Kingdon & Sons Newlyn East, Newquay
1983	P Kingdon & Sons Newlyn East, Newquay	P Kingdon & Sons Newlyn East, Newquay	P Kingdon & Sons Newlyn East, Newquay
1984	Rowe & Co Farms Ltd Tregony, Truro	W L & E Sandercock Delabole	W L & E Sandercock Delabole
1985	P Kingdon & Sons Newlyn East, Newquay	P Kingdon & Sons Newlyn East, Newquay	P Kingdon & Sons Newlyn East, Newquay
1986	P Kingdon & Sons Newlyn East, Newquay	P Kingdon & Sons Newlyn East, Newquay	P Kingdon & Sons Newlyn East, Newquay
1987	P Kingdon & Sons Newlyn East, Newquay	R J Ward & Son Grampound Road, Truro	P Kingdon & Sons Newlyn East, Newquay
1988	P Kingdon & Sons Newlyn East, Newquay	P Kingdon & Sons Newlyn East, Newquay	P Kingdon & Sons Newlyn East, Newquay
1989	P Kingdon & Sons Newlyn East, Newquay	P Kingdon & Sons Newlyn East, Newquay	P Kingdon & Sons Newlyn East, Newquay
1990	P Kingdon & Sons Newlyn East, Newquay	P Kingdon & Sons Newlyn East, Newquay	P Kingdon & Sons Newlyn East, Newquay
1991	P Kingdon & Sons Newlyn East, Newquay	P Kingdon & Sons Newlyn East, Newquay	P Kingdon & Sons Newlyn East, Newquay
1992	R J Ward & Son Besowsa, Grampound Road	P Kingdon & Sons Newlyn East, Newquay	R J Ward & Son Besowsa, Grampound

SUFFOLK SHEEP

	BEST SUFFOLK RAM *Exhibitor*	BEST SUFFOLK EWE *Exhibitor*	CHAMPION SUFFOLK *Exhibitor*
1974		B H Rossiter Kingsbridge, Devon	
1977	B H Rossiter Kingsbridge, Devon		B H Rossiter Kingsbridge, Devon
1978	B W Strout Boyton, Launceston		
1979	D Walter Bude		D Walter Bude
1980	B H Rossiter Kingsbridge, Devon	F G Brewer & Sons (Farms) Ltd Fraddon	B H Rossiter Kingsbridge, Devon
1981	D C W Hampson North Tawton, Devon	F G Brewer & Sons (Farms) Ltd Fraddon	D C W Hampson North Tawton, Devon
1982	D C W Hampson North Tawton, Devon		D C W Hampson North Tawton, Devon
1983	D C W Hampson North Tawton, Devon	N K Allin Winkleigh, Devon	D C W Hampson North Tawton, Devon
1984	D Walter Bude		D Walter Bude

1985	D Walter Bude	F G Brewer & Sons (Farms) Ltd Fraddon	D Walter Bude
1986	N K Allin Winkleigh, Devon		N K Allin Winkleigh, Devon
1987	D Walter Bude	N K Allin Winkleigh, Devon	D Walter Bude
1988	F G Brewer & Sons (Farms) Ltd St Columb	N K Allin Winkleigh, Devon	F G Brewer & Sons (Farms) Ltd St Columb
1989	Mrs R F Pither Taunton, Somerset		Mrs R F Pither Taunton, Somerset
1990	Mrs R F Pither Taunton, Somerset		Mrs R F Pither Taunton, Somerset
1991	R S Robson Bolventor, Launceston	R H & J D Mayo Weymouth, Dorset	R H & J D Mayo Weymouth, Dorset
1992	R H & J D Mayo Weymouth, Dorset	F G Brewer & Sons (Farms) Ltd St Columb	R H & J D Mayo Weymouth, Dorset

DEVON & CORNWALL LONG-WOOL SHEEP

	BEST DEVON & CORNWALL LONG- WOOL RAM	BEST DEVON & CORNWALL LONG- WOOL EWE	CHAMPION DEVON & CORNWALL LONG-WOOL
	Exhibitor	*Exhibitor*	*Exhibitor*
1977	W T R Hawke & Son Newquay	W T R Hawke & Son Newquay	W T R Hawke & Son Newquay
1978	Drake Bros North Tawton, Devon	D Darke Kingsbridge, Devon	Drake Bros North Tawton, Devon
1979	Drake Bros North Tawton, Devon	D Darke Kingsbridge, Devon	Drake Bros North Tawton, Devon
1980	Drake Bros North Tawton, Devon	D Darke Kingsbridge, Devon	Drake Bros North Tawton, Devon
1981	D Darke Kingsbridge, Devon		D Darke Kingsbridge, Devon
1982	Drake Bros North Tawton, Devon	Presingoll Farm Ltd St Agnes	Presingoll Farm Ltd St Agnes
1983	Drake Bros North Tawton, Devon	D Darke Kingsbridge, Devon	Drake Bros North Tawton, Devon
1984	Drake Bros North Tawton, Devon	D Darke Kingsbridge, Devon	Drake Bros North Tawton, Devon
1985	D Darke Kingsbridge, Devon	D Darke Kingsbridge, Devon	D Darke Kingsbridge, Devon
1986	Drake Bros North Tawton, Devon	Presingoll Farm Ltd St Agnes	Drake Bros North Tawton, Devon
1987	W T R Hawke & Son St Columb	Presingoll Farm Ltd St Agnes	W T R Hawke & Son St Columb
1988	Drake Bros North Tawton, Devon	Drake Bros North Tawton, Devon	Drake Bros North Tawton, Devon

| 1989 | Drake Bros
North Tawton, Devon | G S Tancock & Son
Callington | Drake Bros
North Tawton, Devon |
|---|---|---|---|
| 1990 | R E & D Snell
Cullompton, Devon | R E & D Snell
Cullompton, Devon | R E & D Snell
Cullompton, Devon |
| 1991 | M J Britton
Cullompton, Devon | D & J L Wonnacott
Bude | D & J L Wonnacott
Bude |
| 1992 | M J Britton
Cullompton, Devon | D & J L Wonnacott
Bude | M J Britton
Cullompton, Devon |

TEXEL SHEEP

| | BEST TEXEL RAM
Exhibitor | BEST TEXEL EWE
Exhibitor | CHAMPION TEXEL
Exhibitor |
|---|---|---|---|
| 1985 | J Kingsley-Heath
Tregony, Truro | T A S Carlyon
Newquay | J Kingsley-Heath
Tregony, Truro |
| 1986 | Misses C & R Austin
Crediton, Devon | Mr & Mrs G McCallum
Lynton, Devon | Mr & Mrs G McCallum
Lynton, Devon |
| 1987 | E W Quick & Sons
Crediton, Devon | Mrs F M Hellyer
Okehampton, Devon | E W Quick & Sons
Crediton, Devon |
| 1988 | J Kingsley-Heath
Tregony, Truro | Mrs F M Hellyer
Okehampton, Devon | J Kingsley-Heath
Tregony, Truro |
| 1989 | R C L Watts
Taunton, Somerset | R C L Watts
Taunton, Somerset | R C L Watts
Taunton, Somerset |
| 1990 | Mr & Mrs T W Helyer
Salisbury, Wilts | E W Quick & Sons
Crediton, Devon | E W Quick & Sons
Crediton, Devon |
| 1991 | Misses C & R Austin
Crediton, Devon | M C Yeo
Barnstaple, Devon | Misses C & R Austin
Crediton, Devon |
| 1992 | Mr & Mrs T W Helyer
Salisbury, Wilts | Misses C & R Austin
Bude | Mr & Mrs T W Helyer
Salisbury, Wilts |

JACOB SHEEP

| | BEST JACOB RAM
Exhibitor | BEST JACOB EWE
Exhibitor | CHAMPION JACOB
Exhibitor |
|---|---|---|---|
| 1989 | Mrs E R Ruscombe-King
Boscastle | Mrs A M Gilbert
Paignton, Devon | Mrs E R Ruscombe-King
Boscastle |
| 1990 | Mrs A M Gilbert
Paignton, Devon | Mrs A M Gilbert
Paignton, Devon | Mrs A M Gilbert
Paignton, Devon |
| 1991 | Mrs A M Gilbert
Paignton, Devon | Mrs D Ruscombe-King
Boscastle | Mrs A M Gilbert
Paignton, Devon |
| 1992 | Mrs D Ruscombe-King
Boscastle | E R Tucker
Borough, Plymouth | Mrs D Ruscombe-King
Boscastle |

BRITISH CHAROLLAIS SHEEP

| | BEST BRITISH
CHAROLLAIS MALE
Exhibitor | BEST BRITISH
CHAROLLAIS FEMALE
Exhibitor | CHAMPION BRITISH
CHAROLLAIS
Exhibitor |
|---|---|---|---|
| 1991 | | Mr & Mrs D J F Burrough
Honiton, Devon | Mr & Mrs D J F Burrough
Honiton, Devon |
| 1992 | V K Crocker
Yeovil, Somerset | Mr & Mrs D J F Burrough
Honiton, Devon | Mr & Mrs D J F Burrough
Honiton, Devon |

BRITISH BLEU DU MAINE SHEEP

	BEST BRITISH BLEU DU MAINE MALE	BEST BRITISH BLEU DU MAINE FEMALE	CHAMPION BRITISH BLEU DU MAINE
	Exhibitor	Exhibitor	Exhibitor
1992	Mrs M Vile Taunton, Somerset	Mrs M Vile Taunton, Somerset	Mrs M Vile Taunton, Somerset

ANY OTHER PURE BREED SHEEP

CHAMPION ANY OTHER PURE BREED SHEEP

	Exhibitor	Breed
1988	E Richardson Newquay	British Charollais
1989	Mrs M Pringle Totnes, Devon	Vendeen

ANY OTHER PURE BREED SHEEP — Breeds of British Native Origin

CHAMPION ANY OTHER PURE BREED SHEEP—
British Native Origin

	Exhibitor	Breed
1990	A T Guard Torrington, Devon	Dorset Down
1991	C Mortimer Chagford, Devon	Dartmoor
1992	Hemsworth Farms (M S Tory) Wimborne, Dorset	Hampshire Down

ANY OTHER PURE BREED SHEEP — Breeds of Continental Origin

CHAMPION ANY OTHER PURE BREED SHEEP—
Continental Origin

	Exhibitor	Breed
1990	E W Quick & Sons Crediton, Devon	Rouge de L'Ouest
1991	E W Quick & Sons Crediton, Devon	Rouge de L'Ouest
1992	P Kingdon & Sons Newquay	Ile de France

COMMERCIAL SHEEP

CHAMPION COMMERCIAL SHEEP

	Exhibitor	Breed
1987	F G Brewer & Sons (Farms) Ltd St Columb	
1988	F G Brewer & Sons (Farms) Ltd St Columb	Scotch Half Bred Ewe
1989	Mrs A M Gynn Liskeard	Hocken Hybrids
1990	Mrs A M Gynn Liskeard	Hocken's Hybrid X
1991	Mrs A M Gynn Liskeard	Hockens Hybrid
1992	F D Banbury St Ervan, Wadebridge	Ile de France

MAJOR CHAMPIONSHIP AWARDS

		BEST RAM IN THE YARD		BEST PEN OF EWES IN THE YARD	
	Exhibitor		Breed	Exhibitor	Breed
1875	George Turner, Brampford Speke, Exeter		Leicester	Charles Norris, Motion, Exeter	Long Wool
1877	W Tremaine, Polsue, Philleigh		Leicester	Mr Corner	
1878	W Tremaine, Polsue, Philleigh		Leicester	W Tremaine, Polsue, Philleigh	Leicester
1879	W Tremaine, Polsue, Philleigh		Leicester	Sir J Heathcote Amory, Knightshayes Court, Tiverton	Long Wool
1880	Sir J Heathcote Amory, Knightshayes Court, Tiverton		Long Wool	C Norris, Motion, Exeter	Long Wool
1881	W Tremaine, Polsue, Philleigh		Leicester	Sir J Heathcote Amory, Knightshayes Court, Tiverton	Long Wool
1882				Sir J Heathcote Amory, Knightshayes Court, Tiverton	Long Wool
1883	Sir J Heathcote Amory, Knightshayes Court, Tiverton		Long Wool	J N Franklin	Long Wool
1884	P Thompson, Tregony		Leicester	James Stooke, East Sherford, Plympton	South Hams
1885	James Stooke, East Sherford, Plympton		South Hams	Sir J Heathcote Amory, Knightshayes Court, Tiverton	Long Wool
1886	T Potter, Thorverton				
1888	James Stooke, East Sherford, Plympton		South Hams	Sir J Heathcote Amory, Knightshayes Court, Tiverton	Long Wool
1891	E Jacka, Lanner, Sithney			J E Hawkey, Tredore, St Issey	
1892	Sir J Heathcote Amory, Knightshayes Court, Tiverton		Long Wool	J E Hawkey, Tredore, St Issey	
1897	John S Hallett, Sherford Barton		South Hams		
1898	John S Hallett, Sherford Barton		South Hams	W H Treweeke, Chipping Norton	Down
1899	John S Hallett, Sherford Barton		South Hams	E Stooke, Callington	South Hams
1900	W H Treweeke, Chipping Norton		Down	Robert Cook, Chevithorne, Tiverton	Long Wool

1901	F White Williton, Somerset	Long Wool	F White Williton, Somerset	Long Wool
1909	J Stooke Sherford, Brixton			South Hams
1910	E Hoskin Ludgvan, Penzance	Down	R B Trant Menheniot	South Hams
1911	J Stooke Sherford, Brixton	South Hams	P G Brown Tremadart, Duloe	South Hams
1914	J Stooke Sherford, Brixton	South Devon	J S Hallett Sherford, Brixton	South Devon
1915	P G Brown Tremadart, Duloe	South Devon		
1924	E Jacka & Son Lanner, Sithney	Oxford Down		
1925	E Jacka & Son Lanner, Sithney	Oxford Down		

BEST RAM LAMBS, of any breed

1894	N Cook Chevithorne, Tiverton	Long Wool
1895	N Cook Chevithorne, Tiverton	
1896	R J Sobey St Keyne, Liskeard	South Hams
1897	R J Sobey St Keyne, Liskeard	

BEST SHEEP OR PEN OF SHEEP — £10 Special Prize given by
The Rt Hon the Viscount Falmouth

Exhibitor *Breed*
1870 James Tremaine
Polsue, Philleigh

BEST EWE OR RAM IN SHOW
1927 F White   Devon Long-Woolled
     Williton, Somerset

SUPREME CHAMPION SHEEP
Exhibitor   *Breed*

1977	B H Rossiter Kingsbridge, Devon	Suffolk
1978	A T Guard Torrington, Devon	Dorset Down
1979	Drake Bros North Tawton, Devon	Devon & Cornwall Long-Wool
1980	Drake Bros North Tawton, Devon	Devon & Cornwall Long-Wool
1981	D Darke Kingsbridge, Devon	Devon & Cornwall Long-Wool
1982	Presingoll Farm Ltd St Agnes	Devon & Cornwall Long-Wool
1983	D C W Hampson North Tawton, Devon	Suffolk

1984	Drake Bros North Tawton, Devon	Devon & Cornwall Long-Wool
1985	D Darke Kingsbridge, Devon	Devon & Cornwall Long-Wool
1986	Drake Bros North Tawton, Devon	Devon & Cornwall Long-Wool
1987	P Kingdon & Sons Newlyn East, Newquay	Dorset Horn & Poll Dorset
1988	I G Mortimer Esq Chagford, Devon	Scotch Black-Faced
1989	Mrs R F Pither Taunton, Somerset	Suffolk
1990	W J Jordan Chagford, Devon	Scotch Black-Faced
1991	Mr & Mrs D J F Burrough Honiton, Devon	British Charollais
1992	M J Britton Cullompton	Devon & Cornwall Long-Wool

WOOL CLASSES

	HEAVIEST ENTIRE FLEECE OF WOOL *Exhibitor*	MOST VALUABLE FLEECE OF WOOL FROM RAM/EWE *Exhibitor*
1797	Thomas Key St Breock	
1799	Rev Robert Walker St Winnow	
1802	John Bligh (ram) St Mabyn Thomas Trood (ewe) Poughill	
1803	John P Peters (ram) John P Peters (ewe) Philleigh	
1804	Michael Mill (ram) St Columb	
1805	Thomas Key (ram) St Breock	
1806	Thomas Key (ram) St Breock	
1812		Mr Irving (Spanish ram) Mr Irving (Spanish ewe)
1813		Mr Thomson (ram) Cardinham
1814		Rev Robert Walker (ram) St Winnow
1815		Thomas A Lee Boconnoc

	BEST FLEECE OF WOOL	
1858	Robert Nicholls Lostwithiel	

SHEEP SHEARING — 1795 TO 1858

	BEST SHEARER OF SHEEP	BEST SHEARER OF SHEEP, being an apprentice	BEST SHEARER OF SHEEP who shall belong and reside in the County of Cornwall
	Exhibitor	*Exhibitor*	*Exhibitor*
1795	John Wyatt		
1796	John Pile, Kenn, Devon		Thomas Wade, Gorran
1797	Philip Osborne, St Dennis		John Gummoe, Probus
1799	John Vincent, Moorwinstow		
1802	Richard Sawle, Probus		
1803	Daniel Farley, Devon		
1804	James Bridle, Dock		
1805	George Roberts, Newlyn		
1806	John Johns, Mevagissey		
1807	William Rowe, Newlyn		
1808	John Greenslade, Newlyn		
1810	Richard Lawer, Newlyn		
1812	Jeremiah Harris, Spritton, Devon		
1812	Mark Symons, Newlyn		
1813	John Jeffery, Newlyn		
1814	George Turner, Cadbury	Richard Dyer, St Breock	
1814	John Tremayne, Newlyn		
1815	Stephen Bray, Marhamchurch		
1815	Thomas Wills, Egloshayle		
1816	Francis Woolcock, St Erme		
1817	Charles Rogers, Fowey		
1817	Samuel Serpell, Liskeard		
1818	Samuel Serpell, Daureath		
1820		H Symons, Newlyn	

1821	Robert Lees Morval	Michael Cayzer Mawgan
1822	William Dingle Philleigh	Samuel Trevan Sheviock
1825	John Trescott St Stephens	Thomas Oliver Morval
1831	John Andrew	
1832	F Woolcock St Erme (45 mins)	
1833	William Hawkey Newlyn	
1834	John Sawle Probus	
1835	F Woolcock St Erme	
1836	Roger Jackman Lifton, Devon	

BEST SHEEP SHEARER, not having won within the last 4 years the best Prize offered by the Society

1837 Robert Kent
     Newlyn
1838 William James
     Newlyn

BEST SHEARER OF SHEEP

1839 R Kent
     Newlyn (80 mins)
1840 John Sawle
     Probus (64 mins)
1841 Henry Smith
     Probus (75 mins)
1842 R Johns
     St Ewe
1843 R Johns
     St Ewe (65½ mins)
1844 Henry Smith
     Probus (67½ mins)
1845 John Sawle
     Probus (68 mins)
1846 Richard Sawle
     Probus
1847 John Pill
     Gorran (90 mins)
1848 Richard Johns
1849 John Pill
     (74 mins)
1849 John Pill
     (74 mins)

1850 John Pill
     Creed (88 mins)
1851 Richard Johns
     (73 mins)
1852 Richard Sawle
     Probus (84 mins)
1853 Nicholas Pill
     Mevagissey (66 mins)
1854 William Pill
                (65 mins)
1855 Thomas Sawle
     Probus (72 mins)
1856 John Pill
     St Mewan (68 mins)
1857 Nicholas Pill
     Gorran
1858 Thomas Sawle
     Probus

PIG SECTION RESULTS — 1794 to 1972

No Pig classes staged since this date

BEST BOAR
Exhibitor
1794 Thomas Hicks
Lanivet
1795 William Courtney
St Ervan
1802 Richard Cowling
1803 Thomas Dungey
St Ewe
1805 Edward Lawrence
St Martin
1806 Nathaniel Roberts
Manaccan
1807 Thomas Edwards
St Kew
1808 John Williams
St Hilary
1810 Oliver Edwards
St Anthony in Meneage
1811 John Lark
Launcells

	BEST BRED BOAR *Exhibitor*	BEST BRED SOW *Exhibitor*
1828	William Carne Kenwyn	Capt Julian Kea
1829	T Daniel Trelissick	
1830	W Green Truro	Mrs Tippet Kenwyn St, Truro
1831	James Johns St Clement	E H Hill Gerrans
1832	C Daubuz St Clement	James Hendy Ladock
1933	Withheld for want of Merit	G Carlyan Kenwyn
1834	J Trestrail Chevlan	A Huddy St Erme
1835	James Hendy Trethurffe	Thomas Tank Ladock
1836	James Hendy Trethurffe	G W F Gregor Trewarthenick
1837	Mr Trenance Veryan	Thomas Tank Ladock
1838	John Stevens Sancreed	John Stevens Sancreed
1839	W Veale St Columb Major	Thomas Tank Ladock

1840	I Ivey Merther	G G Bullmore Newlyn
1841	M Hotton Probus	R Julian Perranzabuloe
1842	J Permewan St Buryan	H E Bull Kenwyn
1843	J Trevenan Crowan	J Tyack Sithney
1844	G Pearce Kenwyn	Rev W Hocken Endellion
1845	Colonel Scobell Penzance	W Trethewey Probus
1846	G Pearce Kenwyn	J Knight Truro
1847	G Pearce Kenwyn	C Colman Bodmin
1848	Peter Davis Probus	S Tremenhere Paul
1849	R Ivey Sancreed	S Tremenhere Paul
1850	J Knight Kenwyn	J D Gilbert Trelissick
1851	J Lyle Bonython	B Beckerleg Penzance
1852	J R James Botallack, St Just	S Tremenhere Paul
1853	Truro Loan Society	Truro Loan Society
1854	Charles Sanders Truro	Samuel Randall Stythians
1855	Samuel Randall Stythians	Samuel Randall Stythians
1856	Rev H N Barton St Ervan	Samuel Randall Stythians
1857	Joseph Lyle Cury	Edward Skewes Cury

PIGS — LARGE BREED

BEST BOAR

Exhibitor

1858

BEST BREEDING SOW,
in farrow or that has farrowed within 6 months of the meeting

Exhibitor

Robert Bilkey
Tremenbeere, Ludgvan

BEST SOW

1860	Mr Foster Castle, Lostwithiel	Mr Foster Castle, Lostwithiel
1862	John Widdicombe Torrhill, Ivybridge	John Widdicombe Torrhill, Ivybridge
1863	T D Eva Camborne	John Widdicombe Torrhill, Ivybridge

		BEST BREEDING SOW, in farrow or that has farrowed within 6 months of the meeting	BEST PEN OF 2 BREEDING SOWS, (of the same litter), not exceeding 12 months old
1864	John Widdicombe Torrhill, Ivybridge	John Antony Yealmpton, Devon	
1865	George Biddick Trelissick Farm	George Biddick Trelissick Farm	
1867	L Langford Stokeclimsland	John Widdicombe Torrhill, Ivybridge	
1869	Thomas James St Hilary		
1870	Messrs Duckering & Sons Lincolnshire	Messrs Duckering & Sons Lincolnshire	
1871	Sydney Davey Redruth	James Ball Pelynts Barn, Truro	
		Exhibitor	*Exhibitor*
1872	Messrs J Wheeler & Sons Long Compton, Shipston-on-Stour	Messrs J Wheeler & Sons Long Compton, Shipston-on-Stour	James Tremain
1873	Messrs J Wheeler & Sons Long Compton, Shipston-on-Stour	Messrs Duckering & Son Kirton Lindsey, Lincs	Messrs Duckering & Son Kirton Lindsey, Lincs
1874	Jacob Dove Bristol	Jacob Dove Bristol	Mr Duckering Kirton Lindsey, Lincs
1875	Jacob Dove Bristol	Jacob Dove Bristol	Jacob Dove Bristol
1876	Jacob Dove Bristol	W Palmer St Winnow	Jacob Dove Bristol
1878	Messrs R E Duckering & Sons Northorpe	B St John Ackers	
1879	Lord Moreton Torton Court, Glos	Messrs R & T Russell Sithney	Lord Moreton Tortworth Court, Glos
1880	Thomas Salmon St Columb	Dr Bow Lostwithiel	
1881	R P Humphrey Watcombe, Torquay	Thomas Salmon St Columb	Charles Hawke St Columb
1882	R P Humphrey Watcombe, Torquay	Charles Hawke St Columb	Charles Hawke St Columb
1883	N Benjafield Motcombe, Shaftesbury	N Benjafield Motcombe, Shaftesbury	N Benjafield Motcombe, Shaftesbury
1885	N Benjafield Motcombe, Shaftesbury	N Benjafield Motcombe, Shaftesbury	N Benjafield Motcombe, Shaftesbury
1886	N Benjafield Motcombe, Shaftesbury	N Benjafield Motcombe, Shaftesbury	N Benjafield Motcombe, Shaftesbury
1887	William Simmons Kenwyn		
1888	W B Skewis Rowden, Brentor	W B Skewis Rowden, Brentor	Charles Hawke St Columb

1889	Henry Ford Ivybridge, Devon	Henry Ford Ivybridge, Devon		
1890	W B Skewis Rowden, Brenton	W B Skewis Rowden, Brenton		

PIGS — SMALL BREED

	BEST BOAR *Exhibitor*	BEST SOW *Exhibitor*		
1858	J R James St Just			
1860	Ambrose S Toll Quethiock	Robert Nicholls Lostwithiel		
1862	John Cardell Pollywin	Richard Foster Lostwithiel		
1863	John Knight Truro	John Knight Truro		
1864	Sir Massey Lopes Bt MP Maristow	Robert Pront Milton Abbot, Devon		
1865	Edward Skewes Cury	Ambrose S Toll Quethiock		
1867	R Jackman Lifton, Devon	R Jackman Lifton, Devon		

	BEST BOAR, not not exceeding 12 months old *Exhibitor*	BEST BOAR, exceeding 12 months old *Exhibitor*	2 BREEDING SOWS, under 12 months old *Exhibitor*	BEST SOW *Exhibitor*
1869	J Sydney Davey Redruth	Thomas Thomas Sancreed	Sydney Davey Redruth	Sydney Davey Redruth
1870	John Palmer Lewannick	Messrs Duckering & Sons Lincolnshire	Sydney Davey Redruth	R Bicle Lifton, Devon
1871	W M Ware Newham Hse, Helston	W M Ware Newham Hse, Helston	Sir F M Williams Bt MP Goonvrea	Mr Hendy Trenowth, Probus
				BEST BREEDING SOW in farrow or that has farrowed within 6 months of the meeting
1872	W M Ware Newham Hse, Helston	Richard Roskelly St Enoder		Messrs Wheeler & Sons
1873	Messrs Duckering & Son Lincolnshire	Messrs Duckering & Son Lincolnshire		Messrs Duckering & Son Lincolnshire
1874	Mr Duckering Kirton Lindsey, Lincs	Mr Duckering Kirton Lindsey, Lincs	Jacob Dove Bristol	Mr Duckering Kirton Lindsey, Lincs
1875	John Cardell St Erth	Richard Foster Lostwithiel	John Cardell St Erth	Thomas Salmon St Columb

1876	Messrs R & J Russell Sithney	Jacob Dove Bristol	Jacob Dove Bristol	Jacob Dove Bristol
1877	Messrs R & J Russell Sithney	Lord Moreton Tortworth Court, Glos		Thomas Salmon St Columb
1878	W F Collier Horrabridge, Devon	Mr Partridge Bow	Lord Moreton Tortworth Court, Glos	Mr Partridge Bow
1879	Thomas Salmon St Columb	Thomas Salmon St Columb	Messrs R & T Russell Sithney	Thomas Salmon St Columb
1880	Lord Moreton MP Tortworth Court, Glos	J Andrew St Columb	Thomas Salmon St Columb	Lord Moreton MP Tortworth Court, Glos
1881	J Andrew St Columb	J Andrew St Columb	J Andrew St Columb	G Oliver Trehane
1882	W B Northey	Lord Moreton Tortworth Court, Glos	V P Camady	W S Northey Tinhay, Lifton
1883	J Andrew St Columb	N Benjafield Motcombe, Shaftsbury	W S Northey Tinhay, Lifton	Harry Hoblyn St Columb
1884	Harry Hoblyn St Columb	Harry Hoblyn St Columb	Mrs Glanville The Vicarage, Ivybridge	Harry Hoblyn St Columb
1885		W S Northey Tinhay, Lifton	Thomas Roberts Trengothal, St Levan	Thomas Roberts Trengothal, St Levan
1886		W S Northey Tinhay, Lifton	John Andrew Barn, St Columb	W T Merrifield St Columb
1887		W S Northey Tinhay, Lifton	W S Northey Tinhay, Lifton	W S Northey Tinhay, Lifton
1888		W S Northey Tinhay, Lifton	W S Northey Tinhay, Lifton	W S Northey Tinhay, Lifton
1889		W S Northey Tinhay, Lifton		W S Northey Tinhay, Lifton
1890		W S Northey Tinhay, Lifton		W S Northey Tinhay, Lifton

PIGS — of breeds other than those eligible for the breed classes

	BEST BOAR, under 12 months old	BEST BOAR, exceeding 12 months old	BEST SOW, under 12 months old	BEST BREEDING SOW, in farrow or that has farrowed within 6 months of the meeting, exceeding 12 months old
	Exhibitor	*Exhibitor*	*Exhibitor*	*Exhibitor*
1891	J & N Stephens Hendra Park, St Tudy	W S Northey Lifton, Devon	W S Northey Lifton, Devon	Ward & Chowen Brentor, Tavistock
1892	Ward & Chowen Burnville, Tavistock	T D Eva Troon, Camborne	Ward & Chowen Burnville, Tavistock	Ward & Chowen Burnville, Tavistock

| 1895 | William Lean
St Issey | Philip Blare
St Germans | Major Cardell | Nathanial Stephens
Bodmin |
| --- | --- | --- | --- | --- |
| 1896 | Frank Allmand
Wrexham | Frank Allmand
Wrexham | Frank Allmand
Wrexham | William Lean
St Issey |
| 1897 | William Lean
St Issey | Frank Allmand
Wrexham | John Webber
St Columb | John Bastard
Tinten, St Tudy |
| 1898 | A Hiscock Jnr
Motcombe,
Shaftesbury | A Hiscock Jnr
Motcombe,
Shaftesbury | A Hiscock Jnr
Motcombe,
Shaftesbury | A Hiscock Jnr
Motcombe,
Shaftesbury |

BERKSHIRE PIGS

BEST BOAR
Exhibitor

BEST SOW
Exhibitor

1889 A Carlyon
Trenithon, St Keverne

J C Williams
Werrington, Launceston

1890 J C Williams
Werrington, Launceston

YORKSHIRE WHITE PIGS

BEST BOAR
Exhibitor

BEST SOW
Exhibitor

1892 Sanders Spencer
St Ives, Huntingdonshire

Sanders Spencer
St Ives, Huntingdonshire

LARGE BLACK PIGS

BEST LARGE BLACK
BOAR
Exhibitor

BEST LARGE BLACK
SOW
Exhibitor

CHAMPION LARGE
BLACK
Exhibitor

1899 John Frayn
St Stephens, Launceston

E Hosking
Ludgvan, Penzance

1914
    J Oscar
    Heathcot, Yelverton
    'Heathcoat Excelsior'

1915
    Messrs W & H Whitley
    Primley Farm, Paignton
    'Primley Godiva'

1919
    J H Glover
    Cornwood, Devon

1920
    H E Bastard
    St Tudy
    'Trevisquite Padstonian'

1921
    John Warne
    Grampound Road

1922
    H E Bastard
    St Tudy, Bodmin
    J H Glover
    Cornwood, Devon

1924	J Olver Wooland Valley, Ladock 'Valley General'	J H Glover Cornwood, Devon	
1925	J Warne Tregonhayne, Tregony	R Gynn Treslay, Camelford	
1926		J H Glover Cornwood, Devon	John C Olver Wooland Valley, Ladock
1927	W J Warren Staplegrove, Taunton	J Warne & Son Tregonhayne, Tregony	
1928	Messrs R Gynn & Son Treslay, Camelford 'Hendra Sunstar'	Messrs R Gynn & Son Treslay, Camelford 'Westpetherwin Sunbeam'	
1929	Messrs J Warne & Son Tregonhayne, Tregony 'Trewithen Bounder'	Messrs J Warne & Son Tregonhayne, Tregony 'Banns Ruth 3rd'	
1930	John Warne & Son Tregonhayne, Tregony	John Warne & Son Tregonhayne, Tregony	
1931	H E Bastard Tinten, St Tudy 'West Petherwin Leader 2nd'	R Gynn & Son Treslay, Camelford 'Treslay Belle 19th'	
1932	H E Bastard Tinten, St Tudy 'West Petherwin Leader 2nd'	W Truscott Trewithen, St Austell 'Treveglos Biddy 2nd'	
1933	R Gynn & Son Treslay, Camelford 'Treslay Bumper 4th'	H E Bastard Tinten, St Tudy	
1934	T H Luscombe & Son Lower Parks, Crediton 'Treslay Bumper 4th'	H E Bastard Lemail, Wadebridge	
1935	H E Bastard Lemail, Wadebridge 'Tinten Tailor 1st'	H E Bastard Lemail, Wadebridge	
1936	Truscott Bros Trewithen, St Austell 'Trewithen Token'	H E Bastard Lemail, Wadebridge 'Tinten Princess 39th'	
1937	Truscott Bros Trewithen, St Austell 'Trewithen Token'	J Laity Bosfranken, St Buryan	J Laity Bosfranken, St Buryan
1938	J Laity Bosfranken, St Buryan	Truscott Bros Trewithen, St Austell 'Trewithen Moonbeam 3rd'	
1939	W T Stephens Whitley, Liskeard 'Kirlington Punch 4th'	J Laity Bosfranken, St Buryan	J Laity Bosfranken, St Buryan
1947	A J Baker & Sons Comeythowe, Taunton 'Thaluston King John XI'	A J Baker & Sons Comeythowe, Taunton 'Hartington Nocturne 19th'	A J Baker & Sons Comeythowe, Taunton
1948	J Laity Bosfranken, St Buryan	A J Baker & Sons Comeythowe, Taunton	A J Baker & Sons Comeythowe, Taunton
1949			Truscott Bros Trewithen, St Austell

1950	A J Baker & Sons Comeythowe, Taunton 'Downhills King John 12th'	R A White Trevillyan, Bugle 'Luxulyan Constance 6th'		
1951	T H Luscombe & Son Lower Parks, Crediton	A J Baker & Sons Comeythowe, Taunton		
1952	R Gynn & Son Boscastle	R A White Trevillyan, Bugle		
1953	A Raymont Bodrigan, Blisland 'Royston Enterprise'	Truscott Bros Trewithen, St Austell 'Trewithen Fancy 95th'		
1954	A Raymont Bodrigan, Blisland	Truscott Bros Trewithen, St Austell		
1956	J H Menhenitt Tredustan, Wadebridge 'Treslay Play Boy 4th'	J Laity Bosfranken, St Buryan 'Bosfranken Queen 2nd'		
1957	J H Menhenitt Tredustan, Wadebridge	J Laity Bosfranken, St Buryan		
1958	J H Menhenitt Tredustan, Wadebridge	R Gynn & Son Treslay, Boscastle		
1959	A Raymont & Son Bodrigan, Blisland	R Gynn & Son Treslay, Boscastle	R Gynn & Son Boscastle 'Treslay Larkspur'	
1960	E D Snell & Son Yeovil, Somerset	John C Gynn Treslay, Boscastle	John C Gynn Treslay, Boscastle	
1961	J Laity Raventor, Lelant	John C Gynn Treslay, Boscastle	John C Gynn Treslay, Boscastle	
1962	R H Trenouth Tresallyn, St Merryn 'Treslay Knockout 3rd'	A Raymont & Son Bodrigan, Blisland 'Bodrigan Bess 140th'	A Raymont & Son Bodrigan, Blisland 'Bodrigan Bess 140th'	
1964	John C Gynn Treslay, Boscastle	John C Gynn Treslay, Boscastle	John C Gynn Treslay, Boscastle	
1965	John C Gynn Treslay, Boscastle 'Treslay Mainspring 3rd'	John C Gynn Treslay, Boscastle 'Treslay Royal Lady 15th'	John C Gynn Treslay, Boscastle 'Treslay Royal Lady 15th'	
1966	John C Gynn Treslay, Boscastle 'Langland Chieftain 9th'	M J Hick Hugus, Truro 'Bixley Jane 75th'	M J Hick Hugus, Truro 'Bixley Jane 75th'	
1967	M T Burrow Uffculme, Devon	M T Burrow Uffculme, Devon	M T Burrow Uffculme, Devon	
1968	M J Hick Hugus, Truro	John C Gynn Treslay, Boscastle	M J Hick Hugus, Truro	
1969	M J Hick Hugus, Truro 'Tolcross Prospector 3rd'	E D Snell & Son Mudford Sock, Yeovil 'Sock Dorothy 79th'	M J Hick Hugus, Truro 'Tolcross Prospector 3rd'	
1970	P G Snell Mudford Sock, Yeovil 'Thelveton Malcolm 128th'	P G Snell Mudford Sock, Yeovil 'Sock Dorothy 94th'	P G Snell Mudford Sock, Yeovil 'Sock Dorothy 94th'	
1971	P G Snell Mudford Sock, Yeovil 'Sock Black Boy 4th'	P G Snell Mudford Sock, Yeovil 'Sock Dorothy 103rd'	P G Snell Mudford Sock, Yeovil 'Sock Dorothy 103rd'	
1972	T J Trembath Rospavean, Ludgvan	P G Snell Mudford Sock, Yeovil	P G Snell Mudford Sock, Yeovil	

LARGE WHITE PIGS

	BEST LARGE WHITE BOAR *Exhibitor*	BEST LARGE WHITE SOW *Exhibitor*	CHAMPION LARGE WHITE *Exhibitor*
1899	James Balsdon Launceston	Thomas Manuell Trevorva, Probus	
1925			W White & Sons Pool Farm, Taunton
1926			E Hosking Jnr Pulsack, Hayle
1928			W White & Son Pool Farm, Taunton
1929			W White & Son Pool Farm, Taunton
1930			W White & Son Pool Farm, Taunton
1931			W White & Son Pool Farm, Taunton
1932			W White & Son Pool Farm, Taunton 'Taunton King David 3rd'
1933			Sir G A Cooper Bt Hursley Park, Winchester 'Farley Bonetta 27th'
1934			W White & Son Pool Farm, Taunton
1935	G H Johnstone Trewithen, Grampound Road	T P P Kent Ashford, Barnstaple 'Laybrook Jay Queen'	
1936	G A Garceau Trebears, St Merryn 'Taunton Bradbury 37th'	G H Johnstone Trewithen, Grampound Road	
1937	G H Johnstone Trewithen, Grampound Road	G H Johnstone Trewithen, Grampound Road	G H Johnstone Trewithen Grampound Road
1938	W A Whidden Brampford Speke, Exeter 'Uffculme Boy 8th'	G H Johnstone Trewithen, Grampound Road	
1939	G H Johnstone Trewithen, Grampound Road	G H Johnstone Trewithen, Grampound Road	G H Johnstone Trewithen, Grampound Road
1947	G P Day Winnianton, Gunwalloe 'Arden Cote Bradbury'	Hammetts Dairies Ltd Pinhoe, Exeter 'Former Maple Leaf 15th'	Hammetts Dairies Ltd Pinhoe, Exeter 'Former Maple Leaf 15th'
1948	G H & Miss E A Johnstone Trewithen, Grampound Road	A J Baker & Son Comeythowe, Taunton	G H & Miss E A Johnstone Trewithen, Grampound Road
1949	G H & Miss E A Johnstone Trewithen, Grampound Road	G H & Miss E A Johnstone Trewithen, Grampound Road	

1950	A Gruzelier West Dean, Salisbury 'Wall Field Marshall 83rd'	F W & T M Trewhella Trenerth, Leedstown	
1951	H J B Bartlett Dartmouth, Devon	Trewithen Pedigree Farms Trewithen, Grampound Road	
1952	Messrs F W & T M Trewhella Leedstown, Hayle	N J Hosking St Buryan	
1953	H W Long Henley, Oxon 'Flowercote King David 7th'	E E Geary Ringwood, Hants 'Hightown East Lass 5th'	
1954	Trewithen Pedigree Farms Trewithen, Grampound Road	F W & T M Trewhella Trenerth, Leedstown	
1956	W B Gubbin Penhale, Coads Green 'Ardencote Royal Turk 88th'	Trewithen Pedigree Farms Trewithen, Grampound Road 'Trewithen Diana 3rd'	
1957	W J B Bartlett Dartmouth, Devon	Trewithen Pedigree Farms Trewithen, Grampound Road	
1958	H M Martyn & Co Bridgerule, Holsworthy	W B Gubbin Launceston	
1959	H M Martyn & Co Bridgerule, Holsworthy	R Roach & Sons Wellparks, Crediton	
1960	Mrs D P Suthrell Tiverton, Devon	R J Roach Wellparks, Crediton	
1961	R J Roach Wellparks, Crediton	Mr & Mrs C Martin Shepherds Hse, Newlyn East	
1962	Mrs D P Suthrell Tiverton, Devon 'Cowleymore Laddie 82nd'	R J Roach Wellparks, Crediton 'Kyrtonian Madam Matilda 158th'	
1964	P N Cullen Marldon, Paignton	Antony Pedigree Farm Torpoint	P N Cullen Marldon, Paignton
1965	John Roach (Crediton) Ltd Wellparks, Crediton 'Kyrtonian King David 10th'	Mr & Mrs F I Harvey Shepherds, Newlyn East 'Newlyndene Lassie 2nd'	John Roach (Crediton) Ltd Crediton, Devon 'Kyrtoninan King David 10th'
1966	Mrs D P Suthrell Tiverton, Devon 'Cowleymore King David 46th'	H M Martyn & Co Bridgerule, Holsworthy 'Boroughfarm Blackberry 2nd'	H M Martyn & Co Bridgerule, Holsworthy 'Boroughfarm Blackberry 2nd'
1967	H M Martyn & Co Bridgerule, Holsworthy	H M Martyn & Co Bridgerule, Holsworthy	H M Martyn & Co Bridgerule, Holsworthy
1968	H M Martyn & Co Bridgerule, Holsworthy	H M Martyn & Co Bridgerule, Holsworthy	H M Martyn & Co Bridgerule, Holsworthy
1969	A R Clements Bury St Edmunds, Suffolk 'Shimpling Field Marshall 46th'	A R Clements Bury St Edmunds 'Shimpling Beautiful 15th'	A R Clements Bury St Edmunds, Suffolk 'Shimpling Field Marshall 46th'

1970	P N Cullen Marldon, Paignton 'Battleaxe Field Marshall 58th'	A R Clements Bury St Edmunds 'Shimpling Beautiful 15th'	A R Clements Bury St Edmunds, Suffolk 'Shimpling Beautiful 15th'
1971	A R Clements Bury St Edmunds, Suffolk 'Shimpling Field Marshall 55th'	A R Clements Bury St Edmunds 'Shimpling Beautiful 15th'	A R Clements Bury St Edmunds, Suffolk 'Shimpling Field Marshall 55th'
1972	F Fern Westleigh, Chacewater	F Fern Westleigh, Chacewater	F Fern Westleigh, Chacewater

GLOUCESTERSHIRE OLD SPOT PIGS

	BEST GLOUCESTERSHIRE OLD SPOT BOAR *Exhibitor*	BEST GLOUCESTERSHIRE OLD SPOT SOW *Exhibitor*
1920	Ralph Mitchell Trewithian, Grampound Road	J Douglas Kingswood, Bristol
1921	Charles Osborne Grampound Road	J Douglas Kingswood, Bristol
1922	Ralph Mitchell Jnr Trewithian, Grampound Road	Charles Osborne Grampound Road

MIDDLE WHITE PIGS

CHAMPION MIDDLE WHITE
Exhibitor

1925 Bicton Pig Farm
St Ive, Liskeard

1926 W F S Hodgson & Son
Morebath, Bampton

1928 E M Jowitt
Bridport, Dorset

1929 W C Badcock
Middle Colenso, Marazion
'Shawlands Prince Palatine'

1930 W C Badcock
Middle Colenso, Marazion

1931 Messrs J W White & Son
Poole Farm, Taunton
'Amport Bella 68th'

1932 W C Badcock
Middle Colenso, Marazion

1933 W C Badcock
Middle Colenso, Marazion

1934 W White & Son
Poole Farm, Taunton

NATIONAL LONG WHITE LOP-EARED PIGS

	BEST NATIONAL LONG WHITE LOP-EARED BOAR *Exhibitor*	BEST NATIONAL LONG WHITE LOP-EARED SOW *Exhibitor*	CHAMPION NATIONAL LONG WHITE LOP-EARED *Exhibitor*
1927			J H Bickell Lumburn, Tavistock 'Lumburn Lily 5th'
1928			F Richards Lidcutt, Bodmin 'Lidcutt Longsides'
1929	B J Hoopell Bigbury On Sea, Devon 'Folly Merryman'	W J Westlake & Son Godwell, Ivybridge 'Godwell Princess 13th'	
1930	G H Eustice Bezurrell, Hayle	Marshall Bros Ivybridge	
1931	G Eustice Bezurrell, Hayle 'Afton Gay Boy'	Blight Bros Trolvis, Stithians 'Trolvis Ruby 50th'	
1932	G H Eustice Bezurrell, Hayle 'Afton Gay Boy'	Blight Bros Trolvis, Stithians	
1933	H R Jasper East Petherwin, Launceston 'Trolvis Ben 11th'	F Richards Lidcutt, Bodmin	
1934	A H Johns & Son Treringey, Crantock 'Treringey Longsides 2nd'	H R Jasper East Petherwin, Launceston 'Petherwin No 21'	
1935	W H Neal Walreddon, Tavistock	G H Eustice Bezurrell, Hayle	
1936	W H Neal Walreddon, Tavistock	G H Eustice Bezurrell, Hayle 'Bezurrell Ruby 2nd'	
1937	G H Eustice Bezurrell, Hayle 'Bezurrell Ben'	W H Neal Walreddon, Tavistock 'Yealmpstone Dainty 9th'	W H Neal Walreddon, Tavistock 'Yealmpstone Dainty 9th'
1938	W H Neal Walreddon, Tavistock 'Yealmpstone Gay Boy 19th'	Blight Bros Trolvis, Stithians	Blight Bros Trolvis, Stithians
1939	W H Neal Walreddon, Tavistock 'Yealmpstone Gay Boy 19th'	G H Eustice Bezurrell, Hayle	W H Neal Walreddon, Tavistock
1947	W A Johns Treringey, Crantock 'Lidcutt Royalty VI'	G H Eustice Bezurrell, Hayle 'Bezurrell Mary 78th'	W A Johns Treringey, Crantock
1948	F L Collings Pengover, Liskeard	W Whidden Brampford Speke, Exeter	F L Collings Pengover, Liskeard
1949	F L Collings Pengover, Liskeard 'Yealmstone Winston'	W A Whidden Brampford Speke, Exeter 'Treringey Lady Choice'	W A Whidden Brampford Speke, Exeter 'Treringey Lady Choice 6th'

1950		G H Eustice Bezurrell, Hayle	G H Eustice Bezurrell, Hayle
1951	W A Johns Treringey, Crantock	G H Eustice Bezurrell, Hayle	W A Johns Treringey, Crantock
1952	W A Johns Crantock, Newquay	G H Eustice Bezurrell, Hayle	G H Eustice Bezurrell, Hayle
1953	G H Eustice Bezurrell, Gwinear 'Traine Grand Boy'	W L Best Fursdon Farm, Liskeard 'Bezurrell Mona 71st'	W L Best Fursdon Farm, Liskeard 'Bezurrell Mona 71st'
1954	G H Eustice Bezurrell, Gwinear	H R Jasper	H R Jasper
1956	F H Doidge Lamerhooe, Tavistock 'Tamar Donald'	William A Johns Kenwyn, Truro 'Treringey Lady Choice 31st'	William A Johns Kenwyn, Truro 'Treringey Lady Choice 31st'
1957	G H & D Eustice Bezurrell, Gwinear	G H & D Eustice Bezurrell, Gwinear	
1958	F L Collings Trewolland, Liskeard	William A Johns Kenwyn, Truro	
1959	G H & D Eustice Bezurrell, Gwinear	J H Pearse & Sons Berry Pomeroy, Totnes	
1960	G H & D Eustice Bezurrell, Gwinear	F L Collings Trewolland, Liskeard	
1961	G H & D Eustice Bezurrell, Gwinear		
1964	G H & D Eustice Bezurrell, Gwinear	G H & D Eustice Bezurrell, Gwinear	
1965	G H & D Eustice Bezurrell, Gwinear 'Bezurrell Mark 5th'	F L Collings Trewolland, Liskeard 'Trewolland Sunshine 85th'	
1966	F L Collings Trewolland, Liskeard 'Trewolland Mike 2nd'	G H & D Eustice Bezurrell, Gwinear 'Bezurrell Actress 55th'	
1967	F L Collings Trewolland, Liskeard 'Bezurrell Mark 6th'	G H & D Eustice Bezurrell, Gwinear	
1968	G H & D Eustice Bezurrell, Gwinear	W A Johns Kenwyn, Truro	

BRITISH LOP PIGS

	BEST BRITISH LOP BOAR *Exhibitor*	BEST BRITISH LOP SOW OR GILT *Exhibitor*
1969	F L Collings Trewolland, Liskeard 'Trewolland Mike 16th'	F L Collings Trewolland, Liskeard 'Trewolland Sunshine 122nd'
1970	T G Richards Porthleven, Helston 'Methleigh Hero 38th'	F L Collings Trewolland, Liskeard 'Trewolland Sunsine 132nd'

1971	T G Richards Porthleven, Helston 'Methleigh Hero 40th'	F L Collings Trewolland, Liskeard 'Trewolland Sunshine 132nd'
1972	G & J Collings Liskeard	G H & D Eustice Bezurrell, Gwinear

WESSEX SADDLEBACK PIGS

	BEST WESSEX SADDLEBACK BOAR	BEST WESSEX SADDLEBACK SOW	CHAMPION WESSEX SADDLEBACK
	Exhibitor	*Exhibitor*	*Exhibitor*
1937			G S Bray Truscott, Launceston
1938			G A Cole Sidbury Mills, Sidmouth 'Sid Vale Novice'
1939			G A Cole Sidbury Mills, Sidmouth
1947			Trewithen Pedigree Farms Trewithen, Grampound Road
1948			Hammetts Dairies Ltd Pinhoe, Exeter
1949			A J Dibley Alton, Hants
1950			A W Quance Wootton, Shebbear 'Plumley Dinah 25th'
1951	W G Williams & Son Coleshill, Swindon	W G Williams & Son Coleshill, Swindon	
1952	Trewithen Pedigree Farms Trewithen, Grampound Road	Trewithen Pedigree Farms Trewithen, Grampound Road	
1953	W R Jackson Redhill, Wrington 'Chancellor's Viscount 22nd'	W R Jackson Redhill, Wrington 'Chancellor's Spot Love 95th'	
1954	W R Jackson Redhill, Wrington	W R Jackson Redhill, Wrington	
1956	W R Jackson Redhill, Wrington 'Chancellor's Viscount 27th'	W R Jackson Redhill, Wrington 'Chancellor's Spot Love 95th'	
1957	Lt Cmdr L P H Bott RN Craddock, Cullompton	W R Jackson Redhill, Wrington	
1958	W R Jackson Redhill, Wrington	Lt Cmdr J P H Bott, RN Craddock, Cullompton	
1959	Lt Cmdr J P H Bott RN Craddock, Cullompton	Lt Cmdr J P H Bott RN Craddock, Cullompton	
1960	G S Bray Ercildoune, Launceston	W R Jackson Redhill, Wrington	

1961	W R Jackson Redhill, Wrington	Lt Cmdr J P H Bott RN Craddock, Cullompton
1962	J A Church Stokegabriel, Totnes 'Merrywood Renown'	Lt Cmdr J P H Bott RN Craddock, Cullompton 'Garth Sunbeam 81st'
1964	Lt Cmdr J P H Bott RN Craddock, Cullompton	W R Jackson Redhill, Wrington
1965	Whiteways Cyder Co Ltd Whimple, Exeter 'Garth Viscount 210th'	W R Jackson Redhill, Wrington 'Chancellor's Babble 97th'
1966	Whiteways Cyder Co Ltd Whimple, Exeter 'Whimple Viscount 4th'	Whiteways Cyder Co Ltd Whimple, Exeter 'Whimple Violet 22nd'
1967	Lt Cmdr J P H Bott RN Craddock, Cullompton	Lt Cmdr J P H Bott RN Craddock, Cullompton

LANDRACE PIGS

	BEST LANDRACE BOAR *Exhibitor*	BEST LANDRACE SOW *Exhibitor*	CHAMPION LANDRACE *Exhibitor*
1957			Alan H Rose Bordon, Hants 'Lindford Crava Dansy'
1958			Douglas A Pratt Lewdown, Devon 'Bluegate Aurora 13th'
1959			W F J Bell Callington 'Penlittle Ace 24th'
1960			W H Bond & Sons Saltash
1961			A J & G J Stephens Heathfield Lodge, Tavistock
1962			W H Bond & Sons Saltash 'Trerule Celia 48th'
1964			A W Stephens & Son Callington
1965			A W Stephens & Son Callington 'Pencreber Belinda 61st'
1966			M J Tancock Callington 'Hingston Dragon 8th'
1967			J F & M J Tancock Metherell, Callington
1968			J G H & E M Laity Kea, Truro
1969			J G H & E M Laity Kea, Truro 'Carlyon Vega 3rd'

1970		A W Stephens & Son Callington 'Pencreber Belinda 118th'
1971		J G H & E M Laity Truro 'Carlyon Bishop 35th'
1972		A W Stephens Callington

BRITISH SADDLEBACK PIGS

	BEST BRITISH SADDLEBACK BOAR Exhibitor	BEST BRITISH SADDLEBACK FEMALE Exhibitor	CHAMPION BRITISH SADDLEBACK Exhibitor
1968	Lt Cmdr J P H Bott Craddock, Cullompton	D P Paynter Penhayes, Grampound	
1969	W F Matthews Pengelly, Blisland 'Highgelly Golden Arrow 5th'	Whiteways Cyder Co Ltd Whimple, Exeter 'Heather 24th'	
1970	W F Matthews Pengelly, Blisland 'Highgelly Golden Arrow 5th'	D Paynter Penhayes, Grampound 'Merrywood Rosette 120th'	
1971			Whiteways Cyder Co Ltd Whimple, Devon 'Whimple Heather 44th'

PIGS — SPECIAL AWARDS

FAT PIGS OF ANY BREED, which on being slaughtered yields the best bacon
Exhibitor

1892	George R Oliver Trehane, Probus	

BEST BOAR, calculated to produce pigs most suitable for bacon curing purposes

	Exhibitor	Breed
1924	W J Thomas Treskerby, Redruth	Large White 'Treskerby Baron 2nd'
1925	W White & Sons Pool Farm, Taunton	
1932	Sir G A Cooper Bt Hursley Park, Winchester	Large White 'Farley Signal 59th'

BEST SOW, calculated to produce pigs most suitable for bacon curing purposes

	Exhibitor	Breed
1924	George Blight & Son Tregonning, Breage	Berkshire
1925	R Gynn Davidstow, Camelford	
1932	W White & Son Poole Farm, Taunton	Large White 'Taunton Champion Bonetta'

BEST PIG or PEN of PIGS
in the Showyard

	Exhibitor	Breed
1870	Sydney Davey Redruth	Small Breed 'Sugar and Salt'
1875	Thomas Salmon St Columb	Small Breed
1877	Messrs R & J Russell Sithney	
1878	Mr Partridge Bow	Small Breed
1879	Thomas Salmon St Columb	Small Breed
1880	Lord Moreton Whitfield Farm, Falfield	Small Breed
1881	J Andrew St Columb	Small Breed
1882	W S Northey Tinhay, Lifton	Small Breed
1883	Harry Hoblyn St Columb	Small Breed
1885	Thomas Roberts Trengothal, St Levan	Small Breed
1891	Ward & Chowen Brentor, Tavistock	
1892	Ward & Chowen Burnville, Tavistock	
1900	James Frayne Pipers Pool, Egloskerry	
1904	R S Olver Trescowe, Bodmin	
1905	N Stephens Bodmin	
1907	J Warne Treveglos, St Mabyn	
1910	T Warne St Mabyn	
1911	J C Olver Woodland Valley, Ladock 'Queen of the Valley II'	
1912	J Warne St Mabyn	
1914	John Warne Treveglos, St Mabyn	
1915	T Warne Trevisquite, St Mabyn	
1919	J Warne Tregony, Truro	

PRESIDENT'S TROPHY—
EXHIBITOR GAINING MOST POINTS IN THE PIG AND SHEEP SECTIONS.
From 1973, Sheep Section Only

Year	Exhibitor	Year	Exhibitor
1961	John C Gynn, Boscastle	1978	P Kingdon & Sons, Newquay
1962	J F Julyan, Truro	1979	A T Guard, Torrington, Devon
1963	G H Tripp, Portscatho, Truro	1980	A T Guard, Torrington, Devon
1965	W R Jackson, Wrington, Somerset	1981	J A Darke Ltd, Kingsbridge, Devon
1966	W F Matthews, Blisland, Bodmin	1982	A T Guard, Torrington, Devon
1967	K W Watts, Okehampton	1983	A T Guard, Torrington, Devon
1968	Morley Hicks, Truro	1984	Drake Bros, North Tawton, Devon
1969	F L Collings, Liskeard	1985	D Darke, Kingsbridge, Devon
1970	F R Cameron, Gorran, St Austell	1986	I G Mortimer, Chagford, Devon
1971	F L Collings, Liskeard	1987	Presingoll Farm Ltd, St Agnes
1972	A T Guard, Torrington, Devon	1988	Drake Bros, North Tawton, Devon
1973	J F Julyan, Ruan High Lanes	1989	I G Mortimer, Chagford, Devon
1974	A T Guard, Torrington, Devon	1990	E W Quick & Sons, Crediton, Devon
1975	A T Guard, Torrington, Devon	1991	E W Quick & Son, Crediton, Devon
1976	Drake Bros, North Tawton, Devon	1992	C Mortimer, Chagford, Devon
1977	W T R Hawke & Son, Newquay		

HORSES SECTION RESULTS – 1794 TO 1992

BEST STALLION

Exhibitor
- 1794 John Tyeth
  Launceston
- 1795 John Howard
  Bechym
- 1796 Richard Grose
  St Kew, Bodmin
- 1797 Thomas Hallett
  Mawgan in Pyder

BEST STALLION, for Pack & Cart
- 1799 William Eddy
  St Michael Carhayse

BEST STALLION
- 1802 William Cayzer
- 1803 Nicholas Stephens
  Bodmin

HORSES – NON AGRICULTURAL

BEST ENTIRE HORSE,
calculated to improve the
breed of horses for the saddle
in this county

Exhibitor
- 1828 C Peters
  Probus
- 1829 J Taunton
  Liskeard
- 1830 Mr Lakeman
  Launceston
- 1831 Mr Donnithorne
  Creed
- 1832 C Trelawney
- 1833 Mr Fry
  St Teath
- 1834 W Moore
  Grampound
- 1835 J B Trevanion
  Carhayes

- 1836 W Winsor
  Torbryan, Devon
- 1837 William Tremain
  Newlyn
- 1838 Joseph Fawcett
  St Erme
- 1839 J Hendy
  Ladock

BEST MARE,
calculated to improve the
breed of horses for the saddle
in this county

Exhibitor
Thomas Daniell
  Feock
Mr Bullmore
  Newlyn
T Bate
  St Clement
Mr Kendall
  Probus
Thomas Daniell
  Feock
Mr Glasson
  Godolphin
G Bullmore
  Newlyn
James Hendy
  Trethurffe, Ladock

BEST BRED MARE, (hitherto
kept or used for this season,
to be used as a brood mare)
calculated to improve the breed of
horses for saddle in this county

James Pernewan Jnr
  Sennen
James Hendy
  Trethurffe, Ladock
William Hodge
  Perranzabuloe
W Veale
  St Columb Major

		BEST BROOD MARE, calculated to improve the breed of horses for the saddle in this county and bona fide kept for breeding purposes	
		Exhibitor	
1840	J H Tilly Tremough	J Magor Cosworth	
1841	J H Tilly Tremough	G W F Gregor Trewarthenick	
1842	H Tilly Tremough	James Hendy Trethurffe	
1843	J H Tilly Tremough	T Harvey Falmouth	
1844	J H Tilly Tremough		
1845	John Hitchens Penryn		
1846	John Hitchens Penryn	J Tremain Trerice, Newlyn	
1847	J D Gilbert Trelissick	J D Gilbert Trelissick	
1848	Mr Puddicombe Devon	G W Bevan	
			BEST 3 YEAR OLD COLT or FILLY, calculated for carriage purposes (stallions serving during the season not qualified to contend)
			Exhibitor
1849	John Cole Buckworthy, Devon	J D Gilbert Trelissick	T H Tilly Tremough
1850	Charles Trelawny Plymouth		J D Gilbert Trelissick
			BEST 3 YEAR OLD COLT or FILLY, calculated for the saddle, or single or double harness (agricultural horses and riding stallions serving during the season not qualified to contend)
			Exhibitor
1851	Robert Dunn St Austell	T H Tilly Tremough	J D Gilbert Trelissick
1852	R Sparks Truro	James Hendy Trethurffe, Ladock	J D Gilbert Trelissick
1853	James Hendy Trethurffe, Ladock	William Drew Creed	
1854	Richard Sparks Truro	Francis Stocker Carhayes	James Tremain Newlyn

			BEST 3 YEAR OLD COLT or FILLY, calculated for the saddle
			Exhibitor
1855	John Tremayne Heligan	William Michell Gwennap	William Michell Gwennap
1856	Sampson Taylor Laneast	James Tremain Newlyn	John Tremain Newlyn
1857	Samuel Harvey Sennen	John Tremain St Columb	Henry Michael Williams Perran Wharf
1858			James Tremain Newlyn

THOROUGHBRED STALLION

1865 Brydges Willyams
    Nanskeval
1869 R Yeo
    Bodmin
1874 H Laity
    Crowan
1875 H Laity
    Crowan
1876 Representative of the
    Late Lord Glasgow
1877 Col Ballard
    South Wales
1878 T K Bickle
    Lamerton
1880 H Laity
    Crowan

THOROUGHBRED STALLIONS,
to be used exclusively in
Cornwall in the year of the
show

1884 Yeo Bros
    Bodmin & Plymouth
1885 Henry Laity
    Praze, Camborne
1886 Charles Laity
    Townshend, Hayle
1887 Charles Laity
    Townshend, Hayle
1888 Richard Cardell
    Trebelsue,
    St Columb Minor
1888 Henry Laity
    Praze, Camborne
1890 Richard Cardell
    Trebelsue, St Columb
    Minor
1892 Henry Laity
    Praze Stud Farm,
    Camborne
1894 Henry Laity
    Praze Stud Farm,
    Camborne

Year	Name	Location	Horse
1895	Henry Laity	Praze	
1896	T K Bickle	Lamerton	
1897	Henry Laity	Praze, Camborne	
1898	J K Bickle	Lamerton	
1899	T K Bickell	Lamerton	
1900	Henry Laity	Praze, Camborne	
1901	Henry Laity	Praze, Camborne	'Lord Molynoo'
1902	T K Bickell	Lamerton, Tavistock	'Taget'
1903	The Moor Grove Stud Co	Lelant	'Button Park'
1904	The Moor Grove Stud Co	Lelant	'Button Park'
1905	The Moor Grove Stud Co	Lelant	'Button Park'
1906	The Moor Grove Stud Co	Lelant	
1907	The Moor Grove Stud Co	Lelant	'Button Park'
1908	The Moor Grove Stud Co	Lelant	'Squire Darling'
1909	Execs of the late Gen Jago	Trelawney, Coldrenick	'Kans'
1910	John Williams	Scorrier House, Scorrier	'Squire Darling'
1911	Martyn Taylor	Ermington, Ivybridge	'Flaxby'
1912	Charles Laity Jnr	Camborne	'Irish Linen'
1914	T K Bickell	Lamerton, Tavistock	'Tryanny II'
1915	William Craze	Penzance	'Kano'

1919 J Warne
       Tregonhayne, Tregony
       'Bexhill'
1920 T K Bickell
       Lamerton, Tavistock
       'Neyland'
1921 T K Bickell
       Lamerton, Tavistock
       'Neyland'
1922 Alfred Renfree
       Bicton Farm, Hatt
1923 F J Parsons
       Venn Barton Stud, Beaworthy
       'Longboat'
1924 A E G Renfree
       Bicton Stud, Hatt
       'Suspiro'

HACKS OR HUNTERS

BEST STALLION	BEST MARE and foal or in foal
Exhibitor	*Exhibitor*
1858 James Northy Borough Kelly, Devon	Frederick Williams Tregullow
1859 James Northy Borough Kelly, Devon	James Tremain Newlyn
1860 Richard Quick St Just	Thomas P Tyacke Helston
1862 Richard Sparks Truro	Frederick Smale Lewannick
1863 Mr Trelawny Coldrenick	
1864 William Barrett Puddavean, Totnes	Major Carlyon Tregrehan, St Austell
1870	Richard Cleave Advent
1871 J K Bickell Tavistock	
1872 J K Bickell Tavistock	Capt Holder Jetwell Hse, Camelford
1873 J E Bickell Tavistock	Viscount Falmouth Tregothnan
1874	Major Carlyon Par Station
1875	Viscount Falmouth Tregothnan
1876	Viscount Falmouth Tregothnan
1878	W Stephens Wadebridge
1879	Wesley Stephens Hendra, Wadebridge
1880	W Marrick Liskeard

BEST STALLION,
capable of getting roadsters
Exhibitor

1881	H Laity Praze, Camborne	W Stephens Hendra, Wadebridge
1882	Henry Chapman MP Lincolnshire	Samuel Hicks Bodmin
1883	J K Bickell Lamerton	S Hicks Bodmin
1884	Henry Laity Praze, Camborne	Richard Cardell Trebelsue, St Columb Minor
1886		W J Treneer Degrembris, Newlyn East

RIDDEN HUNTERS

CHAMPION RIDDEN   BEST LADIES
HUNTER   HUNTER
Exhibitor   *Exhibitor*

1924 Mrs Savile Petch
     Milborne Port, Somerset
     'Astrologer'
1925 Mrs J Pooley
     Rosewarne, Gwinear
1926 Constance, Duchess of
     Westminster
     Sparkford, Somerset
1927 Mrs J Pooley
     Rosewarne, Gwinear
     'Red Flame'
1928 Mrs J Pooley
     Rosewarne, Gwinear
1929 J K Stevenson
     Welland, Malvern
     'Blue Train'
1930 Mrs Savile Petch
     Milborne Port, Sherborne
1931 J Pooley
     Rosewarne, Hayle
     'Eclipse II'
1932 J Pooley
     Rosewarne, Hayle
1933 Mrs Saville Petch
     Milborne Port, Sherborne
     'Trusty Knight'
1934 Mrs Savile Petch
     Milborne Port, Sherborne
     'Trusty Knight'
1935 Savile Petch
     Milborne Port, Sherborne
     'Viceroy'

1936	Miss K S Hutchinson Stoney Bridge, Par 'Foxhunter'	
1937	Savile Petch Milborne Port, Sherborne	
1938	Savile Petch Milborne Port, Sherborne	
1939	Savile Petch Milborne Port, Sherborne	
1947	N H Partridge Castle Horneck, Penzance	
1948	J F Cann Newland, Cullompton	
1949	J F Cann Newland, Cullompton	
1950	R J Lobb Norton, Bodmin 'Airmail'	
1951	J F Cann Newland, Cullompton	
1953	R Powning Camborne 'Royal Sovereign'	
1954	R J Lobb Norton, Bodmin 'Golden Slipper'	
1956	Lt Col & Mrs A B Coote Ashton, Exeter 'Counter Attack'	
1958	Mrs V W B Smyth MFH Stock Rivers, Barnstaple 'Whiskey'	
1959	P J H Murray Smith Kingsbridge, Devon 'Sandyman'	
1961	P J H Murray Smith South Milton, Kingsbridge 'Free Lance'	
1962	F M Lawrey Long Rock, Penzance 'Marlborough'	
1963	Miss B Murphy Winford, Bristol 'Boy Blue'	
1965	P J H Murray-Smith Kingsbridge, Devon 'Fine Art'	Miss J S Talbot-Clayton Cirencester, Glos 'Island Gem'
1966	Miss B Murphy Winford, Bristol 'Boy Blue'	Miss B Murphy Winford, Bristol 'Boy Blue'
1967		Miss B Murphy Winford, Bristol 'Boy Blue'

1968		C N Evans Newton Abbot 'Myritus'
1969		Mrs R Chichester Wiscombe, Colyton 'What A Game'
1970		Miss M Scott Trecorner, Altarnun 'Stud Poker'
1971		Miss B & Miss A Record Alresford, Hants 'Crown Land'
1972	A H Newbury Oaklands, Exeter 'Game Fair'	
1973	M Westaway Mardon, Paignton 'Carver Doone'	
1974	Witney Horse Blanket Co Witney, Oxon 'Game Fair'	Witney Horse Blanket Co Witney, Oxon 'Game Fair'
1975	Miss M L Steavenson Darlington, Co Durham 'King of Zenda'	R Singleton Bovey Tracey, Devon 'Nubian'
1976	L G Fox Newent, Glos 'Master Kempley'	
1977	Miss M L Steavenson Darlington, Co Durham 'Overflow'	Mrs N Harley Newent, Glos 'Lord and Master'
1978	P White Newent, Glos 'Dual Gold'	Mr & Mrs P B Warren Kernick, Penryn 'Better Times'
1979	Mr & Mrs P B Warren Kernick, Penryn 'Frosty Court'	Mr & Mrs P B Warren Kernick, Penryn 'Better Times'
1980	Mobility Ltd Newent, Glos 'The Consort'	Mrs M Priest Newent, Glos 'Brewster'
1981	C Sandison Newent, Glos 'Glenstawl'	Mr & Mrs J A Crofts Albury, Surrey 'Princely Way'
1982	Gem's Signet Bloodstock Ltd Newent, Glos 'Brewster'	Gem's Signet Bloodstock Ltd Newent, Glos 'Brewster'
1983	Miss R M Dudley Cardiff, S Wales 'Freckles'	Gem's Signet Bloodstock Ltd Newent Glos 'Brewster'
1984	Mr & Mrs J A Crofts Albury, Surrey 'Periglen'	Mrs E Kelly Lanivet, Bodmin 'Royal Crown'
1985	Gem's Signet Bloodstock Ltd Upleadon, Newent 'Dancin'	Mrs E Kelly Lanivet, Bodmin 'Royal Crown'

| 1986 | Mrs D Whiteman
'Iceman' | Mrs & Mr D K E Mainwaring
Chippenham, Wilts
'Niatyks' | **BEST WORKING HUNTER** |
|---|---|---|---|
| 1987 | Mr & Mrs J A Crofts
Albury, Surrey
'Periglen' | P Hobbs
Pershore, Worcs
'Micky Springfield' | Mr & Mrs J A Crofts
Albury, Surrey
'Boley Hill' |
| 1988 | Miss R V O White
Market Harborough
'Classic Tales' | Sir Richard Cooper Bt
Kirthington, Oxford
'Leicester' | Miss Z E Dunstan
Penryn
'Crackrattle' |
| 1989 | Miss R V O White
Oakham, Rutland
'Classic Tales' | Miss R V O White
Oakham, Rutland
'Life Guard' | Mrs W J & Miss E Gibson
Bristol
'Mickey Springfield' |
| 1990 | Mr & Mrs R J Claydon
Silverley, Newmarket
'Carnival Time' | S R Elford
Whitestone, Exeter
'Bubbling Lord' | Mr & Mrs R J Claydon
Silverley, Newmarket
'Carnival Time' |
| 1991 | Mr & Mrs D Curtis
Market Harborough
'Sudden Flight' | Mrs R J Claydon
Silverley, Newmarket
'Carnival Time' | Mrs J Doble
Ilminster, Somerset
'Clonmore Silver' |
| 1992 | Mrs P Caldwell
Milton Keynes, Bucks
'Sir Harry Stokracy' | M J Doidge
Horrabridge, Yelverton
'March Wind' | Mrs T F J Ball
St Columb
'Robert The Tinner' |

BEST RIDDEN COB

Exhibitor

1963 Mrs Z S Clark
Southampton
'Sport'

1965 T H Lobb
Bodmin
'Lady Fern'

1966 E D Richards
Helston
'Marciano'

1991 Mrs R J Claydon
Silverley, Newmarket
'Mr Toad'

HUNTER BREEDING

BEST EXHIBIT—
HUNTER YOUNG STOCK

Exhibitor

1968 J B Stevens
Twyford, Banbury

1969 Lt Col & Mrs A B Coote
Ashton, Exeter, Devon
'Budget'

1970 Mr & Mrs W C Skinner
Exeter, Devon
'Rougette'

1971 Mr & Mrs N C Skinner
      Farringden, Exeter
      'Ladram Bay'
1973 D Oliver
      Bodmin
1974 S Luxton
      Okehampton, Devon
      'Patrick'
1975 Mrs A Ferguson
      St Peter Port, Guernsey
      'Clipston'
1976 Mr & Mrs J N T May
      Morwenstow
      'Killin'
1977 Mr & Mrs R J Burrington
      Exeter, Devon
      'Another Fortune'
1978 S Luxton
      Okehampton, Devon
      'Colonel'
1979 S Luxton
      Okehampton, Devon
      'The Doper'
1980 S Luxton
      Okehampton, Devon
      'The Doper'

BEST EXHIBIT—
HUNTER BROOD MARES & YOUNG STOCK

1987 W D Kellow
      Bodmin
      'Morgan's Boy'
1988 W D Kellow
      Bodmin
      'Newton Annabla'
1989 W D Kellow
      Bodmin
      'Howard's Way'
1990 W J Jordan
      Chagford, Devon
      'Pop's Birthday'
1991 W J Jordan
      Chagford, Devon
      'Pop's Birthday'
1992 W D Kellow
      Bodmin
      'Top Notch'

HACKS

	BEST EXHIBIT— HACK BROOD MARES & YOUNG STOCK	BEST HACK, Mare or Gelding any height or age, to be ridden by a lady
	Exhibitor	*Exhibitor*
1914		W H Yeo Durnford Mews, Stonehouse 'Rainbow'
1915		W E Arthur Burgotha, Grampound Road 'Rasper'
1919		A J Jackman Plymouth
1920		A J Jackman Plymouth
1921		Mrs J A S Pooley Rosewarne, Gwinear 'Beechwood'
1922		John Richards Porthleven 'Roseberry Despatch'
1923		Mrs J Oscar Muntz Foxhams, Horrabridge
1924		Miss V Wellesley Chard, Somerset
1925		Mrs T Oscar Muntz Foxhams, Horrabridge
1926		Mrs T Oscar Muntz Foxhams, Horrabridge

	BEST EXHIBIT— HACK YOUNG STOCK	BEST RIDDEN HACK
	Exhibitor	*Exhibitor*
1965	Miss M Deane Minehead, Somerset 'Homerule'	
1966	Miss P A Clark Newquay 'Carousel'	M A J Sexon Barnstaple, Devon 'Olympian'
1967	Miss M De Beaumont Bodmin 'Shalbourne April-Love'	Mrs W Mitchell Patching, Sussex 'Love Melody'
1968		A H Newbery Oaklands, Exeter 'Good News'
1969	Mr & Mrs N C Skinner Exeter, Devon 'Red Shoes'	Miss De Beaumont & Miss D Kellow Bodmin 'Shalbourne April-Love'
1970	Mrs R Pritchard Sherborne, Dorset 'Rush Hour'	Miss Du Maurier & Miss N Welsh Manaton 'Belle De Nuit'

BEST EXHIBIT—HACK BROOD MARES & YOUNG STOCK

	Exhibitor	Exhibitor
1971	Mrs H Roberts Trinity, Jersey 'Superstar'	M A J Sexon Barnstaple 'Roebisk'
1972		Mrs B Routledge Morwenstow 'Ridgewood Venture'
1973	Mr & Mrs N A Rees Padstow	
1974	D Kellow Bodmin 'Shalbourne Honeymoon'	Mrs H Roberts Closworth, Yeovil 'Super Shell'
1975	W J Hawke Liskeard	
1976	D Kellow Bodmin 'Honeysuckle Rose'	
1977	Mr & Mrs N A Rees Padstow 'Vanity Fair'	
1978	Mrs D Kellow Woodlands, Bodmin 'Shalbourne New Romance'	
1979	Miss J Leedham Fordingbridge, Hants 'Starlyte Candora'	
1980	Mr & Mrs J R Wilton Padstow 'St Lawrence Orbella'	
1981	Mr & Mrs K J Burton Beausale, Nr Warwick 'Cuckoo Waltz'	
1982	M A J Sexon Barnstaple, Devon 'Tawstock Clarissa'	
1983	M A J Sexon Barnstaple, Devon 'Tawstock Clarissa'	
1984	Miss J E Moore Barnstaple, Devon 'Moorland Mig'	
1985	M A J Sexon Barnstaple, Devon 'Tawstock Clarissa'	
1986	G T Tetsill Chapel, Newquay 'Trembleath Saranrose'	
1987	Mrs P Duff Bodmin 'Honeysuckle Rose'	

1988 Mrs D L Harvey
     Kingskerswell, Devon
     'Celtic Theme'
1989 Mrs A G Loriston-Clarke
     Brockenhurst, Hants
     'May Dancer'
1990 Mrs C Rodrigues   Mr & Mrs L A Whitehall
     Cheshunt, Herts   Milton Keynes
     'Cratfield Fairy Doll'   'Mexican Mill'
1991 R E D Pritchard
     Sturminster Newton, Dorset
     'Horsehill Starlight Express'
1992 R E D Pritchard   Mr & Mrs M J Jerram
     Sturminster Newton, Dorset   Great Dunmow, Essex
     'Horsehill Starlight Express'   'Piran Pyca'

ARABS

	CHAMPION PURE BRED ARAB	CHAMPION ANGLO or PART-BRED ARAB
	Exhibitor	*Exhibitor*
1947		Mrs M Masters St Lawrence Stud, Bodmin
1948	Mrs N Heath Horsham, Surrey	Miss M Rowland, Plympton, Plymouth & Mrs M Masters, Bodmin
1949		Mrs Markby Bokelly, St Kew
1950		J H Hawkey Lanteglos-by-Fowey
1951		J H Hawkey Lanteglos-by-Fowey
1953	Miss Scard Morgan Venton Veorm, Liskeard 'Arcturus'	Mrs & Miss R Markby Bokelly, St Kew 'Toyba'
1954	Mr & Mrs A L Masters Bodmin	A E Matthews Bodmin
1956	Mrs E M Thomas Ottery St Mary, Devon 'Lakme'	Mrs F M Moore Trewen, Camelford 'Mikado'
1987		Mrs A Holmes Ottery St Mary, Devon 'Abbas Sparkle My Lovely'
1988	M Harris Honiton, Devon 'Sheruggi'	Mrs A Holmes Ottery St Mary, Devon 'Abbas Racing Silk'
1989	Mrs M L Yates Camelford 'Lucian Sunset'	Mrs A Holmes Ottery St Mary, Devon 'Abbas Sparkle My Lovely'

1990	Mrs F M Robinson, Mr K B Bernard, Mr M B Jones, Redruth 'Jaritah Sbeyel'	Mr & Mrs J R Wilton Padstow 'St Lawrence Serenade'	BEST RIDDEN ARAB— PURE BRED, ANGLO or PART BRED
1991	Mrs P Evans Lostwithiel 'Tarib'	Mrs A Holmes Ottery St Mary, Devon 'Abbas Sparkle My Lovely'	Miss W Lane Kingsteignton, Devon 'Diptford Spring Tide'
1992	Mrs S Talken-Sinclair Fairseat, Tintagel 'Lyndham Tarac'	Mrs A Holmes Ottery St Mary, Devon 'Abbas Sparkle My Lovely'	Mrs A Holmes Ottery St Mary, Devon 'Abbas Sparkle My Lovely'

APPALOOSA

CHAMPION APPALOOSA

Exhibitor

1989 Mrs Y Du Laney-Brown
South Molton, Devon
'Rivaz Soventus'
1990 Mrs R Townsend
Yelverton, Devon
'Bold Lancer'
1991 Mrs R Townsend
Yelverton, Devon
'Bold Lancer'
1992 N J Bellenie
Woodstock, Oxon
'Myriad Beesknees'

PONIES — RIDDEN

CHAMPION
CHILDREN'S RIDING
PONY

Exhibitor

1925 W J C Johns
West Trewirgie, Redruth
1926 Miss S Calmady Hambyn
Buckfast, Devon
1927 A Partridge & Son
Mordref, Plympton
1928 A Partridge & Son
Mordref, Plympton
1929 Mrs J Vinson Thomas
Little Lyndridge,
Okehampton
1931 F A Bate
Launceston
1932 A M Williams
Werrington Park,
Launceston
1933 J P Warren
Gweek, Helston

1934 Miss E Williams
    Pressland House,
    Hatherleigh
1935 Mrs C R Stephens
    Restronguet, Falmouth
1947 Susan Fletcher
    Nynehead, Wellington
1953 Mrs Matthews
    Blackdown, Duloe
    'Miranda'
1954 Lt Col & Mrs J F S Bullen
    Charmouth
1956 David Kellow
    Sweetshouse, Bodmin
    'Gaiety-Girly'
1957 Miss Joy Kestell
    Padstow
    'Mifari Pretty Polly'
1958 Miss P M & B A Currin
    Totnes, Devon
    'Wise Penny'
1962 Miss M De Beaumont
    Bodmin
    'Shalbourne Mayflower'
1963 Miss D Kellow
    Bodmin
    'Welland Valley Love Wire'
1965 A H Newbery
    Exeter, Devon
    'China Tea'
1966 Hall & Newbery
    Exeter, Devon
    'Creden Lucky Charm'
1968

1969 Mr & Mrs M W Kempe
    Minehead, Somerset
    'Winaway Springtime'
1970 Miss T Bell
    Callington
    'Lemington Carousel'
1971 Dr & Mrs M G Scott
    Cookham, Berks
    'Bennochy Ailsa Craig'
1974 J S Hine
    Tewkesbury, Glos
    'Minuette of Cusop'
1975 Mrs J H Maynard
    Waley Bridge via
    Stockport, Cheshire
    'Weston Shade Oak'

BEST CHILD'S PONY
ON LEAD REIN
Exhibitor
Mrs E G French
    Chacewater
    'Rudgeway Tiny Tim'
Miss L Bedford
    Merthyr Tydfil
    'Westwell Jewel'
Mrs B G M Wooff
    Newton Abbot, Devon
    'Cusop Comrade'
Mrs A G Harris
    Chippenham, Wilts
    'Maidford Serena'
Mr & Mrs C M C Hancock
    Truro
    'Knockie Adam'
Miss D Opie
    Truro
    'Polaris Seisyll'

		BEST FIRST RIDDEN PONY *Exhibitor*	
1976	Mrs J Haigh Maynard Waley Bridge via Stockport, Cheshire		Mrs P Buckingham Camborne
1977	Mrs E Kelly Bodmin 'Smalland Otto'	Mrs J Nicholls Stithians, Truro 'Swan Lake'	Mrs S D Rogers Helston 'Cwmpennant Beau'
1978	Mrs E Kelly Bodmin 'Smalland Otto'	J & S Curbishley Macclesfield 'Elmwick Romeo'	Mrs S D Rogers Helston 'Cwmpennant Beau'
1979	Miss L Randall Berkeley, Glos 'Tomatin Forest Wind'	H Sinclair Tintagel 'Valley Lake Eve'	Mrs S D Rogers Helston 'Cwmpennant Beau'
1980	Mr & Mrs C M Brooks Reading, Berkshire 'Kilbees Cranbourne Tower'	Mr & Mrs C M Brooks Reading, Berkshire 'Chilmington Magician'	Mr & Mrs R J Anstey Wootton-Under-Edge, Glos. 'Bladon Stardust'
1981	Mrs J A Link Fairwood, Swansea 'Minette'	Mrs C M Buckingham Redruth 'Bleachgreen Heulwen'	Mrs C M Buckingham Redruth 'Thumb of Kilmorie'
1982	Mrs R M Boyatt Newton Abbot, Devon 'Worth Mindaw'	Mrs C M Buckingham Redruth 'Sinton Medley'	Mrs M Stephens Plymouth 'Bengad Dryas'
1983	Mr & Mrs C R Sandison Worcester, Worcs 'Harmony Bubbling Champagne'	Mr & Mrs H W Prouse Crediton, Devon 'Bengad Nodding Greenhood'	Mrs G L Lee Newton Abbot, Devon 'Marsden Hamara'
1984	Mr & Mrs C R Sandison Worcester, Worcs 'Harmony Bubbling Champagne'	Mr & Mrs H W Prouse Crediton, Devon 'Bengad Nodding Greenhood'	Mr & Mrs H W Prouse Crediton, Devon 'Bengad Nodding Greenhood'
1985	Dr & Mrs M Gilbert-Scott Cookham, Berks 'Bennochy Golden Star'	J Newbery Alphington, Exeter 'Carvolth Fairytale'	Mr & Mrs H W Prouse Crediton, Devon 'Bengad Nodding Greenhood'
1986	Mrs T Lister Coventry 'Keston New Day'	Mrs P K Ross Looe 'Smalland Picoleno'	Miss J Newbery Alphington, Exeter 'Gwyndra Pollyanna'
1987	Dr & Mrs M Gilbert-Scott Cookham, Berks 'Bennochy Royal Star'	Mrs S Berry Warminster, Wilts 'Cusop Sponsor'	Miss J Newbery Alphington, Exeter 'Downderry Man Friday'
1988	G Flower Abergavenny, Wales 'Chinook Tweed'	Mrs J Pascoe & Mrs L Trembath Newquay 'Douthwaite Melody'	Mrs P K Ross & Mrs R M Brown Looe 'Toya Tweetie Pie'
1989	Dr & Mrs M Gilbert-Scott Cookham, Berks 'Bennochy Royal Star'	Mrs C Wood Wells, Somerset 'Oakley Bunbury'	Mr & Mrs M French South Brent, Devon 'Sunrising Merle'
1990	Mrs A Alison Leamington Spa 'Cusop Dalton'	Mrs V M Gwennap Camborne 'Silver Cloud'	Mrs J E Brookshaw Berkeley, Glos 'Menai Prince Charming'
1991	Mrs J Comerford Circencester, Glos 'Sandbourne Royal Emblem'	Mrs J E Brookshaw Berkeley, Glos 'Twylands Query'	Mrs J E Brookshaw Berkeley, Glos 'Menai Prince Charming'
1992	Mrs J Comerford Cirencester, Glos 'Sandbourne Royal Emblem'	Mrs J E Brookshaw Berkeley, Glos 'Twylands Query'	Mrs C Roberts Cwmbran, Gwent 'Rosedale Marguerite'

PONIES — RIDDEN

	CHAMPION WORKING HUNTER PONY *Exhibitor*	CHAMPION SHOW HUNTER PONY *Exhibitor*
1978	Mrs J Lello Hayle 'Fair Valentine'	
1979	Miss S Richards Hayle 'Little Mistake'	
1980	Mrs J M Bates Tiverton, Devon 'Ivetsey Jason'	
1981	Mrs J M Bates Tiverton, Devon 'Grey Cygnet'	
1982	Mrs J Ramsay St Agnes 'Willow Valley Jenifer'	
1983	Miss A Bucknell & Mrs K Begley Newbury, Berks 'Page Boy'	
1984	Mrs M Begley & Miss A Bucknell Newbury, Berks 'Page Boy'	
1985	Mrs M Begley & Miss A Bucknell Newbury, Berks 'Page Boy'	
1986	Mrs M & Miss A Begley Newbury, Berks 'Little Diamond'	
1987	Mr & Mrs H W Prouse Crediton, Devon 'Treverva Lilac Time'	Mrs S Rowe & Miss L Blum Kingston Vale, London 'Sarnau Castaway'
1988	Mrs B Rowland Gunnislake 'Atstan Welsh Magician'	Mrs T A Smith Wells, Somerset 'Peter Rabbit'
1989	Mrs G Blackaller Newick, Sussex 'Huckleberry Finn'	Mrs R M Clarke Salisbury, Wilts 'Springfield Legend'
1990	Mrs T A Smith Wells, Somerset 'Peter Rabbit'	Mrs T A Smith Wells, Somerset 'Peter Rabbit'
1991	Mrs T A Smith Wells, Somerset 'Bright Valentalla'	Mrs R M Clarke Salisbury, Wilts 'Morning Melody'
1992	Mrs S Schutte South Brent, Devon 'Mynach Sixpence'	Mrs V Fenton Preston, Lancs 'Mount Caufield Outlaw'

PONY BREEDING

BEST EXHIBIT — PONY BREEDING
Exhibitor

- 1965 Mr & Mrs A L Masters
  St Lawrence Stud Farm, Bodmin
  'Shabre'
- 1969 Miss E Sollis
  Newquay
  'Rustum Cha-Cha'
- 1970 Mr & Mrs A L Masters
  Bodmin
  'Mullaghmore Corneselia'
- 1971 Mr & Mrs A L Masters
  Bodmin
  'Ridgwardine Mandy'
- 1972 Mr & Mrs A L Masters
  Bodmin
  'Mullaghmore Corneselia'
- 1973 Lynn Jeffery
  Treburgett, St Teath
  'Moon Legend'
- 1974 Mr & Mrs J A Turner
  St Agnes
  'Piran Camilla'
- 1975 Mr & Mrs J A Turner
  St Agnes
  'Piran John Halifax'
- 1976 Mr & Mrs C Rose
  Malmesbury, Wilts
- 1977 Mr & Mrs J A Turner &
  Miss A Rusden
  Perranporth
  'Piran Happy Days'
- 1978 Mr & Mrs M J Austin
  Torquay, Devon
  'Maidencombe Elmina'
- 1979 Mr & Mrs A L Masters
  Bodmin
  'Twylands Fiesta'
- 1980 Mr & Mrs A L Masters
  Bodmin
  'Twylands Fiesta'
- 1981 Mr & Mrs A L Masters
  Bodmin
  'Twylands Fiesta'
- 1982 Mr & Mrs J A Turner
  St Agnes
  'Piran Hesper'
- 1983 Mr & Mrs A J Geake
  St Austell
  'Rhodd Courtesy'
- 1984 Mr & Mrs J A Turner
  St Agnes
  'Piran Fair Dawn'

1985 Mrs C J Grain
    Chettisham, Ely
    'Twinwood Bright Eyes'
1986 Miss J Newbery
    Alphington, Exeter
    'Necta Super Star'
1987 Mrs A Holmes
    Ottery St Mary, Devon
    'Abbas Sparkle My Lovely'
1988 Mrs J Comerford
    Cirencester
    'Sandbourne Royal Emblem'
1989 Mrs M Roper-Caldbeck
    Minehead, Somerset
    'Ashlands Sugar Baby'
1990 Mr & Mrs B P Rennocks
    Cheshunt, Herts
    'Rendene Stormy Affair'
1991 Mrs P M Chance
    Exeter, Devon
    'Spinningdale Arabella'
1992 Mr & Mrs B P Rennocks &
    Mrs V Richardson
    Wormley, Herts
    'Oakley Pets Fun'

DARTMOOR PONIES

CHAMPION DARTMOOR PONY
Exhibitor
1965 J H Stephens
    Ivybridge, Devon
    'Hele Judith'
1966 Mr & Mrs T Reep
    Widecombe-in-the-Moor, Devon
    'Shilstone Rocks Windswept'
1967 Miss H McMillan
    Chard, Somerset
    'Whitmore Honeymoon'
1969 Miss J McMillan
    Chard, Somerset
    'Whitmore Honeymoon'
1970 Mrs S E Jones &
    Miss P M Roberts
    Newton Abbot, Devon
    'Penny II'
1971 Mrs S E Jones &
    Miss P M Roberts
    Newton Abbot, Devon
    'Hisley Pedlar'
1974 Mrs Jones & Miss Roberts
    Newton Abbot, Devon
    'Hisley Candlelight'
1975 Miss B & Mr J Holman
    Okehampton, Devon
    'Cosdon May'

1976 E P E Reep
    Widecombe-in-the-Moor, Devon
1977 Mrs Jones & Miss Roberts
    Lustleigh, Devon
    'Sweetie Pie'
1978 Mrs M I Gould
    Newton Abbot, Devon
    'Boveycombe Bunting'
1979 Mrs J Montgomery
    Yelverton, Devon
    'White Willows Sorceress'
1980 Miss B E & Mr J Holman
    Okehampton, Devon
    'Cosdon May'
1981 Miss B E & Mr J Holman
    Okehampton, Devon
1982 Mrs J Montgomery
    Yelverton, Devon
    'White Willows Banquo'
1983 Mrs J Montgomery
    Yelverton, Devon
    'White Willows Halloween III'
1984 Mrs W E Robinson
    Cullompton, Devon
    'Vean Mary Rose'
1985 Mrs P E Robinson
    Cullompton, Devon
    'Vean Mary Rose'
1986 J Holman
    Okehampton, Devon
    'Cosdon May'
1987 Mrs B L Challenor-Tucker
    Kingsbridge, Devon
    'Shilstone Rock Witching Hour'
1988 Mrs J Montgomery
    Yelverton, Devon
    'White Willows Bewitching'
1989 Mrs D M Alford
    Okehampton, Devon
    'Shelly Pool'
1990 Mrs J Montgomery
    Yelverton, Devon
    'White Willows Charmer'
1991 Mrs J Montgomery
    Yelverton, Devon
    'White Willows Myth'
1992 Mrs E P E Newbolt-Young
    Widecombe-in-the-Moor
    'Shilston Rock's Windswept II'

EXMOOR PONIES

CHAMPION EXMOOR PONY
Exhibitor
1966 Mrs S Griffey
    Northam, N Devon

SHETLAND PONIES
CHAMPION SHETLAND PONY
Exhibitor
1969 Mrs C J Jeffery
 Ilfracombe, Devon
 'Lee Prima Donna'
1970 Mr & Mrs W G Shillibeer
 Yelverton, Devon
 'Lakehead Emma'
1971 Mr & Mrs W G Shillibeer
 Yelverton, Devon
 'Lakehead Emma'
1974 W Waite
 Yelverton, Devon
 'Holwell Jenny'
1975 Mrs E W House
 Bridgwater, Somerset
 'Donnachaioh Elsa'
1976 Mrs C J Jeffery
 Ilfracombe, Devon
 'Blossom of Luckdon'
1977 Mrs E W House
 Bridgwater, Somerset
 'Bincombe Posy'
1978 Mr & Mrs W G Shillibeer
 Yelverton, Devon
 'Lakehead Emerald'
1979 Miss A Runnalls
 Bodmin
 'Tawna Meg'
1980 Miss A Runnalls
 Bodmin
 'Tawna Glynn'
1981 Mrs E W House
 Bridgwater, Somerset
1982 Mrs P Lory
 Helston
 'Southley Jenny Wren'
1983 Mr & Mrs J Lory
 Helston
 'Southley Jenny Wren'
1984 Mr & Mrs W G Shillibeer
 Yelverton, Devon
 'Lakehead Emerald'
1985 Mr & Mrs W G Shillibeer
 Yelverton, Devon
 'Lakehead Emerald'
1986 Mrs E W House
 Bridgwater, Somerset
 'Bincombe Volunteer'
1987 Mrs E W House
 Bridgwater, Somerset
 'Bincombe Volunteer'

1988 Mrs E W House
Bridgwater, Somerset
'Bincombe Volunteer'

1989 Mrs M G Rae
Exeter, Devon
'Hannibal of Hinton'

1990 Mrs A White
Seaton, Devon
'Lakehead Double Diamond'

1991 Mrs E W House
Bridgwater, Somerset
'Bincombe Volunteer'

1992 Mrs E W House
Bridgwater, Somerset
'Bincombe Volunteer'

WELSH PONIES AND COBS

CHAMPION WELSH PONY OR COB
Exhibitor

1974 Mr & Mrs J Carter
Dawlish, Devon
'Sinton Whirligig'

1975 Mr & Mrs J Carter
Dawlish, Devon
'Sinton Whirligig'

1976 The Countess of Dysart
South Molton, Devon

1977 Mr & Mrs J Carter
Dawlish, Devon
'Millcroft Suzuya'

1978 Mr & Mrs J Carter
Dawlish, Devon
'Millcroft Suzuya'

1979 Mr & Mrs J Carter
Dawlish, Devon
'Lydstep Ladies Slipper'

1980 Mr & Mrs R A Lutey
Newlyn East, Newquay
'Bayford Bond'

1981 Mr & Mrs R A Lutey
Newlyn East, Newquay

1982 Mrs T Taylor &
Mrs A G Eggar
South Petherton, Somerset
'Straits Spartan Saviour'

1983 J E Enders
Falmouth
'Argal Dulcimo'

1984 Mr & Mrs R A Lutey &
Mr & Mrs P Ravenshear
Newquay
'Bucklesham Prince Arthur'
1985 Mr & Mrs J Carter
Dawlish, Devon
'Millcroft Suzuya'
1986 Mrs K M Whitley
Tavistock, Devon
'Bengad Falcon'
1987 Mr & Mrs R A Lutey &
Mr S D Morgan
Newquay
'Parc Sir Ivor'
1988 Mr & Mrs J Carter
Dawlish, Devon
'Paddock Fairy Lustre'
1989 Mr & Mrs L E Bigley
Escley, Hereford
'Twyford Signal'
1990 Mr D Wray
Dewsbury, W Yorks
'Tiernon Confidence'
1991 Mr & Mrs J Carter
Dawlish, Devon
'Thornberry Royal Gem'
1992 Mr & Mrs C J & Miss P D Seward
Yeoford, Crediton
'Cefnfedw Gwenda'

MOUNTAIN AND MOORLAND PONIES

CHAMPION MOUNTAIN AND MOORLAND PONY
Exhibitor
1990 Mr & Mrs S W Bridges
Beaworthy, Devon
'Heather Jock of Tower' — Highland
1991 Miss M G Longsdon
Wincanton, Somerset
'Bew Castle Bonny' — Fell
1992 Mrs J A Webb
Praze, Camborne
'Townend Sandra' — Fell

DONKEYS

BEST MARE or GELDING
To be tested in harness
Exhibitor
1890 Simon Lawer
Trevilson, Newlyn East
1891 Simon Lawer
Trevilson, Newlyn East
1892 James Harris
Trelay Arms, Redruth

CHAMPION DONKEY
Exhibitor

1974 The Bramshill Stud
 Basingstoke, Hants
 'Bramshill Erin'
1975 Mrs J Pritchard
 Kingsbridge, Devon
 'Bandit of Merrylees'
1976 Mrs J A Hilliard
 St Agnes
1977 Mrs J L Harper
 Chard, Somerset
 'Synderford Victoria'
1978 Mrs J A Hilliard
 St Agnes
 'Grove Hill L'Antonius'
1979 R A Hoare
 Grampound
 'Bimbo of Cranbrook'
1980 Mrs D J Demus
 Bristol, Avon
 'Cedarshade Merlin'
1981 Miss T K Ford
 Coulsdon, Surrey
1982 Mrs T J Mathews
 Bridgend, Glamorgan
 'Daniel of Westra'
1983 Mrs D J Demus
 Winkleigh, Devon
 'Cedarshade Merlin'
1984 Miss S A Sperring
 Truro
 'Sandon Colombo'
1985 Mrs T J Mathews
 Bridgend, Glamorgan
 'Daniel of Westra'
1986 Mrs C Holliday
 Hope Cove, Devon
 'Bolberry Clarabelle'
1987 Mrs C Holliday
 Hope Cove, Devon
 'Bolberry Clarabelle'
1988 Mrs J Price
 Totnes, Devon
 'Happy Valley Kulio'
1989 Mrs J Price
 Totnes, Devon
 'Happy Valley Kulio'
1990 Mrs J Price
 Totnes, Devon
 'Happy Valley Kulio'
1991 Mrs J Price
 Totnes, Devon
 'Happy Valley Kulio'
1992 Mrs J E Price
 Totnes, Devon
 'Happy Valley Kulio'

AGRICULTURAL HORSES
BEST ENTIRE CART HORSE
Exhibitor
1834 R Helson
Plymouth

	BEST ENTIRE HORSE, calculated for the general purpose of husbandry, to be used in the county during the season	BEST MARE, calculated for the general purpose of husbandry and bona fide kept for breeding purposes
1840	C Peters Probus	John Hawkins Trewithen
1841	Mr Searle Lanreath	C Whitford St Erme
1842	G Holloway Fremington	W Tremaine Newlyn
1843	W Fry St Mabyn	
1844	Mr Lenyon St Allen	Mr Marrack St Just
1845	Richard Lanyon St Allen	Mr Magor Coswarth

	BEST HORSE, calculated for the general purposes of husbandry	
1846	O Hocking Helston	J D Gilbert Trelissick
1847		J D Gilbert Trelissick
1848	J K Bickell	James Tremain
1849	J Finnimore St Teath	J Tremain Trevarton, Newlyn
1850	John Guard Devon	James Tremain Trevarton, Newlyn
1851	John Guard High Bickington, Devon	John Cardell Jnr Lower St Columb
1852	Peter Davis Probus	John Cardell Jnr Lower St Columb
1853	John Finnimore St Teath	Thomas Julyan Creed
1854	William Coomb North Devon	Thomas Harris St Agnes
1855	William Parnell Altarnun	Peter Oliver St Clement
1856	W Hawke Budock	Viscount Falmouth Tregothnan
1857	John Finnimore St Teath	John Tremain Newlyn

	BEST STALLION, for agricultural purposes	BEST MARE & foal or in foal for agricultural purposes
	Exhibitor	*Exhibitor*
1858	Richard Sparks Truro	R H Vincent St Ewe
1859	John Finnimore St Teath	Viscount Falmouth Tregothnan

		BEST MARE, for agricultural purposes
1860	Francis Rogers Fowey	James Tremain Newlyn
1862	Francis Rogers Fowey	R H Vincent St Ewe
1863	William Hawke Budock	James Tremain Newlyn
1864	James Horswell North Milton	William Derry Plymouth
1865	W Hawke Budock	James Tremain Trevarthian
1867	William Jackman Bradstone, Milton Abbot	George Elliott Swilley, Plymouth
1869	H Laity Crowan	R N Branwell & Sons Penzance
1870	George Jeffrey Lidford, Devon	T Palmer & Sons Borough Kelly
1871	G Jeffery Lidford, Devon	Edwin Crago Kea

		BEST MARE & foal or in foal, for agricultural purposes
1872	J Rowe Lamorran	T Pellow Okehampton
1873	Henry Laity Crowan	James Tremain Polsue
1874	Henry Laity Praze, Crowan	Mr James Eglos Merther
1875	T B Yeo Pool, Camborne	Mr Cragoe Kea
1876	Mr Laity Praze, Camborne	W H Symons Mawgan
1877	North Cornwall Stud Co	Thomas Pellow Jnr Okehampton
1878	Henry Laity Praze, Crowan	J Pethick Plymouth
1879	Henry Laity Praze, Crowan	S T Tregaskis St Issey
1880	Henry Laity Praze, Crowan	W Hawkey St Columb
1881	Messrs Yeo Bodmin	H Browne St Austell
1882	Henry Laity Praze, Crowan	Edward Mucklow Holsworthy, Devon
1883	Henry Brown St Austell	John N Davies Gweleath, Mawgan

BEST STALLION, to be used exclusively in Cornwall in the year of the show

1884	Charles Laity St Hilary	E Bridges Wylliams Carnanton
1885	Henry Laity Praze, Camborne	
1886	Henry Laity Praze, Camborne	A D Brewer Grampound Road
1887	Henry Laity Praze, Crowan	H V Newton Polstrong, Camborne
1888	Henry Laity Praze, Camborne	H W Higman St Austell
1889	Angove Laity Townshend, Hayle	John N Davies Gweleath, Cury
1890	Angove Laity Townshend, Hayle	W J Treneer Degembris, Newlyn
1892		Richard Adams Rosevallan, Bodmin

BEST STALLION, for cart or agricultural purposes to be used exclusively in Cornwall in the year of the show

1894	Angove Laity Townshend, Hayle	John Kemp Tregaminion, St Keverne
1895	Angove Laity Townshend, Hayle	C O Peter Hoblyn Colquite
1896	Edwin Jacka Helston	H G & E P Peter Hoblyn St Mabyn
1897	J K Bickell St Johns, Lamerton	John Lintern & Son Blackabroon, Bridestow
1898	G Jeffery Dousland	J Lesberil Liskeard
1899	H G & E P Peter-Hoblyn Colquite, St Mabyn	G & O Nickell Helland
1900	H G & E P Peter-Hoblyn Colquite, St Mabyn	John H Lesbirel Liskeard
1901	Henry Laity Penzance	Walter Hicks St Austell
1902	F Jeffery Yelverton, Devon	T F James Marazion
1903	T K Bickell Hexworthy 'Hereford'	John Raby Hendra, St Germans 'Bedwell Goldflake'
1904	T K Bickell Launceston	H Percy Taunton 'Redlynch Pansy'
1905	T K Bickell Lawhitton, Launceston	J W Olver Duloe
1906	T K Bickell Lawhitton, Launceston	W F Sobey Menheniot
1907	The St Columb Stud Co St Columb	Lord Winterstoke Bristol

1908	The St Columb Stud Co St Columb	W & H Whitley Paignton, Devon
1909	The St Columb Stud Co St Columb	W & H Whitley Paignton, Devon
1910	The St Columb Stud Co St Columb	Richard Gynn Davidstow
1911	The St Columb Stud Co St Columb 'Hoole Honest Tom'	W Littleton Treffry, Bodmin 'Treffry Queen'
1912	The St Columb Stud Co St Columb	F J Olver Lanreath
1914	H Bickell Parswell Farm, Tavistock 'Langton Dray King'	W Littleton Treffry, Bodmin 'Bishopthorpe Fashion'
1915	Brooking Trant Liskeard	W & H Whitley Paignton, Devon
1920	T Balderston Lincoln	E J Maddever Liskeard
1921	Thomas Simpson Bedford 'Haynes Conqueror'	Herbert Whitley Paignton, Devon
1922	Arthur Barber Betford, Notts	Frank Pearce Gulval
1923	Simeon Rose Rosewin, Summercourt	L B Beauchamp Norton Hall, Bath
1924	Simeon Rose Rosewin, Summercourt	L B Beauchamp Norton Hall, Bath

BEST AGRICULTURAL HORSE

Exhibitor

1914	Messrs W & H Whitley Primley Farm, Paignton 'Belchford Queen'
1921	Thomas Simpson Bedford
1922	Arthur Barker Retford, Notts 'Milestone Briton'
1925	W H Uren & Son Henfor, Marazion
1926	W J Littleton Helland Barton, Bodmin
1952	H Williams Kehelland, Camborne 'Trescowe Festival Beauty'
1953	Messrs Leigh Bros Crewkerne, Somerset 'North Cote Grey Queen'
1954	W P Mingo Whimple, Exeter
1956	W P Mingo Whimple, Exeter, Devon 'Strete Melody'

1957 W P Mingo
Whimple, Exeter, Devon
'Strete Melody'
1959 A Pascoe
Fraddam, Hayle
1961 A Pascoe
Fraddam, Hayle
1962 H Pollard
Penryn
'Melody'
1963 H Pollard
Penryn
'Melody'
1964 W J Mills & Son
Newquay
1988 A L Flower
Yealmpton, Devon
'Rhyd-Y-Groes Catrin'

BEST AGRICULTURAL HORSE—
of a Breed other than Shire
1989 D J Thomas & Son
Treguddick, Launceston
'South-Win Anita'
1990 D J Thomas & Son
Treguddick, Launceston
'South-Win Tabor'
1991 D J Thomas & Son
Treguddick, Launceston
'South-Win Anita'
1992 D J Thomas & Son
Treguddick, Launceston
'South-Win Anita'

SHIRE HORSES

BEST SHIRE MARE, FILLY
or FILLY FOAL
Shire Horse Society Silver Medal
Exhibitor
1920 E J Maddever
Liskeard
'Clicker Ruby'
1921 S H Jenks
Charmouth
1922 F Pearce
Gulval
'Norbury Melba'
1923 L B Beauchamp
Norton Hall, Bath
'Chilcampton Day Dream'
1924 W J Littleton
Helland Barton, Bodmin
'Theale Crocus'

1925 W H Uren & Son
Henfor, Marazion
1926 W J Littleton
Helland Barton, Bodmin
1927 F W Parsons & Sons
Speckington, Ilchester
'Wootton Manners 3rd'
1928 W J Littleton
Helland Barton, Bodmin
1929 W J Littleton
Helland Barton, Bodmin
'Stapleton Wild Rose'
1930 F W Parsons & Son
Ilchester
1931 R James
Treisaac, St Columb Minor
'Treisaac Permanent Wave'
1932 F W Parsons & Sons
Speckington, Ilchester
1933 E S Ellacott & Sons
Tredragon, Delabole
'Speckington Abbess'
1934 J C Menhinick
Penvose, St Tudy
'Penvose Flower'
1935 J Pearce
Trevaylor, Gulval
'Silver'
1936 F W Parsons & Sons
Speckington, Ilchester
1937 F W Parsons & Sons
Speckington, Ilchester
1938 F Pascoe
Carnhell, Gwinear
'Peggy'
1939 C J Hill
Payhemburg, Honiton
'Bower Winsome Girl'
1947 W F R Bice
Nanswhyden, St Columb
'Nanswhyden Misty Morn'
1948 F Jelbert
Chy-An-Daunce, Gulval
1949 W Rowe
Pleming, Long Rock
'Ellerton Sweet Briar'
1950 W Coakes
Becton, Barton-on-Sea
1951 W Rowe
Plemming, Long Rock
1952 H Williams
Kehelland

1953	A Pascoe Fraddam, Hayle 'Fraddam Liberty'	
1954	W P Mingo Whimple, Exeter	
1956	William Rowe Pleming, Long Rock 'Pleming Wilwyn'	
1957	Hawke Bros Gilley Farm, Roche	
1959	A Pascoe Fraddam, Hayle	
1962	H Pollard Green Lane, Penryn 'Melody'	
1963	A Pascoe Fraddam, Hayle 'Fraddam Grey Lady'	
1964	H Pollard Penryn	CHAMPION SHIRE HORSE *Exhibitor*
1976		G C Pellow Camborne 'Weston Ladybird'
1977		H Pollard Penryn 'Bay View Shirley'
1978		H Pollard Penryn 'Bay View Shirley'
1979		H Pollard Penryn 'Bay View Shirley'
1980		A Pascoe Hayle 'Fraddam Selena'
1981		H Pollard Penryn
1982		A Pascoe Hayle 'Fraddam Selena'
1983		A Pascoe Hayle 'Fraddam Selena'
1984		H Pollard Penryn 'Ribwood Blossom'
1985		A Pascoe Hayle 'Fraddam Duchess'
1986		A Pascoe Hayle 'Fraddam Duchess'

1987	H Pollard Penryn 'Ribwood Blossom'
1988	A L Flower Yealmpton, Devon 'Rhyd-Y-Groes Catrin'
1989	A Pascoe Hayle 'Weather Oak April Charm'
1990	A L Flower Yealmpton, Devon 'Rhyd-Y-Groes Catrin'
1991	B D Banham Norwich, Norfolk 'Acle Pride'
1992	B D Banham Norwich, Norfolk 'Acle Belle'

PRIVATE DRIVING

	OPEN SECTION	CORNISH SECTION	CONCOURS D'ELEGANCE
	Exhibitor	*Exhibitor*	*Exhibitor*
1927	Mrs W Harding Brislington, Bristol 'Leading Article'	Mrs Webber Killigrew St, Falmouth 'Flashlight'	
1928	T Darton Deeble 10 Market Avenue, Plymouth		
1929	F Ray Carter Cot Manor, Barnstaple		
1930	Miss K M Hassard East Sussex		
1931	F Ray Carter Cot Manor, Barnstaple		

CHAMPION HACKNEY

1947	M L Oats Tregannick, Sancreed
1948	M L Oats Tregannick, Sancreed
1949	R Teague Glenwyne, Scorrier
1950	M L Oats Tregannick, Sancreed
1951	M L Oats Tregannick, Sancreed 'Cornishman'
1952	M L Oats Tregannick, Sancreed

1953 M L Oats
Tregannick, Sancreed
'Cornishman'
1956 E S Phillips
New Malden, Surrey
'Cornishman'
1957 E S Phillips
New Malden, Surrey
'Cornishman'
1958 E S Phillips
New Malden, Surrey
'Cornishman'

OPEN SECTION

1979 T B M Newbery | C Berryman
Plymouth, Devon | Falmouth
'Hanover' | 'Trelew Viking'

1980 A Maunder
Saltash
'Hurstwood Senator'

1981 Mr & Mrs W Isaac | A Maunder | Mrs P Baker
Melin, Neath, W Glam | Saltash | Nailsea, Avon
'Rhos-Y-Meirch Fly Away' | 'Brook Acres Silver Tan' | 'Cinnamon'

1982 R Warren | A Maunder | Mrs J R & Mr D Selby
Newton Abbot, Devon | Saltash | Blisland
'Grants Isla' | 'Hurstwood Senator & Brook Acres Silver Tan' | 'Domino'

1983 Mr & Mrs D & Miss R Folland | Mrs J R & Mr D C Selby | Mr & Mrs D & Miss R Folland
Devon | Blisland | Totnes, Devon
'Manmoel Mostyn' | 'Domino' | 'Manmoel Mostyn'

1984 R Warren | Mrs J R & D C Selby & Miss D Rich | Mrs P Baker
Newton Abbot, Devon | Blisland | Nailsea, Avon
'Grants Isla' | 'Lockeridge Jill' | 'Theydon Councillor'

1985 A Maunder | A Maunder | Swains of Stretton
Saltash | Saltash | Neath, Glamorgan
'Grants Alladin' | 'Grants Alladin' | 'Bucklands Perfectionist'

1986 Miss R Folland | A Maunder | Mrs P Baker
Totnes, Devon | Saltash | Nailsea, Avon
'Manmoel Mostyn' | 'Grants Alladin' | 'The Dancer'

1987 A Maunder | A Maunder | Mrs C A Wigmore
Saltash | Saltash | Salisbury, Wilts
'Grants Alladin' | 'Grants Alladin' | 'Winewood Jet'

1988 Mr A P Noble & Mr M E Gould | S H Williams | D Clixby
Southampton | Camborne | Gainsborough, Lincs
'Portsdown Knight Errant' | 'Holy Park Spartacus' | 'Woodside Aramis'

1989 S Murrell | S H Williams | Mrs P Baker
Kensworth, Beds | Camborne | Nailsea, Avon
'Wentworth Prince Regent' | 'Holy Park Spartacus' | 'Theydon Councillor'

1990	Mrs K A Hill Wellington, Somerset 'Neptunes Bounty'	A Maunder Saltash 'Alladin'	Bailey's Horse Feeds Kensworth, Beds 'Wentworth Prince Regent'
1991	Mr & Mrs S Murrell Dunstable, Beds 'Wentworth Prince Regent'	S H Williams Camborne 'Woodside Priscilla'	Mrs P Baker Nailsea, Avon 'Theydon Councillor'
1992	Mr & Mrs J Barrass Honiton, Devon 'Embleton Ash'	W T Warren & Son Penzance, Cornwall 'William'	Mrs D Wray Dewsbury, W Yorks 'Welsh Pageant'

COACHING MARATHON

	Exhibitor	Names of Horses in Team	Coach
1984	Mrs J R Dick Dunstable, Beds	'Venture', 'Viking', 'Vivid' & 'Maximillian'	
1985	Mr & Mrs J R Dick Dunstable, Beds	'Venture', 'Viking', 'Vivid', 'Victory' & 'Maximillian'	
1986	Pro-Mil Engineering Tamworth, Staffs	'Spider', 'Bobby', 'Victor' & 'Rocket'	'The Wonder'
1987	N D B Smith Ascot, Berks	'Bromly', 'Ulco', 'Aadrof' & 'Zorro'	
1988	N D B Smith Ascot, Berks	'Zorro', 'Aador', 'Bromly', 'Binko' & 'Ulco'	Private Coach
1989	N D B Smith Ascot, Berks	'Zorro', 'Aadorp', 'Bromly', 'Woldkonig', & 'Binko'	
1990	G C Mossman Luton, Beds	'Duncan', 'Drummer', 'Drifter' & 'Dragon'	
1991	G Mossman Luton, Beds	'Duncan', 'Drummer', 'Drifter' & 'Dragon'	
1992	P Munt & N Smith Bulmer, Sudbury	'Russel', 'Duke', 'King', 'Earl' & 'Bromley'	Private Coach

MAJOR CHAMPIONSHIP AWARDS

BEST HORSE IN THE CLASSES of Hunters, Hacks or Harness

Exhibitor

1905	James Penna & Sons Mawgan
1906	J Tremain St Columb Minor
1907	Maj Gen Jago-Trelawney Menheniot, Liskeard
1908	Gen Jago-Trelawney Coldrenick
1909	Execs of the Late Gen Jago-Trelawney Coldrenick
1910	Edwin Jacka Lanner, Helston

1912 Edwin Day
 Cury Cross Lanes, Helston
1914 The Hon T C Agar-Robartes MP
 Lanhydrock, Bodmin
1915 A J Slade
 St Budeaux
1920 H Copplestone
 Lostwithiel
1921 J F Hocking
 St Cleer
1922 R P Cawsey
 Torrington, Devon
1923 Harry Rowe
 Treeza House, Porthleven
1925 W H Benney
 Boderloggan, Wendron
1926 Capt G Babington
 Woodtown, Bideford

BEST EXHIBIT IN THE LIGHT HORSE CLASSES
— Bred and Exhibited by a bona fide resident Cornish Exhibitor
Exhibitor
1949 Miss N Eddy
 Treloweth, St Erth
 'Edna'
1953 M L Oats
 Tregannick, Sancreed
 'Cornishman'

HER MAJESTY THE QUEEN'S CUP
— **BEST EXHIBIT IN THE LIGHT HORSE CLASSES**
Exhibitor
1961 P J H Murray Smith
 Kingsbridge, Devon
 'Freelance'
1962 Mr & Mrs N C Skinner
 Exeter, Devon
 'Rouge Croix'
1963 Miss B Murphy
 Winford, Bristol
 'Boy Blue'
1964 F M Lawrey
 Long Rock, Penzance
1965 P J H Murray-Smith
 Kingsbridge, Devon
 'Fine Art'
1966 Miss B Murphy
 Winford, Bristol
 'Boy Blue'

1967 A H Newbery
      Alphington, Exeter
1968 Miss H Hunt
      Blackwater, Truro
      'Macdonald'
1969 J E Treffry
      Truro
      'Benbow'
1970 Mr & Mrs W C Skinner
      Exeter, Devon
      'Ocean Cheer'
1971 Mr & Mrs N C Skinner
      Farringden, Exeter, Devon
      'Ladram Bay'
1972 A H Newbery
      Alphington, Exeter
      'Game Fair'
1973 M Westaway
      Marldon, Paignton
      'Carver Doone'
1974 Mr & Mrs N C Skinner
      Honiton, Devon
      'Ocean Cheer'
1975 Mrs A Ferguson
      St Peter Port, Guernsey C.I.
      'Clipston'
1976 Mr & Mrs J N T May
      Morwenstow
      'Killin'
1977 Mr & Mrs R J Burrington
      Exeter, Devon
      'Gay Boy'
1978 S Luxton
      Okehampton, Devon
      'Colonel'
1979 S Luxton
      Okehampton, Devon
      'The Doper'
1980 S Luxton
      Okehampton, Devon
      'The Doper'
1981 S Luxton
      Okehampton, Devon
      'The Doper'
1982 R A Shuck
      Newent, Glos
      'Celtic Gold'
1983 D Kellow
      Bodmin
      'Manor Court'
1984 Gem's Signet Bloodstock
      Upleadon, Newent
      'The Candyman'
1985 Gem's Signet Bloodstock
      Upleadon, Newent
      'Dancin'

1986 Mrs D Whiteman
'Iceman'
1987 Mr & Mrs J A Crofts
Albury, Surrey
'Periglen'
1988 Miss R V O White
Market Harborough
'Classic Tales'
1989 Miss R V O White
Oakham, Rutland
'Classic Tales'
1990 Mr & Mrs R J Claydon
Silverley, Newmarket
'Carnival Time'
1991 Miss R V O White
Oakham, Rutland
'Corbally'
1992 Mrs P Caldwell
Milton Keynes, Bucks
'Sir Harry Stokracy'

IN-HAND CHAMPIONSHIP

Exhibitor

1974 Mr & Mrs N C Skinner
Honiton, Devon
'Ocean Cheer'
1975 Mrs A Ferguson
St Peter Port, Guernsey C.I.
'Clipston'
1977 Mr & Mrs R J Burrington
Exeter, Devon
'Gay Boy'
1978 S Luxton
Okehampton, Devon
'Colonel'
1979 S Luxton
Okehampton, Devon
'The Doper'
1980 D Kellow
Bodmin
1981 Mr & Mrs R J Burrington
Kenton, Exeter
1982 R A Shuck
Newent, Glos
'Celtic Gold'
1983 D Kellow
Bodmin
'Manor Court'
1984 Gem's Signet Bloodstock
Upleadon, Newent
'The Candyman'
1985 Mrs C J Grain
Chettisham, Ely
'Twinwood Bright Eyes'

1986 Miss J Newbery
      Alphington, Exeter
      'Necta Super Star'
1987 W D Kellow
      Bodmin
      'Morgan's Boy'
1988 Mrs J Comerford
      Cirencester
      'Sandbourne Royal Emblem'
1989 Mrs M Roper-Caldbeck
      Minehead, Somerset
      'Ashlands Sugar Baby'
1990 W J Jordan
      Chagford, Devon
      'Pop's Birthday'
1991 Mrs P M Chance
      Exeter, Devon
      'Spinningdale Arabella'
1992 Mrs A Holmes
      Ottery St Mary, Devon
      'Abbas Sparkle My Lovely'

HORSE SHOEING COMPETITIONS

HORSE SHOEING
Exhibitor
1896 F P Frayn, Egloskerry &
      James Frayn, Launceston

HORSE SHOEING—   HORSE SHOEING—
Cart   Hacks
1900 J H Daniel   Walter Hoskin
      St Austell   Launceston

HORSE SHOEING—   HORSE SHOEING—
Agricultural   Roadsters
1902 C Hill   E Hoskin
      Redruth   Launceston
1904 J Johns
      Breage
1905 Edgar C Yelland
      Tregony
1907 James Frayn
      Druckham, Launceston
1908 R Dunstan
      Penryn
1909 Reuben Hoskin
      Launceston
1910 E Brown
      St Austell
1912 W Bennetts
      Godolphin Cross, Breage
1915 J Pollard
      St Erth

		HORSE SHOEING— Smiths Under 21	HORSE SHOEING— Open	
1919	J H Dowrick Ladock	J Oliver St Austell	E Brown St Austell	
1920	T E Masters Week St Mary			
1921	F Parkin St Columb Major			
1922	J Maunder Plymouth			
		HORSE SHOEING— Roadsters Smiths Under 25	HORSE SHOEING— Roadsters—Open	
1923	T S Beswetherick Mawgan, St Columb	C Dowrick Grampound	G Trout Ugborough, South Brent	
		HORSE SHOEING— Agricultural Smiths Under 21		
1924	J Hill Broadbury, Beaworthy	C Dowrick Tregony (Gold Medal Winner)	T E Masters Week St Mary	
				HORSE SHOEING— Cart Horses
1925	S J Daniel High Cross St, St Austell	W E F Pollard Fraddam, Hayle (Gold Medal Winner)	T E Masters Week St Mary	F Parkin Parkland, St Columb Minor
1926	G Gale Okehampton	C J Dowrick Tregony	J Hill Broadbury, Beaworthy	F Parkin Parkland, St Columb Minor
1927	J H Baker Barnstaple, Devon	C J Dowrick Tregony	J H Baker Barnstaple, Devon	T H Cundy Grampound (Centenary Cup)
1928	T R S Glidden Ashwater, Beaworthy	E Warne St Columb	T E Masters Week St Mary	S Beswetherick Mawgan, St Columb
1929	J Vanstone Okehampton	J B Williams Prospidnick, Sithney	J Vanstone Okehampton	T E Masters Week St Mary
1930	J Baker Barnstaple,Devon			T H Cundy Grampound
1931	J Vanstone Okehampton	P J Cundy Grampound	T H Cundy Grampound	S Beswetherick Mawgan, St Columb
1932	H F Gliddon Ashwater, Beaworthy	F Stapleton Bideford	T E Masters Week St Mary	W Higman St Columb
1933	J Vanstone Okehampton	G Masters Week St Mary	T H Cundy Grampound	T E Masters Week St Mary

HORSE SHOEING—
Agricultural Cornish Only, never having won a 1st at any R.C. Show

1934	A Thomas Shaftsbury, Dorset	W H Trevaskis Shop Hill, Marazion	H F Gliddon Ashwater, Beaworthy	T E Masters Week St Mary
1935	H F Gliddon Ashwater, Beaworthy	J C Dyer Cubert, Newquay	H F Gliddon Ashwater, Beaworthy	T E Masters Week St Mary
1936	J Vanstone Okehampton	G Masters Week St Mary	H F Gliddon Ashwater, Beaworthy	T E Masters Week St Mary
1937	J Vanstone Okehampton	J C Dunstan Trenance, St Austell		T H Cundy Grampound

HORSE SHOEING—
Agricultural Cornish Only

HORSE SHOEING—
Agricultural Cornish Only Under 25

1938	H Vanstone Okehampton	G Masters Week St Mary	T H Gilbert Camborne
1939	J Vanstone Okehampton	T E Masters Week St Mary	
1949	T E Masters Week St Mary		
1950	W G C Masters Week St Mary		
1951	W G C Masters Week St Mary		

HORSE SHOEING—
Light Horse

1953	A J Palmer AFCL Chagford, Devon	A J Palmer AFCL Chagford, Devon
1956	J C Dunstan Trenaneerd, St Austell	J C Dunstan Trenaneerd, St Austell

MISCELLANEOUS RESULTS — 1793 ONWARDS

FOUR MEN SERVANTS IN HUSBANDRY, WHO HAVE LIVED THE GREATEST NUMBER OF YEARS IN THE SAME SERVICE

1796 John Rounceval, Padstow — 51½ years
John May, St Issey — 34 years
John Dawe, Quethiock — 37 years
Charles Higgings, Liskeard — 33 years

FOUR WOMEN SERVANTS IN HUSBANDRY, WHO HAVE LIVED THE GREATEST NUMBER OF YEARS IN THE SAME SERVICE

1796 Sibilla Row, Lanteglos by Fowey — 35 years
Joan Williams, Merther — 32 years
Julian James, Wendron — 30 years
Agnes Piper, Anthony — 33 years

TO THE SERVANT OR LABOURER IN HUSBANDRY, WHO SHALL HAVE LIVED THE LONGEST PERIOD IN ONE CONTINUED SERVICE (BEING OF GOOD MORAL CHARACTER)

			In Service To
1835	Peter Buddle, St Erme	— 50 years	Edward Collins, Truthan & his late father
1837	Jonathan Webber		
1840	Joseph Penrow, St Clement	— 58 years	John Vivian, Pencalenick & his pre-decessors
1842	P Eustace, Perranzabuloe	— 54 years	
1843	Charles Lilly, Feock	— 55 years	
1844	J Richards, Bodrean	— 50 years	H P Andrews
1845	William Roberts, Illogan	— 51 years	Mr Paul
1846	Richard Keast, Ladock	— 54 years	Mr Bone, Ladock
1847	John Benny	— 52 years	Walter Andrew, St Enoder
1848	William Saundry	— 49 years	Mr Hoblyn, Gloweth
1849	John Kent	— 47 years	S Symons, Legonna, St Columb
1850	Thomas Scoble	— 58 years	Mr Hoskin, St Allen
1851	Richard Musgrove	— 54 years	Mr Gatley, Polsue
1852	James Truscott	— 48 years	J P Peters, Philleigh
1853	Thomas Hill	— 50 years	Treveor Estate, Gorran
1854	John Osborne, Rosewaste, St Columb	— 50 years	
1855	William Carne	— 51 years	H Pearce, Kenwyn
1856	William Long	— 55 years	Mark Guy, Endellion
1857	Samuel Blake	— 56 years	Jonathan George, Trefreock
1858	John Phillips	— 58 years	Mark Guy, Endellion

FOUR HUSBANDRY LABOURERS, WHO HAVE BROUGHT UP THE GREATEST NUMBER OF CHILDREN, IN HABITS OF HONEST INDUSTRY, WITHOUT RELIEF FROM ANY PARISH—13 CLAIMANTS

1796 James Short, St Ives — 8 children
James Rundle, St Austell — 8 children
Richard Cook, St Columb Minor — 8 children
Charles West, St Mawgan in Pyder — 10 children

TO THE LABOURING MAN IN HUSBANDRY THAT HAS REARED THE LARGEST FAMILY WITHOUT RECEIVING PAROCHIAL RELIEF

1829	Richard Stevens, St Enoder	13 children
1830	James Dunstan, Kea	14 children
1831	Thomas May, Newlyn	11 children
1832	Thomas Michell, Kenwyn	13 children
1833	Stephen Spargue, Constantine	11 children

TO THE COTTAGER IN KENWYN, KEA, FEOCK, ST CLEMENT, MERTHER OR PROBUS WHO SHALL HAVE BROUGHT UP THE LARGEST FAMILY, WITH THE SMALLEST MEANS, AND IN THE BEST MANNER, WITHOUT PAROCHIAL RELIEF, BEING OF GOOD MORAL CHARACTER
£5 Prize given by The Rt Hon The Earl of Falmouth

1833	James Dunstan, Kea	14 children
1834	Edward Merrifield, Kenwyn	14 children, 12 living
1835	Thomas Michell, Kenwyn	17 children, 12 living
1837	William Teague, St Clement	
1840	Thomas Michell, Kenwyn	20 children, 14 living
1842	R Jeffery, Miner, Kenwyn	12 children
1843	Thomas Francis, Kenwyn	10 children
1844	Martin Davey, Kea	21 children, 15 living
1845	Thomas Oats, Kenwyn	17 children
1846	John Whitford, Kenwyn	17 children
1847	John Uren, Kenwyn	17 children, 10 living
1848	Thomas Gray, Kea	14 children
1849	John Skewes, Kea	15 children
1850	Paul Rowe, Kenwyn	12 children
1851	Stephen Kestle, Kenwyn	14 children
1852	James Lobb, Kea	15 children
1853	William Davey, Kenwyn	15 children
1854	Thomas Michell, Carnon Downs	16 children
1855	Thomas Moyle, Kenwyn	14 children
1856	Matthew Tresize, Kenwyn	17 children, 14 living
1857	Edward Matthews, Kea	19 children, 10 living

BEST ONE HORSE CART
| 1799 | Alexander Menhennick |
| 1803 | Edward Harvey, Launceston |

BEST CROP OF POTATOES IN THE YEAR 1798 (as a Laborer)
| 1799 | Thomas Crap, St Columb |

BEST CROP OF TURNIPS — *Not less than 3 acres; §Not less than 5 acres
1796	Joseph Hambly, St Germans
1799	Charles Peter, Padstow
1801*	William Kendall, Padstow
1801§	John Coryton, Penetillie Castle
1802*	John Wherry, St Breock
1802§	William Kendall, Padstow

TURNIP HOEING
1799 Richard Pope, Laborer, Egloshayle, Wadebridge
1802 Jacob Hawken, Bodmin and Oliver Veal, St Wenn

BEST CROP OF RUTA BAGA
1802 John Slyman, St Germans and Rev Jeremiah Trist, Veryan

GREATEST NUMBER OF STOCKS OF BEES
1801 Emanuel Hollman, Illogan
1802 Francis Pearce, Breage, Helston
1803 William Welsh, St Mabyn

EXPERIMENTS ON WHEAT
1804 Revd Robert Walker, St Winnow

BEST CROP OF SPRING WHEAT — in year previous to show
1814 E W Stackhouse, Pendarves — Average of 41 Winchester bushels

BEST CROP OF HORSE BEANS
1813 Mr Trood, Poughill, Bude

SUNDRY IMPLEMENT AWARDS
NEW OR IMPROVED HUSBANDRY IMPLEMENT superior to any in common use in this County

1806	George B Worgan, Glynn	
1841	R Doble	Turnip Hoe
	S Anstey	Corn Drill
	H Tilley	Randsome's Ridge Plough
	E Hennah	Chaff Cutter
1844	Mr Harvey	Liquid Manure Cart & Pump
	Mr Harvey	Skim Coulter Plough
	Messrs Fox	Randsome's Chaff Cutter
	Mr Harvey	Swing & Wheel Ploughs
1845	W Gerrans, Tregony	Turnip & Manure Drill
	Messrs Buckingham, North Hill Launceston	Turnip & Manure Drill
	Mr Watts, Charlestown	Turnip & Manure Drill
1846	W H Vingo, Penzance	Corn Drill
	Mr Gerrans, Tregony	Turnip Drill
	Mr Trethewy, Probus	Horse Rake
	Mr Holman, St Just	Steam Apparatus
	J D Gilbert, Treliske	Clod Crusher
	Mr Holman, St Just	Rake
	Mr Manley, Helston	Winnowing Machine
	Peter Dungey	Iron Harrows
	Mr Truscott, St Stephens	Iron Harrows
1847	W Gerrans, Tregony	Broadcast Sowing Machine
	W Gerrans, Tregony	Corn Drill & Turnip Drill
	J Holman, Camborne	Steaming Apparatus
	C Puckey, Ladock	Double Skim Plough
	James Magor, Colan	Norwegian Harrow
1848	G Bullmore, Newlyn	Ransom's Single Wheel Plough
	J Truscott, St Stephens	Shaver & Skim Coulter Plough
	C Puckey, Ladock	Double Plough & Skim Coulter
	M Burley, Tregony	Two Single Ploughs

1848	Mr Holman, Camborne	Steaming Apparatus, Horse Rake, 2 Skim Coulter Ploughs, Banking Plough, Lawn Rake, Cultivator
	T Jeffery, Camborne	Double & Single Furrow Ploughs
	S & R Davey, Redruth	Norwegian Harrow & Cultivator
	W Gerrans, Tregony	Three & Four Rows Seed Drills & Broad Cast Drill
1850	Mr Gerrans, Tregony	Manure Distributor
	L Truscott, St Stephens	Improved Turnip Drill
	J Truscott, St Stephens	Improved Plough
	Mr Burley, Tregoney	Improved Skim Plough
1851	Mr Jeffery	Plough & Norwegian Harrow
	Mr Tilly	Winnowing Machine
	Broad Sharp	Plough & Wain
	Mr Burley	Horse Rake
	Mr Truscott	Corn & Seed Machine
	Mr Reed, St Allen	Shaver
	Mr Gerrans	Turnip Drill
1852	Abraham Reed, St Allen	Iron Single Plough
	G Bullmore, Newlyn	Straw Shaker
	Andrew Stephens, Phillack	Wheel Plough & Hay Making Machine
	Mr Stanbury, Rosemerryn	Drill & Horse Hoe
	James Davis, Cornelly	Winnowing Machine
	William James, Merther	Horse Rake
	John Hotten, Cornelly	Bentall's Cultivator
	Probus Farmers Club	American Reaping Machine by Messrs Garrett & Co
	John Randle	American Reaping Machine by Dean & Dray
1853	James Williams, St Enoder	Corn, Seed & Manure Drill

AWARDS MADE AT THE WINTER SHOWS STAGED IN TRURO ON THE OCCASION OF THE ASSOCIATION'S ANNUAL GENERAL MEETING DURING DECEMBER — 1841 to 1851

LIVESTOCK SECTION

Year	BEST FAT OX OR STEER Any Breed	BEST FAT COW OR HEIFER Any Breed	BEST 10 FATTED WETHERS	BEST COTTAGERS FATTED PIG
1848	Thomas Julyan Creed	Thomas Julyan Creed		
1849	Josiah Stephens Probus	S Tresawna Probus	Mr Tremain Bodrean	William Staple Kenwyn
1850	John Cardell Lower St Columb	Messrs Doble, Kendall & Co Probus		John Whitborne Newlyn
1851	Thomas Julyan Creed	Messrs M A Doble & Co Probus		James Lawer Probus

GRAIN SECTION

	WHITE WHEAT	WHEAT, of other varieties	RED WHEAT	BARLEY	OATS
1841	J Hotten Cornelly 35 bushels/acre	J Lawry St Just, Roseland	J Lawry St Just, Roseland 45 bushels/acre	P Davis Probus 51 bushels/acre	T Treloar 40 bushels/acre
1842	R Doble Philleigh			J Cardell Jnr Lr St Columb	W Pellow St Allen
1843	Stephen Doble Mylor			Peter Davis Probus	William Benny Little Colan
1844	Stephen Doble Mylor				
1845	R Doble Philleigh		R Whitford St Just	Peter Davis Probus	J Huddy Probus
1847	P Clark Kenwyn		W Whitford St Just	James Huddy Probus	J Harding St Enoder
1848	Stephen Doble Mylor	Mr Stephens Penpoll, Feock		Edward Pascoe Mevagissey	Richard Lanyon St Allen

ROOTS SECTION

	6 SWEDE TURNIPS —Skirvinge's Variety	6 YELLOW TURNIPS	6 WHITE TURNIPS	10 CARROTS	10 PARSHIPS
1841	M A Doble Probus 24 tons/acre	William Corfield Jnr 22 tons/acre			

6 SWEDE TURNIPS

1842	M A Doble Bartilever	J Cardell Jnr Lr St Columb	J Huddy Probus	T H Tilly Tremough	
1843	H Middlecoat Veryan	The Earl of Falmouth Tregothnan	J Harding St Enoder	J Gwatkin Veryan	
1844	William Corfield Feock	John Aver Cuby	John Hotten Cornelly	J Gwatkin Veryan	
1845	The Earl of Falmouth Tregothnan	J Lawry Gorran	J Huddy Probus	J Gwatkin Veryan	
1847	Mr Doble Probus	Mr Doble Probus	J Harding St Enoder	J Gwatkin Veryan	J Gwatkin Veryan

	12 SWEDE TURNIPS	12 YELLOW TURNIPS	12 WHITE TURNIPS		
1848	Paul Clark Perranzabuloe	John Haddy Probus	J Cardell Lr St Columb	Mr Hicks Trevellan	

	6 SWEDE TURNIPS	6 YELLOW TURNIPS	6 WHITE TURNIPS		
1850	Samuel Simmons Lagona, St Columb	Samuel Simmons Lagona, St Columb			
1851	Richard Cowling Perranzabuloe	James Tremain Newlyn		Joseph Roberts St Clement	Rev T Phillpotts Feock

	6 MANGEL WURTZELS Long Variety, Yellow or Red	6 MANGEL WURTZELS Globe Variety, Yellow or Red	3 CABBAGE, CALCUL- ATED FOR THE AGRICULTURALIST
1842	Lady Basset Tehidy Park		C H T Hawkins Trewithen, Probus
1843	J H Tilly Tremough	Lady Basset Tehidy Park	William Gill St Clement
1844	J Gwatkin Veryan	The Earl of Falmouth Tregothnan	R Rosewarne Phillack
1845	J Gwatkin Veryan	The Earl of Falmouth Tregothnan	
1847	J Gwatkin Veryan	The Earl of Falmouth Tregothnan	J Paul Camborne

	12 MANGEL WURTZELS Long Variety, Yellow or Red	12 MANGEL WURTZELS Globe Variety, Yellow or Red	
1848	J D Gilbert Trelissick	J D Gilbert Trelissick	Rev Thomas Phillpotts Feock

	6 MANGEL WURTZELS Long Variety, Yellow or Red	6 MANGEL WURTZELS Globe Variety, Yellow or Red	
1851	J D Gilbert Trelissick	William Paul Illogan	W Trethewy Probus

PLOUGHING MATCH RESULTS   * Without a driver;   § With a driver

	Venue	Best Ploughman	His Boy/Driver	Best Apprentice	His Boy/Driver
1793	Truro	Matthew Toman St Merryn	Thomas Toman		
	Bodmin	William Burt St Minver	Richard Blake		
1794	Liskeard	John Vosper Coldrinick	John Pole		
	Helston	Francis Ferris Helston	Joseph Martin		
1796	Tregony	George Gummoe Probus	Thomas Woolcock	John Ford Probus	John Sparks
	St Columb	John Julian St Columb Major	Richard Carne		
1797	Callington	William Rickard Penatillie	John Rickard		
1798	Stratton	John Trace Moorwinstow	Thomas Shephard		
1799	Truro	John Sloggett Servant to John Enys	Thomas Jennings		
	Wadebridge	Thomas Burt St Minver	George Hicks	William Hawke St Breock	John Bonear

1801	St Austell	John Pearce Bodrigan	William Husband		
1802	Camelford	Joseph Burt St Minver	Thomas Hicks	John Blake St Endellian	John Harris
	Bodmin	Samuel Bastard Michaelstow	Stephen Hawken	Thomas Hill Bodmin	Francis Wills
1803	Marazion	John Nicholls Trelowarren	Edward Cook	William Bant St Hilary	William Pearce
	Mitchell	John Knight Truan	George Hicks	George Sweet St Columb Major	John Hicks
1804	Launceston	Christopher Spear South Petherwin	William Cornelius	John Downing St Stephens	John Fisher
	Padstow	Thomas Hawkey St Issey	John Mitchell	George Johns St Breock	Richard Johns
1805	Probus	John Barrett St Columb	George Rough	William Peneliggan Probus	Joseph Manuel
1819	Bodmin	William Dunn* Fowey			
		Thomas Hawkey§ St Breock	Thomas Carn		

361